Fire Prevention Applications

Second Edition

By Brett Lacey and Paul Valentine

FIRE PROTECTION PUBLICATIONS
OKLAHOMA STATE UNIVERSITY

RECYCLABLE

Design/Layout: Missy Hannan
Editors: Cindy Brakhage, Keith Wolfe
Indexer: Nancy Kopper
Cover Photo: Courtesy of Lake Zurich Fire Department, Lake Zurich Illinois

10 9 8 7 6 5 4 3 2 1   Printed in the United States of America

ISBN 978-0-87939-621-3
Library of Congress Control Number: 2017941749
Second Edition
First Printing, June, 2017
Printed in the United States of America

Contents

NOTE

Case studies are provided at the start of each chapter. The purpose of the case study is to provide a situation where the material presented in the chapter can be applied. The reader should think about the case study as he or she reads the chapter material. After reading the chapter, reference should be made back to the case study and provide a solution to the case study citing specific chapter material. Most readers will find the case studies will reflect similar situations in his or her community.

About the Authors

Brett T. Lacey

Brett is the Fire Marshal for the Colorado Springs, (CO) Fire Department. He is a graduate of Oklahoma State University and is a registered professional engineer, and a certified safety professional. He has worked in the fire service for over 36 years serving in the ranks through firefighter, paramedic, and fire protection engineer up to his current position. He has also worked for IFSTA/ Fire Protection Publications. Brett has served on various IFSTA technical validation committees and is a current member of the IFSTA Executive Board. Most recently, he served as Chair of the technical validation committee for the IFSTA *Fire Inspection and Code Enforcement,* 8th edition manual. He currently serves on the NFPA Technical Committees for NFPA 1031, *Standard for Professional Qualifications for Fire Inspector and Plan Examiner*; NFPA 1300, *Standard on Community Risk Assessment and Community Risk Reduction Plan Development*; NFPA 1452, *Guide for Training Fire Service Personnel to Conduct Community Risk Reduction*; and NFPA 1730, *Standard on Organization and Deployment of Fire Prevention Inspection and Code Enforcement, Plan Review, Investigation, and Public Education Operations.*

He is co-author of the Fire Protection Publication textbook *Fire Prevention Applications for the Company Officers* (1st edition) and has co-authored articles for Firehouse.com magazine and NFPA *Fire Journal*. Brett has served on the Colorado Fire Marshal's Association Code Committee as well as various other state committees. He has completed various curriculums at the Center for Creative Leadership and served as a member of the City of Colorado Springs Strategic Leadership Team that was responsible for major change and leadership initiatives for the City.

You can contact the author at:
Colorado Springs Fire Department Office of the Fire Marshal
Brett T. Lacey
375 Printers Parkway
Colorado Springs, Colorado 80910
(719) 385-7355
blacey@springsgov.com

Acknowledgements

I would like to thank foremost my wife, Janell, for her patience and support in allowing me the space, time, and periodic frustration in revising this book. I am indebted to so many other people who, throughout my career, provided me with invaluable information, mentoring, opportunities, and ideas and especially all of my staff over the years who put up with my thoughts, opinions and at times, experiments to find what works best in the world of fire prevention.

All of these people have argued with me, struggled with me, put up with me, but in the end helped me learn and work through so many things. I am especially grateful to Paul Valentine who allowed me the opportunity to work with him on this book. Without his patience and ability to cover my back, I couldn't have survived. Thanks also to all the staff at IFSTA/Fire Protection Publications who make our ideas understandable.

Brett

About the Authors
(Continued)

R. Paul Valentine

Paul Valentine has 30 years of fire protection experience in both the municipal and private sectors. His 20-year fire service career began as a volunteer firefighter with the Stillwater, (OK) Fire Department. Paul served as Fire Marshal and fire protection engineer for the Mount Prospect Fire Department in Mount Prospect, Illinois.

After his fire service career, Paul managed a fire protection engineering division in one of the largest fire protection consulting firms. He specialized in risk management, power generation (fossil and nuclear), hazard evaluations, municipal code and management consulting, and review of equivalent means and methods for compliance. His private sector experience also included serving as the site fire protection engineer for a Department of Energy research laboratory and as a Loss Control Consultant in the insurance industry.

Paul presently provides expert fire protection and risk management consulting services to the power industry, complex industrial occupancies, and municipal fire prevention bureaus throughout the United States.

He has a Bachelor of Science Degree in Fire Protection and Safety Engineering Technology from Oklahoma State University and a Master of Science Degree in Management and Organizational Behavior from Benedictine University. He is a graduate from the National Fire Academy's Executive Fire Officer Program. Paul has been an instructor at a local community college and for various fire service certification courses. He has served on many International Fire Service Training Association (IFSTA) committees and as a member and Vice Chair of the IFSTA Executive Board. He is a principal committee member of many NFPA codes and standards as well as Chair of the NFPA Fire Marshal Professional Qualification Standard. Paul is also the coauthor of the Fire Protection Publications textbook *Fire Prevention Applications for the Company Officer* (1st edition) and has coauthored articles for Firehouse.com magazine.

You can contact the author at:
R. Paul Valentine
1515 Fender Road
Naperville, Illinois 60565
rpaulvalentine@gmail.com

Acknowledgments

I would like to thank my wife, Kellie, my kids, Lucas, Rachel, and Sarah, who graciously allowed me to take time away from them while I worked on this project. Thank you to my parents, Richard and Peggy; without them I would not have been provided with so many opportunities. I feel fortunate to have had my father's example of hard work and dedication to the fire service to be the greatest influence on my career.

I am grateful for those people who throughout my career have unconditionally helped me whenever I ask them. Although too many to list, these people include Dennis Compton, John Malcolm, Doug Forsman, Bob Barr, Gary Jensen, Pat Brock, Jim Hansen, Ed Cavello, Abe Froman, Dean Maggos, Tony Huemann, Jerry Howell, Kevin Roche, Joe McElvaney, Chris Truty, Alan Berkowsky and the entire FPP staff.

Lastly, I owe a great deal of gratitude and acknowledgement to my great friend in this project, Brett Lacey. His insight, knowledge and forward fire prevention thinking is commendable. I am fortunate to have had the opportunity to know him, work with him and call him friend.

Paul

Dedication

To our fire service role models, mentors, and best friends... our fathers.
Men who honorably dedicated their time and talents to the service of others.

Lt. Richard Valentine

Champaign Fire Department,
Champaign Illinois

Chief Milton F. Lacey

Paradise Hills
Volunteer Fire Department

BCFD No. 7

Introduction

Introduction

Our service is continually facing challenges that require us to rethink how we do things. For example, you may remember when hazardous materials response was a new issue facing firefighters. Most remember when terrorism was only a threat, and we now face the reality that foreign terrorism on our home soil is more prevalent than ever. Today, government officials at the federal, state, and local levels are responsible for Homeland Security. Amid all of this, the fire department remains one of, if not, *the* most critical player involved in protecting our citizens.

The fire department is among the most critical players in protecting American citizens.

We have recently experienced one of the toughest economic climates in the history of our country, which forced many fire departments to do more with fewer resources at a time when calls for services continued to escalate. The changing roles of the fire service necessitate an ever-increasing demand for fire and injury prevention and disaster mitigation efforts. The reality is that the fire service will likely be unable to continue increasing and maintaining human resources for emergency response the way it has in the past. The prevalent culture found in our country's fire service reflects support for operational functions that far outweigh those of prevention, mitigation, or preparedness efforts. Unbalanced support includes both funding and allocation of personnel.

According to the National Fire Protection Association's (NFPA®) Fire Analysis and Research Division, public fire departments in 2012 received reports of around 1,375,000 fires. This number equates to a fire department responding to a fire somewhere in the United States every twenty-three seconds. Every eighty-five seconds, one of those fires was reported in a home. Every three hours and four minutes a fire death occurred in a home. In fact, as of 2005, fire deaths in the home accounted for over 94% of civilian fire deaths. As reported by NFPA in 2013, over 67% of fire department calls for service were emergency medical incidents. [1]

The odd thing is that, as a country, we appear to be on the cutting edge of technology, but we still suffer multiple-life losses from fires and other disasters **(Figure 1.1, pg. 4)**. That is not accounting for an inordinate number of calls for service, particularly emergency medical service. Unfortunately, these occurrences are largely due to apathy and lack of awareness or education. In recent years, Rhode Island suffered one of the most tragic loss-of-life fires this country has seen. Unfortunately, it was a nightclub fire very similar to many others in

Figure 1.1 One of the important outcomes of fire prevention is the enhancement of fire-fighter safety and reductions of firefighter deaths.

years past in other parts of our country (See Chapter 2). Most, if not all of these incidents could have been stopped or diminished through fire prevention and mitigation programs developed to meet the needs of the community. Many lives could have been saved had people been made aware of the combustibility of interior finishes, the effects of blocked or restricted exits, and other dangerous factors. The fact that we continue to experience these large life-loss fire incidents is clear evidence that we concentrate far more on a reactionary approach to disasters rather than taking a preventative and preparedness approach. This should be a wake-up call to the fire administration ranks. While popular and visible, our unbalanced focus on emergency response, in lieu of prevention and mitigation, could be seen as an unprofessional and cost-prohibitive approach.

The uncertainty of future health care and reductions in patient coverage by the insurance industry will continue to contribute to a major part of our emergency response load. The emergency medical response agencies are becoming our primary means of health care, which is neither cost effective nor efficient in its current operational model. As previously stated, since the majority of our alarm load consists of emergency medical calls, some suggest we should change our name to something other than the fire service or fire department. In a progressive and professional process, we would be aggressively looking at smarter ways of reducing our calls for service and proactively protecting our citizens. This process may involve non-traditional methods, such as fire department clinics for seniors, neonatal and baby wellness, satellite inoculation centers for infectious diseases or pandemics, and educational awareness programs targeting hypertension, diabetes, and other chronic conditions that ultimately lead to emergency calls.

As a basic service consideration, each community must establish a level of fire and injury prevention efforts that meet its needs. The prevention, education, and preparedness efforts needed in one community may differ from those needed in another. Does that make one community safer or more hazardous than another? Not necessarily. It all boils down to determining the level of risk, protection, and consequential service our citizens expect and then developing a program that meets or exceeds those expectations. The costs and available funding for these processes, in conjunction with the perceived and actual benefits and advantages of the service, will determine the type and level of service to be provided. Determining the necessary level of protection and then selling the need for that protection is the fire professional's greatest future challenge.

As the cost of emergency fire response escalates along with the costs that result from the devastation of fire and other community harm, the demand should increase for preventing these types of injuries from occurring or, at the very least, mitigate their effects (**Figure 1.2**). There needs to be a greater focus on built-in fire protection systems to control fires until fire department personnel can arrive. More proactive approaches to community medicine and self-care need to exist, as no quick or easy health care system fixes are available. This does not necessarily mean it is the fire service's problem to fix. However, as we remain the first and principle response agency, we will continue to be pinged by these requests for assistance unless we take a proactive and professional approach to stem the tide.

Each community must establish a level of fire prevention effort that meets its unique needs.

Figure 1.2 Most citizens only associate fire department activities with manual suppression.

As a disproportionate number of fire deaths continue to occur in our country's homes, suppression systems, such as fire sprinklers in residences, need to become more prevalent. We now have national codes requiring residential sprinklers in single family homes. Clearly, residential sprinklers will be the next generation of home fire protection. The evolution of residential home protection can be compared to the time when smoke alarms became a common form of home fire protection. It is evident that prevention is and will continue to be less costly by far than responding to a fire incident after it has occurred.

Providing financially limited citizens a location or method to receive treatment for acute emergent issues as well as chronic conditions will significantly reduce their dependence on us for service. Our overwhelming need to prevent fires and community injuries and deaths from occurring should be obvious. However, fire departments cannot do this alone, and they cannot do this through a predominantly emergency response focused lens.

Fire departments need to make a paradigm shift, moving strongly toward educational and awareness campaigns and processes. They need to emphasize proactive discovery of hazards and fix them. They need to focus much more on eliminating bad days from people's lives rather than trying to fix them after they have already happened. They need to look for opportunities to build coalitions and make the prevention of fires and injuries a combined effort. The following chapters examine some methods to build those coalitions and ways to prevent or mitigate injurious community events.

Unfortunately, obtaining staffing and resources for fire and injury prevention efforts continues to be more difficult than obtaining firefighters or paramedics. We emphasize and train for major emergency medical incidents rather than helping people stay healthy or maintaining safe and health-conscious behavior. Why does this happen? Because keeping an incident from becoming serious or preventing it from occurring at all is not dramatic, it is not newsworthy, and if nothing terrible happens, nobody notices.

The grass-roots level of our fire service needs to become as proud of prevention and mitigation as we are of emergency response.

Our firefighters' visibility sends a clear message that they will be there to "save the day," which they certainly will do whenever they can. However, as a profession we need to work smarter and better by helping people keep themselves safe. It goes back to the old Ben Franklin saying, "An ounce of prevention is worth a pound of cure." A grass-roots level change needs to occur, as our fire service needs to become as proud or more of prevention and mitigation efforts as we are of our quick and robust emergency response.

In this book, we take a different approach toward fire prevention activities. We emphasize a holistic approach to fire protection and community injury that encompasses not only fire suppression and medical response, but community involvement in all injury prevention efforts. Remember that community injury takes many forms. In fact, it takes any form that is injurious, such as fire, explosions, medical emergencies, pandemics, natural disasters, and so on. Many other texts provide detailed explanations and descriptions of dealing with specific hazards or code violations, such as not storing gasoline next to water heaters and the height at which a fire extinguisher must be hung. We, however, wanted to address a more global philosophy of community-based fire and injury prevention and how it can be designed and developed to work within any community.

In an effort to apply the material covered, each of the following chapters begins with a case study. Keep the case study in mind as you read. The information presented in the chapter addresses the case study issues and, hopefully, sparks some conversation or debate on the subject matter.

We begin by examining the history of fire and injury prevention efforts and end with the methods and opportunities available using some of the latest technology to help prevent these risks. We also discuss at length the roles of prevention versus mitigation and the utilization of risk management techniques to focus our prevention efforts in the right direction. Fire and injury prevention efforts involve the engineering of appropriate solutions to eliminate or reduce fire hazards, educating the public on the appropriate behaviors to prevent and mitigate their risk, and enforcing the appropriate fire and life safety codes and standards. This text explores methods for applying each of these principles to produce the community's desired level of prevention and preparedness services. Not only must we deliver this expanded and expected level of service for our community, but we must also have a means of prioritizing what gets done first. This is critical given our current economic climate and the future limitations that exist regarding funding and resources.

The "Three Es" of fire prevention are Engineering, Education, and Enforcement.

We hope this text is informative, but more than that, we hope it challenges you to think "outside of the box." We further propose the notion that if we all work together and make more people aware of the issues, firefighter safety will be enhanced, community safety will be enhanced, and our vocation will rise to a higher professionally capable level than ever before. The community looks to us for leadership and guidance. For this reason alone, we must strive to consistently provide the most professionally advanced solution possible. We cannot lead from behind.

Footnotes

http://www.nfpa.org/research/reports-and-statistics/the-fire-service/fire-department-calls/fire-department-calls

History and Development of Fire Prevention

Table of Contents

Key Points

1. Studying historical fires lays the groundwork for understanding today's field of fire prevention.

2. The emergency services can increase its value at little cost by preparing the community to deal with emergencies.

3. If the fire service is to survive, it must market its services and demonstrate its value.

4. Failure to enact codes and legislation until after incidents occur keeps fire prevention and mitigation efforts at least one step behind the available technology.

5. Most destructive fires are caused by the careless actions of people, largely through apathy or ignorance.

6. Fire service professionals must shift from overwhelmingly emphasizing response functions to stressing prevention functions.

7. If fire prevention work is as important as we believe, then departments must also factor in technical training on those functions.

Learning Objectives

1. Identify laws, codes, ordinances, and regulations as they relate to fire prevention.*

2. Understand code enforcement as it impacts life and property loss.*

3. Define the national fire problem and its role in Fire Prevention.*

4. Identify and describe fire prevention organizations and associations.*

5. Identify and describe the standards for professional qualifications for FIre Marshal, Plans Examiner, FIre Inspector, Fire and Life Safety Educator, and Fire Investigator.*

6. Describe the history and philosophy of fire prevention.*

7. Recognize significant fires that have occurred in the United States and identify similar lessons learned from each fire.

FESHE Objectives

History and Development of Fire Prevention

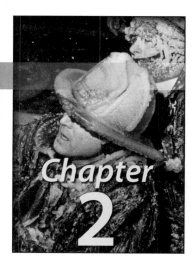

Chapter 2

Case Study

A fire occurs in a three story apartment building resulting in the death of an 85-year-old woman and seriously injuring one firefighter. The fire started when food was left unattended on the stove. The noncombustible building was not protected by automatic sprinklers. Smoke alarms were present in the apartment, but they were not functioning at the time of the incident. In other parts of the building, some residents had to be rescued from balconies, and others were trapped in the main stairwell when the doors to the floors locked behind them.

The deceased victim was the mother of a popular city councilman. Elected officials are demanding that this type of incident should never occur in their community again. Identify what fire prevention measures could be implemented. Include justification for each measure.

Why Study Fire Prevention History?

You may wonder, "What is the point of learning what has happened in the past?" Is it really important for fire prevention professionals to know about fires that occurred before they even began their careers? When will fire prevention professionals ever use historical information? How are historical fire case studies relevant to the daily tasks firefighters perform? In this day and age of technology and all the modern construction features we require, it can be difficult to understand why fires in our country could be a problem.

A knowledge of the relevance of historical fire events lays the groundwork for understanding how we got where we are today in the field of fire prevention **(Figure 2.1, p. 14)**. Most of the national codes adopted by municipalities were created or modified as the result of tragic fire incidents. It is painstakingly obvious that the most cost-effective practice is preventing the incident from occurring by proactively adopting codes. As a society, we typically react to devastating situations and try to prevent them from reoccurring after the fact. The public outcry to these types of events are found in the form of stricter or modified laws, as was the case with many of the devastating fires that have occurred in the United States.

Figure 2.1 Studying historical fires helps identify reasons for code changes and preventing similar tragedies.

Fire prevention professionals will often be asked to explain "Why do we need to do that?" or "What is the purpose or intent of this code requirement?" If they understand what led to the development of fundamental requirements found in most of the model codes, they will be capable of explaining the purpose of or reasoning for code requirements. Most often, achieving code compliance can be easier by presenting facts and logic of why it is needed other than just citing code sections with a "because the code says so" attitude. Case studies of historical fires give fire protection professionals the ability to recognize existing building hazards that are similar to conditions that contributed to historical fires. This in itself can help fire protection professionals prevent or mitigate similar events.

The concept of fire prevention is not new. As early as 24 B.C., large portions of Rome were destroyed by fire. Caesar Augustus stationed an estimated 600 servants at the city gates to fight the fires. After another disastrous fire in 6 A.D., Augustus instituted a Corps of Vigils. The corps was composed of slaves who patrolled the city for fires and alerted the citizens when a fire occurred. Over the next 500 years, this led to dividing the city into districts and developing ranking officers for each district. This is very similar to how fire departments are organized today in many of the larger cities across the United States. The districts in Rome had an estimated 7,000 vigils that were dressed and equipped with buckets and axes. They also took on the duty of fire prevention by enforcing what could be termed fire prevention laws and punishing offenders.

Early Fire Prevention Laws

Laws to prevent fires were issued as early as 1189, when the first lord mayor of London mandated that houses were to be built of stone with slate or burnt tile. These laws also banned thatch roofs to prevent fires from spreading from one building to another. They required a 16-foot by 3-foot party wall to separate buildings from each other. A *party wall* is a noncombustible wall constructed to prevent the spread of fire from one building to the next. Buildings had to have rings, so firefighters could pull them down with fire hooks during a fire to create a firebreak. Early fire apparatus carried these hooks along with ladders and were commonly referred to as *hook and ladder companies.*

Fire prevention laws reached the New World soon after the arrival of Europeans. In 1608, Jamestown was destroyed by fire. Ever since the destruction of Jamestown, Americans have developed tougher codes and increased educational programs to prevent such tragedies. However, as history shows, devastating fires continue to occur. Part of this reason is that we fail to enact codes or legislation quick enough to stem unwanted events. Another reason is that we fail to emphasize sufficient education and code enforcement. Frequently, this keeps our fire prevention and mitigation efforts at least one step behind the available technology. Not until the mid-1970s with the report *America Burning,* did the United States finally begin to take a proactive scientific approach to our fire problem.

The detailed report *America Burning* is a must read (copies of this report can be obtained from the U.S. Fire Administration). Although it is a long document, it explains America's fire problem going into the 1970s and the issues that needed to be addressed. Many of the identified issues still play a significant role in fire service and prevention planning today. The report provides a detailed summary of recommendations for mitigating the hazards we faced and, unfortunately, continue to create.

Most destructive fires are caused by the careless actions of people, largely through apathy or ignorance. Many factors played a part in creating the fire environment that caused these problems. *America Burning* identified a number of issues, some of which remain crucial to implementing successful fire prevention efforts (**Figure 2.2**):

Figure 2.2 Historically, *America Burning* is one of the most significant fire prevention documents.

- Place more emphasis nationally on fire prevention.
- Implement better training and education in the fire service.
- Educate Americans about fire safety.
- Be aware that in both design and materials, the environment in which Americans live and work presents unnecessary hazards.
- Improve the fire protection features of buildings.
- Continue with important areas of research.

The needs identified 30 years ago are still the key elements for furthering our fire prevention efforts today. Although we are making great strides in most of these areas, more must be done to achieve the goals and objectives outlined in this landmark report. *America Burning* identified several causes of our ongoing fire problem, but indifference was and continues to be the biggest. People who have experienced a hostile fire firsthand will never forget it; however, most

Americans think the dangers of fire's destructive powers are, at worst, very remote. The apathetic norm believes "It could never happen to me!" Indifference is prevalent even today when it would seem most inexcusable.

American Fire Tragedies

As the United States evolved into the prosperous country of today, many fires resulted in large life loss. The following are some of the most notable fires that directly influenced many of the codes enforced today throughout the country (**Figure 2.3**):

- Iroquois Theater, 1903
- Triangle Shirtwaist, 1911
- Cocoanut Grove, 1942
- Our Lady of Angels School, 1958
- Beverly Hills Supper Club, 1977
- MGM Grand Hotel, 1980
- Happy Land Social Club, 1990
- Food Processing Plant, 1991
- Station Nightclub, 2003
- Cook County Building, 2003

Code Changing Fires			
Fire	**Date**	**Lives Lost**	**Code Changing**
Triangle Shirtwaist Factory, New York City, NY	3-25-1911	147	Formation of Safety to Life committee and development of the Building Exits Code known today as the *Life Safety Code®*.
Iroquois Theater, Chicago, IL	12-30-1903	602	Exit doors swing outward in places of assembly.
Cocoanut Grove Nightclub, Boston MA	11-28-1942	492	Regulations of combustible interior finishes.
Our Lady of Angels School, Chicago, IL	12-1-1958	93	Two available exits and need to control interior finishes.
Beverly Hills Supper Club, Beverly Hills, CA	5-28-1977	164	Sufficient number of exits must be provided free from obstruction. Alarm notification requirements needed.
The Station Nightclub, Warwick, RI	2-19-2003	100	Sprinklers in places of assembly.
One Meridian Plaza, Philadelphia, PA	2-23-1991	3	Improved standpipe design and need to review pressure reducing values.
MGM Hotel, Las Vegas, NV	1-21-1980	87	The need for automatic sprinklers throughout high-rise buildings. The need for providing stairwell re-entry.

Figure 2.3 Notable fires that directly influenced code changes.

Iroquois Theater (1903), Chicago, Illinois

The Iroquois Theater fire in Chicago, Illinois, resulted in 602 deaths. It was one of the worst theater disasters in American history. According to reports, the building was said to be fireproof. At the time of the fire, the permissible occupant load of the building was 1,602, but an estimated 1,774 persons were in the building.[1] Few of the exits were marked and some were covered with draperies. Some of the doors were locked with levers that the occupants did not know how to operate.

The fire is believed to have started when one of the spotlights ignited the draperies.[2] After the fire broke out, one of the stagehands tried to extinguish it, and only a few of the occupants started to evacuate. In fact, witnesses stated that the orchestra continued to play in an attempt to avert panic and to calm patrons.

As the fire from the draperies spread to the drop scenery, those trying to evacuate had difficulty due to the large number of occupants blocking travel paths, acrid smoke that obscured visibility, and poorly identified and locked or blocked exit doors. The fire grew so rapidly that many of those who perished were in or near their seats. Officials found bodies that were stacked as many as four high near many of the exits.

The fire provided a number of lessons:

- Protect stages with automatic sprinklers. At the time of the fire, the building had none.
- Use fire-resistive draperies, such as a stage curtain.
- Obey permit occupant load limits.
- Train employees in evacuation procedures.

Triangle Shirtwaist (1911), New York, New York

The industrial fire at the Triangle Shirtwaist factory in New York City resulted in 145 deaths. This is still the largest life-loss industrial fire in American history. It is also significant because of the codes that were developed in its aftermath.

The building where the fire occurred was a ten-story loft building occupied by approximately 500 employees. Each floor was approximately 10,000 square feet. The building had wood floors and wood trim. Two exit stairs were available for the occupants. The building had no automatic fire sprinklers installed at the time of the fire.

The fire started near quitting time on the eighth floor of the factory. It is believed to have begun in a rag bin.[3] When the fire started, workers tried to extinguish it. The smoke from the fire traveled rapidly to the other floors through the stairwells. The doors apparently were at least partly open, contributing greatly to the spread of fire and smoke. In a very short time, the fire had spread over the entire fire floor and began moving out of the windows to the other floors. Many workers on the ninth floor became aware of the fire when the flames pierced through the exterior windows.

Workers on the eighth floor tried to exit one of the stairwells and found the door locked. After a time, they managed to open the door, although their difficulty was increased because the door opened inward. Once the door finally was opened, people evidently rushed into the stairway. Some of occupants evacuating in the stairwell fell down to the seventh floor landing, where evacuees began to pile up on the stairs. According to reports, a police officer ran up the stairs to help and assisted in untangling the pile-up. An estimated 125 workers escaped down the stairwell. Other workers on the eighth floor climbed out of the windows onto a narrow fire escape. Some fell from the fire escape, and others were able to reenter the building at the sixth floor and exit the facility.

The workers on the ninth floor tried to escape down a stairwell but that door was also locked. Some workers perished trying to escape down the elevator shafts; others tried to escape the flames by going to the window ledges and jumping to their deaths. The fire department tried to extinguish the fire from the outside with large streams of water, but the floors were too high for the water to penetrate. By the time firefighters extinguished the fire using hoselines supplied from hose outlets in stairwells, the upper three floors were burned, with all of their wood trim and finish consumed by the flames. Many lessons were learned from this incident, including the following:

- A building's means of egress must include a sufficient number of stairs in fire resistant shafts with rated fire doors at each opening. Even today, the model building codes as well as the NFPA® 101, *Life Safety Code®* use this fundamental principle to provide occupants of a building a means of safe escape in an emergency. The concept of providing a protected means of escape through rated exit enclosures is still one of the basic code principles taught to architects and engineers.

- The need to install automatic sprinklers was emphasized as with the other tragic fires with large loss of life. This is especially true in industrial occupancies with many combustibles.

- The need was identified to conduct evacuation drills in industrial occupancies. At the National Fire Protection Association's annual meeting in May 1911, R. H. Newborn presented a paper on exit drills and educating factory workers.[4] A year later, Mr. Newborn's paper became the NFPA's first safety-to-life publication under the title "Exit Drills in Factories, Schools, Department Stores, and Theaters."

Some good did evolve out of one of America's worst industrial fire tragedies. The National Fire Protection Association code requirements for the workplace, particularly exiting provisions, changed substantially as a result of this fire. After a number of meetings and discussions regarding topics such as fire escapes, evacuation, and sprinklers, the NFPA Committee on Safety to Life was formed in 1921. The committee's work began by expanding previous committees' publications regarding exits and related features for all occupancies and then evolved into the first publication of NFPA's Building Exits Code in 1927.[5] This later became what is known today as the *Life Safety Code®*.

Cocoanut Grove (1942), Boston, Massachusetts

In 1942, another fire occurred that would also help to foster future building codes and later editions of the *Life Safety Code®*. In one of Boston's most popular nightclubs, a fire claimed the lives of 491 patrons and injured another 200. The original reinforced-concrete building had undergone several additions over the years. There were no automatic sprinklers. The dance floor on the main floor could accommodate about 500 persons. Adjacent to the large dance floor was a room decorated with paper palm trees and a mix of other combustible decorations including cloth coverings on the walls and ceilings. A number of tables and chairs scattered throughout the space blocked or obstructed the building's six exits. Some of the exit doors were covered with drapes or even locked to control access. A single revolving door was the primary means of egress in the club. An estimated 1,000 patrons were inside at the time of the fire.[6]

According to reports, investigators believed the fire started in the basement lounge area. It spread rapidly because of the large amount of combustible decorations on the ceilings and walls. The rapidly traveling flames led the occupants to panic. The main entrance revolving door became jammed with people trying to escape. The flames and toxic gases produced during the fire killed many of those inside. Firefighters had difficulty entering the building because of the number of bodies stacked at the doors. The fire reportedly lasted only 12 minutes before it claimed the patrons' lives.

Several important findings came from this fire:

- The combustible decorations, combined with both the lack of available exits and the absence of sprinkler protection, contributed substantially to the large loss of life.
- Overcrowding and locked or obstructed exits severely limited the occupants' ability to evacuate safely.
- Adequate enforcement of building codes and employee evacuation training also would have helped save lives.

Our Lady of Angels School (1958), Chicago, Illinois

A fire in a Chicago parochial school in 1958 took the lives of 3 nuns and 87 children. During the following months, 5 more children died from their wounds. The fire originated in a pile of combustibles stored at the base of one of the stairways.[7] All but two of the building's stairways were open to the rest of the building. Many additions to the original building had been made without fire separations that would have limited the fire area. The building contained combustible interior finishes and combustible ceiling tile. It had no automatic fire sprinklers.

Students who indicated they smelled smoke were the first to notice the fire. Alerted teachers then went to find the principal. The principal was filling in for an ill teacher in another classroom and was not readily available. A significant delay followed before someone outside the school noticed the fire and found a phone to summon help. The students on the first floor began to evacuate.

Smoke and hot gases traveled rapidly up the open stairs. Smoke began to fill the halls and enter the classrooms through the transoms above the doors. Soon the fire had spread to a concealed space above the stairs. Shortly thereafter, the roof over the stairs collapsed. In an effort to escape the smoke and flames, some students jumped from the windows. Fire department ground ladders saved many students. Doing as they were instructed, other students perished while sitting at their desks. The lessons learned from this incident included:

- Automatic sprinklers with water flow alarms would have reduced or eliminated the loss of life at this incident.

- The open stairways created vertical passageways for the smoke to travel. Like the other fires, this event again demonstrated the need for enclosed stairs.

- The dangers of transoms over doors and combustible finishes were also identified.

Beverly Hills Supper Club (1977), Southgate, Kentucky

Thirty-five years after the Cocoanut Grove fire, another fire in a place of assembly took the lives of 162 people, injured 100 patrons, and caused injuries to firefighters. Factors contributing to the loss of life in this event were very similar to those in the Cocoanut Grove fire. Sadly, lessons learned several times over had yet to be heeded.

The Beverly Hills Supper Club was an irregular-shaped building, approximately 240-feet by 260 feet. It had doubled in size after a fire destroyed the original building in 1970. The split-level building was divided into 18 large rooms with interconnecting corridors. It had no automatic sprinklers. The interior finishes along the walls and ceiling in many areas of the building were combustible.

According to reports, at the time of the fire an estimated 2,800 persons were present. Approximately 1,200 to 1,300 patrons were in the large Cabaret Room—three times the safe occupant load.[8] An insufficient number of exits were available in some areas of the building, and evidence indicated that some of the available exits were obstructed or locked. Moreover, some of the available exits were not clearly identified.

The fire possibly started from an electrical short in a plastered wall. Investigators believed it burned for some time before an employee discovered it, and an attempt to extinguish the fire delayed alerting the other occupants. When firefighters arrived, they found light smoke showing. They had difficulty entering portions of the building to conduct suppression and rescue operations because bodies near the exits blocked the doors. Many of the victims perished from smoke inhalation because they could not evacuate the building. During the incident, the walls eventually collapsed.

The lessons learned from this fire are very similar to the others we have discussed:

- Automatic sprinklers would have reduced or eliminated the loss of life from fire.

- Exits need to be clearly marked and readily accessible to patrons.

- Exits must be free from obstruction or special locking arrangements.
- Combustible interior finishes contributed to fire spread and the rapid development of smoke.

MGM Grand Hotel (1980), Las Vegas, Nevada

In 1980 a fire took place in Las Vegas, Nevada, at the MGM Grand, a 21-story high-rise hotel occupancy that also contained a large casino and convention facility. The building was only partially protected by sprinklers. The fire claimed 58 lives and caused 600 injuries to patrons and 35 injuries to fire fighters.

The fire is believed to have originated in the hotel's deli area.[9] The flames rolled out of the deli and continued inside a dropped ceiling into the casino area. The smoke rapidly spread to the high-rise portion of the complex, traveling through stairways, seismic joints, elevator shafts, and the building's air handling system. Heavy smoke obscured the exits. During the event, the high-rise evacuation signals did not sound. Victims were found trapped in stairwells, corridors, and even in their guest rooms.

The lessons learned included:

- A need for complete automatic sprinkler protection throughout and adequate fire separation construction in vertical openings, especially stairwells.
- The occupants' need for a means to leave the stairwell once they entered the stair enclosure. This was a significant finding at this event, and today the model building codes as well as the *Life Safety Code*® address requirements for reentry from stairwells.

Happy Land Social Club (1990), Bronx, New York

A decade after the fire at the MGM Grand, a 1990 fire at a popular nightclub in the Bronx claimed the lives of 87 people. The building measured only 24-feet by 60-feet, typical of row-type occupancies found in the Bronx. A second-floor addition had been built on the original structure. When the addition took place, sprinkler protection was not extended to all areas. The building had two unenclosed stairways and combustible paneling along the interior walls.

The exit door contained a rolling security door. At the time of the fire, the rolling door was down and the exit was not readily apparent to the occupants. Other egress doors were locked and all of the exit doors had deadbolts.

Unlike the other case-study fires, this one was set intentionally. It originated in the main entrance area to the facility. An accelerant was poured and ignited. Upon the fire department's arrival, the flames were venting out the front doors. Although fire damage was limited to an area near the front door, the occupants could not escape because their only means of egress, the main entrance, was blocked by fire. Bodies were found near the entrance and on the second floor, where the smoke had traveled through the open stair to the second floor. The lessons learned from this fire were:

- A renewed emphasis on the need for complete automatic sprinklers
- The critical importance of having at least two available exits

Food Processing Plant (1991), Hamlet, North Carolina

Even though similar factors had contributed to previous large life-loss fires in the United States, the lessons learned did not prevent another industrial fire. Unlike the Triangle Shirtwaist Fire in 1911, the Hamlet, North Carolina, food processing plant fire in 1991 occurred when technology was on the rise and many years after the establishment of fire codes. The fire spread rapidly and killed 25 people. Although the number of deaths is not as staggering as in the previous case studies, the event is relevant nonetheless because the factors contributing to the large loss of life were so similar to those in earlier years.

The food processing plant was a one-story, 33,000-square-foot building without automatic fire suppression. There had been several additions to the building. According to reports, when the fire started at 8:15 A.M., 90 employees were present.

The fire started when a hydraulic line ruptured and was ignited by a cooker. The burning hydraulic fluid sent dense smoke throughout the building obscuring visibility within minutes of the fire. The occupants tried to evacuate but encountered difficulty.

The fire investigation revealed that exterior exit doors were inoperable. Some were bolted on the outside, which prevented the occupants' escape. One of the primary means of escape was through a loading dock, but it was blocked by a semi-tractor trailer. Some employees were forced to take refuge in large coolers.

This event was significant not only to fire officials but also gained the attention of the United States Department of Labor. Shortly after the incident, the Department of Labor's Occupational Health and Safety Administration (OSHA) issued statements regarding the need to maintain a work environment safe from fire. The director of OSHA urged all employers to survey their workplace to ensure the availability of adequate exits as well as extinguishers and an evacuation plan.

The contributing factors in the food processing plant fire deaths were identical to those in the industrial fire that led to the development of the *Life Safety Code*®. The lack of sufficient available exits combined with the absence of automatic sprinklers and an evacuation plan led to the deaths.

Station Nightclub (2003), West Warwick, Rhode Island

In West Warwick, Rhode Island, on February 23, 2003, a live band was beginning their performance with pyrotechnics. The pyrotechnics ignited combustible sound dampening foam, which had been installed on the walls and ceiling. The foam ignited and the fire rapidly grew, traveling up the walls and across the ceiling of the unsprinklered night club. Occupants began to evacuate even before the fire alarm was sounded; however, the vast majority migrated to the main front entrance rather than using the alternate and remote exits provided. Consequently, a funnel effect ensued as a result of the mass of occupants being slowed down trying to get through the front door. Occupants quickly became overwhelmed by the smoke and heat and had great difficulty escaping. The failure to evacuate quickly resulted in the deaths of 100 building occupants.[10] The National Institute of Standards and Technology Report states the following:[10]

Ten recommendations to improve model building and fire codes, standards and practices (as they existed in February 2003) resulted from the investigation, including (i) urging state and local jurisdictions to (a) adopt and update building and fire codes covering nightclubs based on one of the model codes and (b) enforce those codes aggressively; (ii) strengthening the requirements for the installation of automatic fire sprinklers; (iii) increasing the factor of safety on the time for occupants to egress; (iv) tightening the restriction on the use of flexible polyurethane foam -- and other materials that ignite as easily and propagate flames as rapidly as non-fire retarded foam -- as an interior finish product; (v) further limiting the use of pyrotechnics; and (vi) conducting research in specific areas to underpin the recommended changes.

Cook County Building (2003), Chicago, Illinois

A fire on the 12th floor of the Cook County Administration Building occurred on October 17, 2003. Building occupants were only notified of the fire through the emergency/voice evacuation system (Reference Note 16, Witt Report). Evacuating occupants encountered fire suppression activities on the 12th floor and were instructed to go back up the flight of stairs.[11] Occupants were unable to re-enter the floors because of locked stair doors and became trapped in the stairwell. Six individuals perished.[11] Significant findings of the incident revealed:

- Inconsistencies in building codes.
- The building was not equipped with an automatic fire sprinkler system.
- Better mitigation and preparedness actions were needed.

Where Is the Next Tragedy?

How can we prevent future tragedies? Should we only focus on the potential large loss of life tragedies, or should our efforts also include life loss in single family homes? The answer is complex, but achievable. Is there a problem with how we administer fire prevention programs? Yes, but we have made more progress in the last 10 years than in the 50 before that. The rest of this chapter explores what has been discussed and proposed to address our future needs.

Where Do We Go From Here?

"Where we go from here?" is a big question. It is a question that needs to be explored further by looking at the Wingspread Conference Reports on fire in America. The initial Wingspread Conference convened in 1966 in the Johnson, or Wingspread, House in Racine, Wisconsin. Five additional conferences have followed, each 10 years apart, with the most recent in 2016. Because of the rapid change taking place in the United States Fire and Emergency Services, the participants of Wingspread VI decided that the next conference will be at a 5-year interval instead of 10 years. A virtual "Who's Who in the American Fire Service" has attended all of these conferences.[12] The purpose of the Wingspread conferences is to discuss issues of emerging interest and importance and to reevaluate our current roles and responsibilities as a fire service. These discussions have continued to advance an awareness of the diverse interests of fire departments, both volunteer and paid. The Wingspread Conference reports provide insight on pressing issues. They are not an exclusive list but a good foundation for future thinking. The conference report not only identifies emerging issues, but includes an action plan to address them. For direct access to the electronic version of the Report, please use this link: http://thenfhc.org/resources/2016Wingspread.pdf

Six basic issues of national importance have emerged:

- *Customer Care.* The emergency services have an opportunity to increase their value at little cost by preparing the community to deal with natural disasters, fires, medical emergencies, and other incidents. These programs are not directly connected with emergency response or operations but in fact are more in line with prevention and mitigation functions such as public information and education.

- *Managed Care.* Escalating health care and insurance costs are driving more people than ever before to use 9-1-1 as their source of medical care. This discussion spawned ideas to radically change the delivery of emergency medical services. Again, this attempt at managed care is an opportunity for fire prevention bureaus or divisions to explore their potential in public education, community wellness awareness education and information, and making the right decision. There is not only a federal fiscal necessity, but a practical local need for working toward reducing the number of 9-1-1 emergency calls. There is a need to teach more people in our communities how to handle their own issues and not use emergency response equipment for issues which are not legitimate emergencies.

- *Competition and Marketing.* Wingspread's intent was to recognize that if the fire service is to survive, it must market its services and demonstrate its value. Professional responsibilities dictate that we should be capable of justifying and marketing our services competitively if we, in fact, need to exist. We cover this in detail in our Public Information Officer chapter.

- *Service Delivery.* This topic speaks to deployment, response times, and service-level objectives of the overall system. This book details risk management and planning, which is crucial to keep costs down and provide sufficient public protection. Increased prevention and planning functions can significantly lower the overhead costs of emergency response. This must be aggressively tackled if departments are going to survive in the future.

- *Wellness.* This is a two-pronged issue with relevance on two fronts. First, it relates to fire service employees (firefighters). This issue addresses the need for fire service employees to make sure they are physically fit, mentally prepared, and emotionally healthy. To provide proper support and service, we must be at our best. The second front has to do with community wellness and education. Regardless of how we slice it, many communities utilize emergency response as a means of addressing its health care needs. This requires correction, as the cost of this type of service to communities is very expensive.

- *Political Realities.* This topic stresses the overall importance of good labor/management and customer relations to maximize our overall impact on our communities. Politics is a reality, and the fire service must be better prepared to deal with it.

The conference also discussed seven "Ongoing Issues of National Importance." These discussions include the following:

- **Leadership.** Expressed a critical need for leadership development to move the fire service into the future, particularly with our dynamic environments and evolving political and fiscal challenges. Again, strong fire prevention and mitigation offers a good venue for strong leaders to emerge by providing high-level protection and service at minimal cost.

- **Prevention and Public Education.** Emphasized the need to expand this resource which, in many locations, has been severely reduced or decimated due to budget shortfalls and fiscal stressors.

- **Training and Education.** Addresses managers increasing their professional and leadership roles to remain credible to policy makers, administration, staff, and the public. The participants expressed the need to accomplish this through nationally recognized standards and certifications.

- **Fire and Life Safety Systems.** Highlights the need for adopting and supporting more codes and standards that mandate these protective systems' use. Again, this emphasizes the important need for fire prevention and mitigation program support.

- **Strategic Partnerships.** Participants explored the need for the fire service to reach out and enlist the support of other individuals and groups in accomplishing the overall mission of fire protection and emergency service response. Fire prevention bureaus are a good connection point because of the need to involve so many different organizations. We discuss this in several chapters in this book and cannot overemphasize its importance. Leveraging as many resources as possible is the most efficient and productive way to reach economies of scale.

- **Data.** As discussed in this chapter, measurable data is crucial to understanding where we have been and where we are going. Fire prevention bureaus and divisions are major players in this role. Quality data and improved analysis are a must.

- **Environmental Issues.** These issues concern the need for the fire service to comply with local and federal laws in both mitigating incidents and providing for the safety and welfare of our employees and partners.

The Wingspread Conference reports provide insight on pressing issues. They are certainly not an exclusive list but a good foundation for future thinking. Ironically, where we are is largely where we have already been.

Vision 20/20

Under the auspices of the Institution of Fire Engineers (U.S. Branch), a group of fire service and related professionals joined together to conduct a national strategic planning process.

The result was the Vision 20/20 project. The preparation for a national strategic planning process began in August of 2007. A web forum was developed and conducted to place the background fire loss information before as wide an audience as possible.

More than 500 people participated in the web forum at 13 different satellite locations around the nation, sponsored by various fire prevention associations.

The major strategies of the plan are:

- Increase advocacy for fire prevention
- Conduct a national fire safety education/social marketing
- Raise the importance of fire prevention within the fire service
- Promote technology to enhance fire and life safety
- Refine and improve the application of codes and standards that enhance public and firefighter safety and preserve community assets

Vision 20/20 continues today. Please visit http://www.strategicfire.org for additional details of the work of the committee and a copy of the final Vision 20/20 Strategic Plan.

Greater Emphasis on Fire Prevention

Drawing attention to emergency operations and response programs is much easier because the Big Red Truck (BRT) is far more visible and evident, suggesting that equipment and methods of fire protection are the only method. This was most prevalent in our country more than ever after the tragedies of September 11, 2001. A great deal of attention is centered on the fire service response, but little to fire prevention. This became more of a harsh reality during the recent recession. Many fire departments were faced with significant budget deficits resulting in the need to eliminate services. Fire prevention cuts and even elimination of the service in its entirety is not uncommon during tough economic conditions. We have made progress as a service. We have decreased our fire incidents and are able to focus on other services and prevention efforts. We now tend to move toward customer service and firefighter safety with health initiatives becoming more prevalent in our industry. However, while progress is being made we, as a fire service, still have not embraced fire prevention as one of our most critical services. These attitudes, beliefs, and group norms are unfortunately the culture of many fire departments. Fire departments ranging from large suburban departments to volunteer fire departments tend to view fire prevention as "busy" work performed by another entity within the fire department. The need for fire prevention is even more critical during tough economic conditions when many departments have reduced their fire suppression staffing level. Preventing an incident from occurring is far more economical than the resources required for mitigation of the incident. How can we change these beliefs and norms and is it really necessary to do so?

Many fire suppression personnel see the value of fire prevention efforts. However, do they see the value of fire prevention just as important as the duties they perform? Many will argue that they did not join the fire service for fire prevention… they want to fight fires or be a medic. As a profession, we need to begin a cultural shift to educate the fire service how fire prevention efforts benefit them and their work environment directly. Fire departments need to incorporate fire prevention efforts as part of the job descriptions for fire suppression personnel. We do not have to wait until the professional qualification standards force us to address this. Each fire department has the opportunity to modify their respective job descriptions to address this. These efforts need to then be linked to the performance reviews of the firefighter. We need to foster a culture that incorporates fire prevention as our premier effort.

Modifying an organization's culture is not easy, let alone modifying an entire entity such as the fire service. However, this is not an impossible task. The first step begins with the chief of the organization, who can foster an environment that rewards fire prevention efforts. The organization can enable a culture where fire suppression personnel understand the role of fire prevention and understand how fire prevention benefits them and their fellow firefighters. One of the key starting points of fire prevention education training is at the recruit level. We do not spend enough time teaching recruits the benefits and importance of fire prevention. Sure, we teach recruits the basics of fire sprinklers, fire alarms, and standpipes, but do we address a total fire prevention program to include fire and life safety education, fire inspections, construction document review, fire suppression system maintenance, etc.? The latest edition of IFSTA's *Essentials of Fire Fighting* manual contains information on basic fire and life safety education techniques for firefighters. The bonus instruction CD that accompanies each manual provides a video of how to conduct a basic classroom presentation for fire safety and conduct a station tour. Using this material in recruit training academies may begin the culture shift.

One of the most often overlooked links to a prevention program is the role of the company officer. This individual has the opportunity to lead a team of trained individuals to mitigate any unforeseen emergency they encounter. He or she also has the opportunity to harness this team's skill to enhance the fire departments fire prevention efforts. An effective fire prevention program not only serves the citizens but also serves fire suppression personnel. Fire prevention and mitigation efforts equate to firefighter safety. We need to ensure that fire officers in a leadership role understand that fire prevention efforts positively impact our fire department emergency operations.

Many fire service professionals now refer to fire prevention efforts as a means to "shape our battlefield." The actions we take prior to the incident not only affect the occupants of the structure, but also affect the ability or inability of emergency response personnel to function at their best. The early efforts of fire prevention and mitigation during the construction and ongoing maintenance of a building will help to "shape the battlefield" where incidents do occur. When suppression personnel are called to a fire in commercial structure at 0300, where is the hydrant? What types of hazards are inside, and is there an available water supply? Where is the lock box? Where is the fire department connection? Where is the fire alarm control panel? This is our battlefield! The answers to these questions are "it depends!"………In fact, one could say it depends on your fire prevention efforts and many times the partnership and collaboration with suppression personnel. The text, *Fire Prevention Applications for the Company Officer* (available at Fire Protection Publications, www.IFSTA.org), is a comprehensive approach to educate and integrate fire prevention at the company officer level. Many fire departments have identified the critical role of fire prevention and the company officer and are incorporating a level of fire prevention knowledge into officer promotional exams.

This book focuses on fire prevention at a company officer level. Throughout the book you will see where fire prevention impacts fire department operations at a company level.

Even if your fire department has fostered a culture of fire prevention, it can still be difficult to have individuals willing to move from fire suppression tasks to fire prevention work or to hold onto the philosophy of fire prevention. Fire chiefs and city managers can help address this by having a clear and defined career path for fire prevention personnel regardless of their status as sworn or civilian. During recent economic conditions many city managers reduced costs by targeting fire prevention functions during budget cuts or consolidated their efforts with other municipal entities such as building or planning departments. Tragically, as a result, many fire departments no longer have fire prevention functions as part of their service or have a "bare bones" staff. It may be ineffective for the delivery of fire department services to have fire prevention functions just transferred to another municipal department without consideration of the overall long-term impact.

Given the mission and role of the department, if the value and importance of the work were understood, why were fire prevention functions removed from the fire department? However, we recognize the stress placed on municipal or organizational leadership and understand the reasoning for examining any type of combination that may seem to work. Consolidation with building departments seems to be most common and while it can certainly work, we have seen significant degradation in some locations and great success stories in others. The success or demise of these situations rests upon the organizational leadership and the ability of staff to be good followers.

Some practical difficulties exist in any of these permutations, but as long as the target is highlighted and focus is provided on the critical parts of the mission, it should work. We should try to stay the course and keep fire prevention as a principal part of our service and move forward to making fire prevention as much of our cultural philosophy as fire suppression and EMS. We regularly tout we are professionals, and yet we frequently act in a very contrary way.

If we truly are consummate professionals in preventing and mitigating injury and loss, then why do we typically place over 95% of our resources on a reactive response program as opposed to a proactive and preventative program? Big Red responds only after injury or loss has occurred. Once we arrive, we are well behind the power curve. True professionals should work to prevent the injury or loss or at least keep what might occur to the bare minimum. Operational response should be a last-ditch effort, yet most policy makers view it as the principal method. People are enthralled with "Big Red" as it comes screaming and honking down the street, while they think of the fire inspector visiting a business merely as a thorn in someone's side, interrupting their busy day. Is it not ironic that one can become a hero responding to a tragedy as it is unfolding, but an inspector who visits a business to prevent or greatly mitigate those tragedies is frequently viewed as a Code Cop? Fire service professionals must do more to shift the paradigm from overwhelmingly emphasizing response functions to stressing more proactive and efficient prevention functions.

Better Training and Education

Shortly after the publication of *America Burning,* the fire service was examined in detail. Considerable new research asked questions like "What other ways can we do things?" These studies identified planning as a major element of fire service organizational management. Attempts to finally identify and attack the "real" target became a focus for the fire service. Drawing much attention was the need for a National Fire Academy to not only teach fire suppression and mitigation techniques, but also increase administrators' competencies in management and planning. The subsequent creation of this academy has fostered great effort to address the educational issues.

While the National Fire Academy's existence has helped fire prevention efforts tremendously, ironically it has at the same time compounded the problem. By its very nature, the fire service's mission is extremely broad. Although the mission varies from municipality to municipality or county to county, it typically involves providing a wide range of services to the people it serves. These services can include EMS, fire fighting, Urban Search and Rescue (USAR), emergency management, wildland fire fighting, hazardous materials response, large scale incident management, and various recovery tasks. Added responsibilities result in the need to train firefighters to perform the tasks involved. Many departments have mandatory training requirements, allotting time for each shift. These requirements must be met between running calls, which is the main reason for their existence and performing numerous other duties.

Fire prevention service is just as important, and fire departments must also factor in technical training for those functions. In reality, it becomes very difficult for a fire protection professional to be really good at so many different things. Fire prevention is our most important responsibility and must be a significant consideration in the fire service's overall mission; however, the methods of addressing this are complex.

Smaller departments can benefit significantly from the use of line personnel in performing inspection activities, but a large municipal department's much more extensive staff allows it to have more technically experienced personnel do proactive and aggressive fire prevention work. As we discussed earlier, the Company Officer has the greatest opportunity to continually positively impact fire prevention efforts. Chapter 5 examines the staffing and structuring of fire prevention bureaus to accomplish these complicated tasks in more detail.

Fire and Life Safety Education

From the comfort of your living room or office, you can watch tragic events unfold in real time halfway across the globe. This information transfer is marvelous in one respect; however, some suggest that it numbs people to reality. So, while we can transmit photos, stories, interviews, and sometimes very graphic photos of tragic fire events, many people see them no differently than the made-for-television movie they watched the night before. They understand it happened and can recite many details, but the "realness," the significance of the event to their life, seems too removed to be credible.

So then, education messages must be interactive and animated to gain the attention of the public. It must be timed appropriately, and it must match the immediate needs and values of the individuals we are trying to reach. Our education efforts must better match the mainstream of communication, as it exists today. This should include social media including Facebook, Twitter, blogs, and, to a lesser extent, e-mail, television spots, highway traffic message signs, and other means **(Figure 2.4)**. Chapter 7 explores effective fire and life safety education techniques in greater detail.

Figure 2.4 In a means to reach all audiences, our education efforts range from social media to old fashion sign boards and banners.

Neglected Areas of Research

Fire service research has actually been improving at an astonishing rate. For many years, the fire service did not do much research. Among the various reasons for this was the lack of funding as well as a lack of an academic focus. Technological limitations also played an important role as laborious and detailed computations could not be performed timely enough to be effective. With the advent of the personal computer and the improvement of minicomputers and high-end computers, such as the Cray Supercomputer, our ability to perform research has increased exponentially, particularly over the last ten years. Many of us can remember running hydraulic programs to calculate sprinkler designs on small PCs that had to run all night; then, right when they were about to finish the iterations, they would crash due to insufficient memory. We have come a long way, in part, thanks to the Internet, which has enabled us to share more information faster and obtain state-of-the-art research tools directly. Research needs to continue, but the fire service has made great strides and will only continue to improve.

An additional finding in the commission's report was the fire service's inability to gather information and correlate nationwide trends or results. As late as the 1970s the fire service was ill equipped to handle this problem. "Time and time again—in listening to testimony, in studying the fire problem, in searching for solutions—this Commission found an appalling gap in data and information that effectively separated us from sure knowledge of various aspects of the fire problem."[14] Fire departments in large municipalities were addressing their local fire problems well; however, many were blind to trends prevalent in adjoining communities or nationwide because there was no good clearinghouse for sharing and comparing information. This was a significant and tragic lack of communication that, in effect, prevented any holistic fire prevention efforts for a number of years. Good answer!

Research Organizations

Today, a number of organizations at the local, state, and federal levels can assist fire prevention bureaus. The number of organizations is too large to discuss each in great detail, so we can note only a few here. The *Fire Protection Handbook®*, 20th Edition, published by the National Fire Protection Association, provides a detailed listing of organizations with fire protection interests. Listed in Appendix A, page 361 are a number of Internet sources that may also provide additional organizational Web links.

Private Organizations

National Fire Protection Association®. The National Fire Protection Association (NFPA®), based in Quincy, Massachusetts, promulgates fire codes and standards. Currently the NFPA publishes more than 210 codes and standards. These documents make a significant impact on all aspects of fire protection because they are considered the "standard of good practice" that forms a basis for legislation at all levels of the government, from the local to the federal.

NFPA® is an independent, voluntary, and nonprofit organization that has over 60,000 members from industry, fire departments, architecture and engineering firms, and others. The revenues generated by their publications, membership dues, and seminars support the organization. The activities of NFPA are either technical or educational. An excellent resource for fire protection professionals is NFPA's one-stop data shop. They can provide summaries of incidents relating to a specific occupancy or type of fire.

Insurance Organizations. The Factory Mutual System, based in Norwood, Massachusetts, is well known for its loss prevention engineering, research, and training expertise. It can provide information on property loss prevention worldwide. It also provides technical reference manuals or *Factory Mutual Global Loss Prevention Data Books* that fire departments can obtain as a useful resource. These books not only provide Factory Mutual's recommended fire safety practices, but also provide information on loss history that can be invaluable in making various risk management decisions.

Insurance Services Office, Inc. (ISO), provides a municipal grading service, based on their ability to perform. Fire departments are assigned a numerical grade of 1 to 10, with 10 as the lowest. The municipality's insurance premiums are then based on the fire department's grade.

Fire Testing Laboratories. Underwriters Laboratories, Inc. (UL), based in North Brook Illinois, is a not-for-profit organization whose purpose is to promote safety through scientific investigation, testing, and study of various materials and products. Fire protection professionals rely on UL and similar organizations to test fire protection equipment, such as fire extinguishers, sprinklers, and fire alarm components. After equipment has been tested according to UL, it will bear the UL label. Many times, fire protection professionals verify through UL that a product is being installed or used in accordance with its UL listing.

Southwest Research Institute (SwRI) based in San Antonio Texas, is another non-profit organization devoted to government and industry. SwRI is divided into four sections focusing on different aspects of fire technology: Standard Testing Services, Fire Performance, Fire Chemistry, and the Applied Environmental Toxicology sections.

Another entity that performs testing and research is the Factory Mutual Research Corporation. This organization conducts research and development for property loss control and operates a third-party certification program commonly known as approvals. Similar to UL, Factory Mutual approves products and materials beneficial to loss prevention for their insured clients.

Professional Organizations. A number of professional organizations have an interest in fire protection, including the International Association of Fire Chiefs (IAFC), the International Association of Arson Investigators (IAAI), the International Association of Black Professional Fire Fighters (IABPFF), the International Association of Fire Fighters (IAFF), the Fire Marshals Association of North America (FMNA), Women in the Fire Service (WFS), and the Society of Fire Protection Engineers (SFPE). The many similar state or local organizations can be located by contacting your state fire marshal's office or other appropriate state agencies.

Many other professional organizations have a more specific mission within the overall interest of fire protection. An excellent resource for all fire protection professionals is the International Fire Service Training Association (IFSTA). The purpose of IFSTA is to validate fire service training materials for publication. These manuals are written by fire service and fire protection experts and reviewed by fire safety professionals from a variety of organizations. A unique validation process ensures the technical quality of this organization's products. Those interested in becoming a part of the IFSTA validation process can complete an application found in the back of any IFSTA training manual. Fire Protection Publications produces the IFSTA validated material and also serves as a means for conducting research.

Fire Protection Publications (FPP). Working to save lives and property is a core mission of Fire Protection Publications at Oklahoma State University. In recent years, FPP has also become a major hub in the university's research mission, as well. FPP established a Research Department and has received a significant amount of funding to research and find solutions to the most pressing fire and life safety problems. This includes research in fire and life safety education, firefighter safety, educational methods, and leadership and management issues. These projects are used for the benefit of the fire service community, as well as focusing on the needs of the public. IFSTA and FPP training materials use the latest research results from these projects in their products. Results of these funded research projects are in turn shared with the public at no charge. Additional information regarding the research endeavors at FPP can be found at http://info.ifsta.org/research.

Federal Fire Protection Organizations

United States Fire Administration. The United States Fire Administration (USFA) administers the federal data and analysis program and serves as the primary agency to coordinate arson control programs at the state and federal levels. This agency also administers a program concerned with firefighter health and safety.

National Fire Academy. Located in Emmitsburg, Maryland, the National Fire Academy is part of the Federal Emergency Management Agency's office of training and the National Emergency Training Center. The National Fire Academy provides training programs ranging from fire service management to fire prevention. It is an excellent resource for fire-prevention-related topics as well as an outstanding educational institution. All fire service professionals should take advantage of courses and services offered by the National Fire Academy.

United States Forest Service. The United States Forest Service (USFS) provides technical and financial assistance to state forestry organizations to improve fire protection efficiency. The USFS also provides fire protection for millions of acres of forests and grasslands.

Bureau of Alcohol, Tobacco, and Firearms. The Bureau of Alcohol, Tobacco, and Firearms (BATF) is a branch of the United States Department of the Treasury. The BATF conducts arson investigations and provides fire investigation training and technical assistance to local and state law enforcement agencies as well as those fire departments responsible for conducting fire investigations.

Consumer Product Safety Commission. Canada's Consumer Product Safety Commission (CPSC) collects data related to product failure and investigates injuries from products.

State and Local Organizations

State and local organizations vary. The most common are local insurance providers, civic groups, and even large corporations. State fire marshal offices are located throughout the United States, with the exception of Colorado and Hawaii. Fire marshal functions vary from state to state but can include:

- Code enforcement
- Fire and arson investigation
- Plan review
- Inspections
- Fire data collection
- Fire data analysis
- Fire legislation development
- Public education
- Fire service training
- Licensing

Every fire protection professional should know what fire prevention related services are provided at the state level.

State functions differ from those of local fire prevention offices since they must take care of the whole state. Typically state functions cover those parts of the state that are unincorporated or outside home-rule cities. For example, training may be provided for small outlying fire districts or departments that cannot afford their own training division or sections. The state facilitates this by providing resident and other courses throughout their state. Also, jurisdictions that do not have fire investigators typically can call upon the state to assist with or conduct an investigation, particularly if arson is suspected.

The staffs at state agencies generally are limited and their workload is fairly intense. Like local jurisdictions, they must wade through the political peaks and valleys to obtain funding and resources; however, the problem of convincing legislatures most of the time is far more difficult and trying than having to deal with local elected officials or policy makers.

Summary

The concepts of stricter fire and life safety codes are not new. They have been in place in some form or another since ancient times. The codes in place today have evolved as a result of many fire tragedies. It is important for the fire protection profession to understand the relevance of historical fire events. This lays the groundwork for how we got where we are today in the field of fire prevention. Many situations in fire prevention professionals' careers will require them to explain "Why do we need to do that?" or "What is the purpose or intent of this code requirement?" If they understand what led to the fundamental requirements in most model codes, they will be capable of explaining the purpose of or reasoning for code requirements.

The document *America Burning* and national conferences such as Wingspread have identified the fire problem in the United States and proposed alternatives to address our fire problem. Many issues identified over 20 years ago still need attention. Today, Vision 20/20 is taking the lead to develop and implement a National Strategic Plan for Loss Prevention. Today's fire protection professionals face the challenge to move forward in addressing the country's fire problem and ensure we do not replicate the conditions that led to the tragic fires identified in this chapter. Because of economic conditions, many fire departments no longer have the robust fire prevention staff they once did. In some jurisdictions, the fire prevention bureau has either been eliminated or the responsibilities transferred to another entity within the jurisdiction.

The company officer has the opportunity to harness this team's skill to enhance the fire departments fire prevention efforts. An effective fire prevention program not only serves the citizens but also serves fire suppression personnel. Fire prevention and mitigation efforts equate to firefighter safety. We need to ensure that fire officers in a leadership role understand that fire prevention efforts positively impact our fire department emergency operations as a means to "shape our battlefield."

Case Study

This chapter's case study is an example of a common event in many jurisdictions. The best course of action would have been to adopt codes and ordinances to reduce the potential for loss of life caused by the fire. The tragic historical fires discussed in this chapter had similar outcomes. The lessons learned from them need to be applied locally. For example, lacking automatic sprinklers and locking stairwell doors trapped the occupants. Applying historic events can be used as a proactive code-adoption approach as well as an after-the-incident approach. The key is to have an understanding of how to apply the lessons learned from other historic tragedies to address your fire problem. Even a tragic event that gains national media attention can be used as part of an educational tool at the local level. The public will have an interest in the story. Having a local authority explain the importance of knowing where your exits are and being safe outside of your home will grab the reader's attention.

Chapter 2 Review Exercises

2.1   How does learning about historical fires help an inspector do his or her job better? _____

_____

_____

2.2   How do tragic fires in the past influence today's fire prevention efforts? _____

_____

_____

2.3   Identify some of the earliest established fire prevention rules?

_____

_____

_____

2.4   What are the lessons learned from the Iroquois Theater fire in 1903?

_____

_____

_____

2.5   What significant finding of the Triangle Shirtwaist Fire has become one of the principal exiting requirements in all of today's model building codes? _____

_____

_____

2.6   What national fire protection publication was produced as a result of the Triangle Shirtwaist Fire? What was this publication later called?

_____

_____

_____

2.7   What lessons were learned from Our Lady of Angels Fire in Chicago?

_____

_____

_____

2.8   What factors led to significant loss of life in the Beverly Hills Supper Club Fire? _____

_____

_____

2.9   Compare and contrast the Triangle Shirtwaist Fire with the Food Processing Plant Fire in Hamlet, North Carolina. _____

_____

_____

2.10   Explain project Vision 20/20. _____

_____

_____

2.11   Identify the major strategies of Vision 20/20. _____

_____

_____

2.12   What is *America Burning*?_____

_____

_____

2.13   Name five key issues identified in *America Burning* and explain why each is important. _____

_____

_____

_____

_____

_____

2.14 What is the Standard Test Fire and what is its use today?

_____

_____

_____

2.15 What are some important areas of research that are being neglected?

_____

_____

2.16 What is the NFPA®? _____

_____

_____

2.17 Name two fire-testing laboratories.
1. _____
2. _____

2.18 Name five functions that state fire marshal offices can provide.

_____

_____

_____

_____

_____

2.19 Discuss how the United States Fire Administration works to better fire protection in America. _____

_____

_____

_____

_____

2.20 Discuss how all tragic fires are related and how the outcomes could be significantly altered. _____

_____

_____

_____

_____

_____

Notes

1. Pages from the Past, "Theater Was 'Fireproof' Like a Stove but 602 Persons Lost Their Lives," Fire Engineering (August 1977).

2. Ibid.

3. Paul E. Teague, "Case Histories: Fires Influencing the Life Safety Code," in Ron Cote, Life Safety Code Handbook (Quincy, Mass.: National Fire Protection Association, 2000), pp. 931–933.

4. Ibid.

5. Ibid.

6. Pages from the Past, "Flammable Decorations, Lack of Exits Create Tragedy at Coconut Grove," Fire Engineering (August 1977).

7. Chester Babcock and Rexford Wilson, "The Chicago School Fire," NFPA Quarterly (January 1959).

8. Richard L. Best, "Tragedy in Kentucky," Fire Journal (January 1978).

9. Richard L. Best, Investigation Report on the MGM Grand Hotel Fire (Quincy, Mass.: National Fire Protection Association, 1982).

10. Report of the Technical Investigation of The Station Nightclub Fire, William Grosshandler, Nelson Bryner ,Daniel Madrzykowski ,Kenneth Kuntz (National Institutes for Science and Technology, 2005)

11. Cook County Administration Fire Review, James Lee Witt and Associations, October 1, 2004

12. International Association of Fire Chiefs, et al., The Fire and Emergency Services in the United States, Wingspread IV, October 23–25, 1996.

13. Ibid.

14. NCFPC, America Burning: Report of the U.S. National Commission on Fire Prevention and Control (Washington, D.C.: U.S. Government Printing Office, 1973).

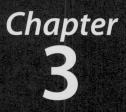

Chapter
3

Codes and Standards

Table of Contents

Key Terms

Learning Objectives

1. Identify laws, codes, ordinances, and regulations as they relate to fire prevention.*

2. Understand code enforcement as it impacts life and property loss.*

3. Define laws, rules, regulations, and codes and identify those relevant to fire prevention of the authority having jurisdiction.*

4. Identify and describe the standards for professional qualifications for Fire Marshal, Plans Examiner, Fire Inspector, Fire and Life Safety Educator, and Fire investigator.*

5. Identify various classifications of building construction.*

6. Understand theoretical concepts of how fire impacts major types of building construction.*

7. Describe building construction as it relates to firefighter safety, building codes, fire prevention, code inspection, firefighting strategy, and tactics.*

8. Explain the different loads and stresses that are placed on a building and their interrelationships.*

9. Classify occupancy designations of the building code.*

10. Understand the difference between a code and a standard.

11. Identify public, federal, state, and private organizations related to fire prevention.

12. Understand performance-based designs and their impact on fire departments.

FESHE Objectives (USFA)

Codes and Standards

Case Study

As a new fire marshal, one of your more experienced fire inspectors is called to a local business, which is housed in a building that was built around the turn of the century. The business is a pub and has been in operation since before 1940. The inspector observed an occupant load of around 320, no sprinkler system, and a kitchen with an old, dry chemical extinguishing system over the cooking equipment. The pub also has an exit through the back of the pub, through the kitchen, and discharges into the alley. The inspector reports back to you that they were cited for the following:

1. Not having a sprinkler system

2. The need for a separate exit, not from the kitchen

3. That the occupant load must be decreased below 50

What do you think about these written comments? What do you suppose the business owner will say about these requirements? What issues would you consider to be immediately dangerous to life safety? Explain some different ways how you could handle this situation.

What Are Codes?

Building and fire codes have existed a long time. Although the early codes enforcement may have been a little stricter by penalty of noncompliance, their detail was not nearly as comprehensive as codes are today. The first well-known use of codes was recorded about 1700 B.C. "King Hammurabi established a law by which a builder could be executed if the house he built collapsed, resulting in the death of the owner." [Section 5, Chapter 13, Revised by Richard E. Stevens, NFPA® *Fire Protection Handbook*, 15th Edition, National Fire Protection Association, Quincy, MA, 1981].

When we use the term **code**, we are referring to a body of law systematically arranged regarding fire issues, to define requirements pertaining to the safety of the general public from fire and similar emergencies. The purpose of codes is to establish *minimum* requirements for fire and life safety. Fire codes provide a means of reducing, controlling, or mitigating hazards by regulating design and construction methods, controlling ignition sources and fuel arrangement, and addressing behavior. Fire codes are a shared responsibility between the public and private sectors as well as the general public.

> **Code** — A body of law systematically arranged to define requirements pertaining to the safety of the general public from fire and other calamities.

The purpose of codes is to establish minimum requirements for life safety.

A major shift in technical advancement started around the turn of the 20th century. Unfortunately, code development during that period stemmed primarily from disasters that had already taken place. After the disasters, changes would be made to prevent or mitigate a reoccurrence of the same or similar situation. For example, after the great Chicago fire of 1871, Lloyd's of London, a large insurance underwriting firm, stopped writing policies in Chicago because buildings were constructed so poorly. Everyone was in such a hurry to rebuild; they often ignored the revised or existing codes and often scoffed at the regulations. "Some of the new buildings were falling down before they were finished." [page 137, Peter Charles Hoffer, Seven Fires, The Urban Infernos that Reshaped America, Public Affairs, Perseus Books Group, 11 Cambridge Center, Cambridge, MA 02142, 2006] Other insurance carriers in the area also had great difficulty selling policies because they had to charge such high rates to cover their forecasted future losses. Throughout the country, the reluctance to consider overall fire risk in building design has led to substantial **community injury**, such as significant loss of property or monetary value, physical injury, or death.

Community Injury — Any significant loss of property or monetary value, as well as physical injury or death.

Prior to the great earthquake in San Francisco, California, in April 1906, the National Board of Fire Underwriters (NBFU), now known as the American Insurance Association (AIA), published the first edition of its *Recommended Building Code*, which later would become the *National Building Code* (NBC)].[1] This document provided uniformity in specifications that reduced the spread of fire by requiring construction of what it termed Class A buildings. These were basically fire-resistive shells that could withstand substantial fire and exposure, likely losing contents but lending themselves to quick renovation and remodeling. These expectations were derived from the conflagration that followed the earthquake. In that fire, 452 people were killed and 28,000 buildings were lost. The few buildings in San Francisco that were designed to the *Recommended Building Code*'s Class A criteria remained mostly intact and could be returned to service relatively quickly.

After various significant fires, larger communities began adopting recommended codes and standards even if only embodied as fragments in local laws and regulations. Insurance companies that were hard pressed to continue paying large amounts in claims created and endorsed many of these regulations, while local community governments forced other requirements. Keep in mind that large companies or corporations that have substantial properties in a community wield a great deal of clout regarding fire protection for their locations and the surrounding areas. These big companies typically had property insurance carriers with a strong vested interest in protecting their losses and, consequently, their clients. Naturally, if a large tire plant was destroyed by fire, the loss would be difficult for a single insurance company to cover. To prevent significant incidents resulting in large loss, the insurers would exert pressure on their clients to use state of the art fire protection systems and administrative controls. In some cases, that pressure extended to the surrounding community, whose infrastructure may have needed improvement to provide water supplies, fire access, public fire protection, and other resources. In other instances, large insured's actually created their own internal or private fire departments or brigades to cover their risk because the surrounding communities did not have sufficient resources. Any number of these locations exist today, many of which provide supplemental protection as a service to their neighboring communities. The net result helped the entire community because everyone reaped the benefits of overall increased fire protection, not only from preventing the direct impact that a fire would have but also in avoiding the loss of jobs and economic instability that would have resulted.

Insurance Services Office

Another practice which significantly affected how fire protection was managed or mandated was **grading**, or **fire suppression rating schedules (FSRS)**. Slightly different than codes, these were basically methods by which the level of fire suppression capabilities was evaluated and credited to individual property fire insurance rates. This process was initially started by the NBFU and later adopted by the Insurance Services Office (ISO). Different revisions have been made over time, with the latest done over the last few years. The rating scales operate in much the same way with some different values that address more appropriately the way the fire service operates today. These schedules worked by providing an analysis of a given community's fire protection capabilities and grading them based on adopted standards.

Grading (Fire Suppression Rating Schedule) — Method of evaluating fire suppression capabilities and crediting them to individual property fire insurance rates.

The new FSR Schedule gives relative weights of:

Water Supply	40%
Fire Department	50%
Fire Alarm and Communication	10%

There is now a Community Risk Reduction section that has a weight of 5.5 points, which can result in 105.5 total points available. This new section now provides extra points to recognize communities that effectively and efficiently use and implement fire prevention practices, without penalizing departments that have not yet adopted or been able to implement these measures.

ISO sends representatives to evaluate the community based on these elements and then determines a grading number of Class 1 through Class 10. Class 1 receives the highest rate recognition, and Class 10 receives no recognition.[2] The assigned classification is factored into the underwriting of various properties. It is important to note that not all insurance companies use this schedule or grading, but it is still a common reference by which fire departments compare its level of service and protection to that of other agencies. Many experts believe that fire service accreditation will surpass the grading method as a means of properly evaluating a community's fire protection capabilities.

Many experts believe that fire service accreditation will surpass the grading method as a means of properly evaluating a community's fire protection capabilities.

Accreditation seems to provide a more accurate reflection of overall fire service delivery as it includes fire, medical, prevention, and other service standards such as response times. While the ISO rating can be an indicator of certain fire defense capabilities relative to standards, accreditation provides a more rigorous test of contemporary expectations and overall service delivery that incorporate standards as well as overall community loss-control needs.

What Are Standards?

When the terms *codes* and *standards* are used, some clarification is needed. Many people incorrectly interchange these terms, but there is a clear difference. Codes are documents that answer the questions who, what, when, and where concerning various requirements and their enforcement **(Figure 3.1)**. **Standards** dictate how something is to be done **(Figure 3.2)**. For example, the 2009 edition of the ICC *International Fire Code®* requires that automatic fire sprinklers be installed in apartment buildings that house over 16 units. Assuming a municipality has adopted this code into law, we can break it down for discussion to see:

Who required it?	The authority having jurisdiction (AHJ) by reason of adopting the code into law
What is required?	Automatic fire sprinklers
When is it required?	Any time there is a newly constructed group R fire area
Where is it required?	Throughout the apartment building

Figure 3.1 This shows a building code that, as stated, is different than a standard.

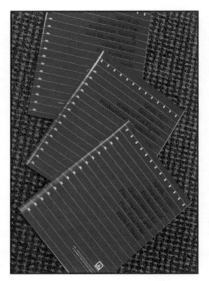

Figure 3.2 This shows copies of NFPA standards that are generally different than codes.

For guidance on how to install the automatic sprinklers, with slight exception, the *International Fire Code®* refers us to NFPA® 13, *Standard for the Installation of Sprinkler Systems* or NFPA® 13R, *Standard for the Installation of Sprinkler Systems in Low-Rise Residential Occupancies*. These standards then answer the final question:

How do we do it?	Standard NFPA® 13 or 13R

Codes are typically adopted by a jurisdiction (district, city, town, county, or state) and then incorporated into the law that governs that jurisdictional responsibility. Standards are then **adopted by reference** through the adopted code.

An easy way to remember the difference between codes and standards is that codes tell us *what* to do, and standards tell us *how* to do it. For example, throughout the 2009 *International Fire Code®* there are references to standards, which are located in Chapter 45. Looking at that chapter, you can see the large number of various standards that the code touches in one way or another. It codifies (arranges in a logical order) these individual standards, lists them specifically, and declares them to be part of the code. However, only specific sections of the listed standards are considered if the chapter is adopted as published. If an entire standard is desired to be adopted and used, jurisdictions must specifically cite or adopt the relevant or desired standards in total if they want to use and to list them in the code or in a separate ordinance or law. A jurisdiction should never assume specific documents are referenced or included when it adopts a model code. Again, only specifically referenced sections in the standards are listed, assuming that specific chapter is adopted and not omitted. The year of the standard is also important, since the requirements within the standard may change from edition to edition.

Adoption by Reference — A local jurisdiction's formal decision to follow state laws exactly as drawn.

Appendices

Another important aspect of code language involves appendices. Appendices in model codes or individual standards contain a great deal of information ranging from background material, history, simplified tables or interpretations, to good fire protection practices. Time after time, jurisdictions have tried to enforce comments or elements contained within the appendix of a various code or standard; however, they cannot do this legally unless the material is specifically cited, included, and adopted. We recommend including the entire appendix or appendices in any state or local ordinance that adopts a code. Additional confusion can enter in when code commentary is cited. Code commentary is very helpful for attempting to interpret or understand various code provisions, however, remember, it is not the code! It can assist you with your argument but should not necessarily crystallize your interpretational duties.

Federal Laws and Properties

Chapter 2 examined many federal agencies that have set forth regulations designed to ensure the safety of the public. These regulations cover a broad spectrum of activities and include such matters as employee safety, transportation of hazardous materials, patient safety in health care facilities, access issues for handicapped citizens, and minimum housing standards. The federal agency that sets the standards in any particular area is also typically responsible for enforcing them. However, in some cases, such as workplace safety laws, the state or county may choose to enforce federal regulations as opposed to having the federal agency do it.

Generally, federal laws can be enacted to provide: (1) that all state laws on the same subject are superseded by the federal law, (2) that state laws not conflicting with the federal law remain valid, or (3) that any state law will prevail if it is more stringent than the federal law.[3] Examples of federal agencies include the Occupational Safety and Health Administration (OSHA), the Department of Health and Human Services (HHS), and the Consumer Product Safety Commission (CPSC).

In most cases, the local fire department is not responsible for enforcing federal regulations; however, we affectionately term those instances when the department is required to enforce federal or state requirements as "Unfunded Mandates." In those instances where we do not have jurisdiction and we find hazards or violations, the department should know how and where to report and refer them to see that they are corrected.

In comparing how local, state, and federal fire prevention laws work, we will contrast their applicability in a make-believe city of 300,000 people. The City of Make-Believe Fire Department has fire code jurisdiction over everything within the city boundaries. This means that the department adopts its own local fire code and standards. It may either adopt one of the two model codes as is or adopt a model code with local amendments or requirements.

Now, the United States Postal Service decides to build a new post office in the middle of downtown. This facility must comply with federal fire code requirements, but it is not required to comply with the City of Make-Believe's local laws because federal property is exempt from local regulation. Out of courtesy, the federal government will likely ask the city organization for input on the plans, but the federal government has no obligation to comply. While the City of Make-Believe has no legal authority to force the federal government to comply, it does have some motivational alternatives to persuade a level of compliance, or at least cooperation, with local laws. Most municipal ordinances do not require the fire department to respond to fires. Typically the language is such that the department has the authority (the right to do so if it chooses) to respond, but it does this at the chief's discretion. So, an interesting discussion that could be had with the Post Office representatives would be to ask them, "who will provide fire protection for them should a fire occur?" The Post Office's answer likely would be, "The City of Make-Believe." To negotiate compliance, the city might respond with something like: "I'm sorry, you are a federal property, and, therefore, we have no legal responsibility to protect your installation. While we would like to provide that service as a good neighbor, there are some issues that must be addressed. For our crews to perform this task properly and safely, we would need you to follow the same codes and standards all of our community follows, so that we can provide the best service and protection possible." This may provide a little incentive for the Post Office to cooperate or at least enter into some discussion of collaboration. Is it "political blackmail?" Yes, but the issue is entirely political anyway. We do not recommend going to this extreme; however, it does provide an example of how jurisdictional discussions might play out among various entities and jurisdictions. We find it difficult to believe that there are jurisdictional or governing entities that are not concerned about the well-being of their citizens. The issues are typically financial. In the federal government's defense, if they had to comply with every single jurisdiction's specific requirements, the cost to all of us as taxpayers could be very significant.

When conflict arises, it usually involves determining what level of protection is "adequate." In utilizing a "systems approach," which is what we encourage throughout this book, this example shows how very important relationships and coalitions are to solving fire protection problems before they become exaggerated or emotional.

Territorial or jurisdictional differences do not exist solely to frustrate regulatory agencies or cause problems. Their intent is to allow certain levels of autonomy regarding various operations and responsibilities. Federal and typically state buildings within a local jurisdiction are not required to comply with local codes. In the past, the agencies that operated these buildings usually enforced their own fire protection regulations with the assistance of groups like the General Services Administration (GSA) in a federal situation or other state agency for state issues. In recent years, however, state or federal government has shown more willingness to follow local codes in resident community state or federal facilities. Relationships with the people responsible for extraterritorial properties should be fostered to ensure that fire protection is maintained at a high level, regardless of which code is followed. This not only benefits the public, but also enhances the safety of our fire suppression crews.

State Laws and Statutes

In addition to enforcing selected federal laws, states are empowered to enforce state laws and statutes. A state government may also regulate specific fire inspection or code activities within its jurisdiction. For example, some states may specifically assign the state fire marshal's office or another state entity to inspect nursing homes, schools, and daycare centers. The local authority may be prohibited from enforcing codes in these locations or be required to enforce a specific code or standard that can be different from what is locally adopted. Other states may assign particular jurisdictional powers to different agencies, such as a department of public safety or department of human services. States that have a state fire marshal typically maintain the responsibility for administration and enforcement of state laws that relate to fire, life safety, or training. This often includes the authority and responsibility to investigate fires and crimes of arson and to enforce fire codes and some building codes.

The variations of duties and responsibilities among state governments are great. Some states have adopted codes that all local jurisdictions must enforce at a minimum. Others have mandated that state codes be used, regardless of local desires. Many others have adopted some combination of the two, where state law will apply only if no other local regulation or home rule exists or if there is a conflict, the more stringent of the two laws shall apply.

State laws can also specify building construction and maintenance details in terms of fire protection and empower agencies to issue regulations. State labor laws, insurance laws, and health laws also have a bearing on fire safety and sometimes encompass fire inspection responsibilities.

Figure 3.3 This photo shows city codes or ordinances that cover all aspects of city governance, which includes adoption of fire codes and other laws.

Enabling Act — Method of adopting state regulations that allows the local jurisdiction to amend them based on local needs or preferences

Keep laws and ordinances current in order to meet growth and other changes in the community.

Maintenance Code — Code that details how to properly safeguard the activities or operations in a building.

Local Laws and Ordinances

Local laws and ordinances, although sometimes based on state laws, are more specific and tailored toward the exact needs of the county, municipality, or fire protection district **(Figure 3.3)**. Local codes should be developed to address the specifics of the community's fire problem. Typically, states allow local jurisdictions to adopt state regulations, either by reference or as enabling acts. To adopt by reference means that the local jurisdiction follows the state laws exactly as written. Adopting them as **enabling acts** allows the local jurisdiction to use state laws as a basis but then amend them based on local needs or preference. Specifically how this is accomplished will vary from community to community and from state to state.

General Principles of Fire Codes and Standards

Keep laws and ordinances current in order to meet growth and other changes in the community. Fire safety regulations generally fall into one of three categories:

- Those that govern the construction and occupancy of a building when it is being planned and constructed
- Those that regulate activities conducted within a building once it has been constructed
- Those that govern the maintenance of building components

Fire departments should be involved in adopting codes that address all three categories; however, their degree of involvement in each will vary across jurisdictions. Typically, the building code specifies how to construct a building to prevent the spread of fire by construction features or hazard arrangement, and fire prevention codes regulate the activities and operations in a finished structure. The fire code then, has historically been the code that details how to properly safeguard the activities or operations in the building, basically referred to as the **maintenance code**. While this particular argument is still made in certain circles, we believe it does not apply in that context today. Fire codes and standards are becoming extremely complex and technical with requirements that are far more relevant to the initial construction of facilities or processes rather than simply to their maintenance **(Figure 3.4)**.

Figure 3.4 It is more important than ever that fire prevention personnel and fire codes are integrated into new construction, not just maintenance inspections.

Tremendous amounts of information are being integrated into fire codes. This is a result of many factors:

- Highly skilled fire prevention staff (fire protection engineers)
- Improved technology
- Computer applications
- Web-based communication
- Mission focus
- Budget constraints

This improved availability of information has allowed a variety of information to find its way into fire codes and standards as never before. While fire codes of years past concentrated primarily on a building's use and occupancy classification, codes today are integrating more and more detail into sections on hazardous materials, manufacturing or system processes, system operations, and the like. This makes the fire code integral to the initial construction of a facility. A good example of this is how the building code identifies the requirements for the storage of flammable liquids through the establishment of Maximum Allowable Quantities (MAQs). Fire code requirements do not apply retroactively to every occupancy. If that were the case, every time a fire code edition was adopted, every building and occupancy would have to come into immediate compliance. If a building was currently operating as it was 50 years ago, then the code that was in effect 50 years ago would apply. Only instances of immediate or grave threats to life safety can be retroactively applied and that must meet the approval and direction of the fire chief. Typically, those retroactive requirements will be some type of derivation from the strict letter of the adopted code to apply to the risks actually present. This is an infrequent occurrence and must be handled with great care.

Appeals Procedures

Code requirements from most jurisdictions may have a method for them to be appealed. The process is fairly straightforward but must be clearly and specifically spelled out. Basically, if an adopted code or standard requirement is imposed upon an individual or company, the individual or company needs to comply with the regulation. However, in some instances the requirement may be viewed as overly restrictive or impractical, or there may be a difference of opinion on how the code was interpreted. Most permit the AHJ to allow certain exceptions to the specific codified requirements. Granting an exception generally requires proof of need, substantial evidence that the lack of compliance will not cause more harm than the original requirement, and possibly an alternate means to accomplish the original requirement's goal. The preferred method to address this issue is to use an equivalent means of compliance that clearly achieves the intent of the code requirement while providing an equal or greater level of protection for the occupants. For those instances where an alternate means or substantial justification for relief from the code cannot be provided, a variance through appeal is justified.

Individuals or businesses that feel they are incorrectly being required to do something should have the right to apply for a variance. To receive consideration for a code modification or variance, the applicant must generally make a formal written request to the AHJ within a specific time frame from receiving initial notification of the violation. This is typically viewed as seven to ten days; however, the time frame is strictly dependent upon the jurisdiction. Once the application for appeal is received by the department, it is placed on a meeting agenda. Most localities have posting requirements that are required to be followed, such as in the classified legal ads. This is typically followed, as these meetings are open to the public, and anyone who wishes to attend and speak for or against, or simply view the proceedings, must be notified of the event within a specified number of days. The board of appeals then hears the case, generally after an introduction of the issue by the fire department. The board may ask questions of the appellant, the department, or any other witnesses. Action on the floor follows standard meeting guidelines, such as Roberts Rules or similar standards. A motion is then made, and the item is voted on rending a decision to accept or deny the appeal. The appellant then receives a signed copy of the board's decision. There may be one additional appeal process for the appellant if they do not like the board's decision. This may go to a town or city council or even district court. Detailed records of the decision are generally kept available in the fire prevention office (**Figure 3.5**). Figure 3.5 shows an example of a request for the modification of a fire code.

Most codes establish an appeals procedure with a board of appeals or other body empowered to interpret the code and issue a ruling. The board of appeals usually consists of three to seven members who have previous experience in the fields of fire prevention or building construction. The exact number of members and their professional qualifications or areas of expertise are specified by the adopted code.

Fire inspectors should understand the appeals process and the workings of the board of appeals.

- Can the fire inspector continue to enforce codes on the property during the appeals process?
- Can a decision on behalf of the property owner affect the way the fire inspector enforces the code or ordinance for other properties in the future?
- Is further action required? An example might be asking the board to clarify whether it has granted a general variance or a one-time variance.

```
CITY OF_____
                    FIRE DEPARTMENT

        APPEAL for Modification of Fire Code

Address all communications to Department of Building

Meeting Date: _____

Petition must be on file at the office of the Board of Appeals _____
one week before meeting date.

Petitioner(s) must be present at meeting.

Approvals on proposed construction are null and void unless permit is obtained within 6 months.

                                        Legal
Address of Job _____  Description: Lot ____ Block ____ Tract ____
                No.        Street
Between Cross Streets _____ and _____

Owner's Name _____

Petitioner's Name _____

Petitioner is: _____ Owner _____ Contractor _____ Architect _____ Engineer

Address of Petitioner _____
                      No.                              Street

Phone No. _____ City _____ P.O. Zone _____
Status of Job: _____ Not Started _____ Under Construction _____ Finished

Permit No. _____ Plan Check No. _____
                                        (If permit has not been issued)
Specific ordinance modification desired: _____

_____

_____

(Additional sheets or data may be attached)
Date: _____ Owner: _____

(It is understood that only those points specifically mentioned are affected by action taken on this appeal.)

        Plot on Reverse Side Required by Yard Modification
```

Figure 3.5 Most jurisdictions have a form for citizens to request a modification of a code.

A one-time variance is binding only for the particular circumstance under review and may not be directly applied to other similar situations. When the board grants a general variance by reason of equivalency, the fire inspector will have to apply this ruling to all future code enforcements at the location. If the board of appeal rules that the code is too vague for enforcement, the fire inspector, or preferably the fire marshal, must take steps to ensure that the code is clarified.

As previously stated, the appeals process is not open-ended. Adopted code regulations usually specify a time limit within which the property owner must submit any appeals. Seven to ten days from the time of the inspection is common, but this varies in different jurisdictions. This requirement is frequently overlooked or omitted but is key to regulate and legitimize the appeal process. Rules and regulations used by the board during its hearing are generally made public. A schematic of a typical appeals process is provided in **Figure 3.6**.

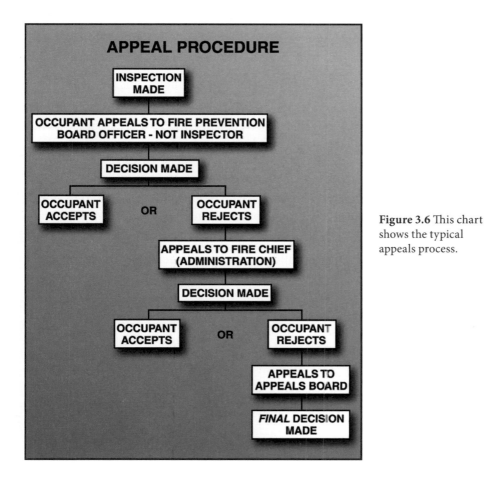

Figure 3.6 This chart shows the typical appeals process.

Building Codes

Initially, building codes were developed to prevent collapse or other failures of construction. Later, changes added protection from ravages by fire or other natural disasters. Today, building codes dictate how we construct buildings to resist wind, earthquakes, floods, fires, snow, and ice, as well as to avoid problems with sanitation, ventilation, loss of energy, and contamination. Not long ago, up to eighty percent of code content dealt with fire and life safety. Today, more than 50 percent of a building code usually refers in some way to fire protection.[4] Due to this large fire protection component in building codes, some communities have designated the fire department to manage the building department along with its usual fire prevention functions. Considering the intent of the building codes and their impact on a community's fire problem, more communities might be well served to consider this approach. This merger of building and fire prevention divisions is becoming more prevalent during tough economic conditions. Refer to Chapter 5 for further discussion of fire prevention bureau structures.

Building Departments

Local building departments or agencies typically enforce building codes that designate a building official as the manager. The building official is the AHJ with regard to building requirements, just as the fire marshal is the AHJ with regard to fire prevention requirements. Building departments are organized much like fire prevention divisions in that they likely staff the following positions:

- Plans examiners
- Inspectors
- Support staff (who issue permits, licenses, enumerations, etc.)

Depending upon the jurisdiction, building departments may be staffed with public works staff (engineers, inspectors, etc.), planning staff (planners, zoning inspectors, hearing officers, etc.), flood plain managers, and a host of others involved in the development and construction process. In some jurisdictions, these interrelated functions may all be housed in a central location or in different offices with plans and meetings moved from office to office. Each community generally tries to provide the most customer-oriented process and organization it can with regard to performing the various tasks efficiently.

More often than not, it is the building official who recommends which model code to use. The building official is responsible for reviewing and inspecting every construction-related plan that comes through his/her jurisdiction. For this reason, he or she must provide the best recommendation for code compliance, use, and consistency. While the building code is the principal code regarding construction, others (fire, mechanical, plumbing, etc.) are viewed as companion documents. To maintain consistency, the appropriate companion documents or codes should accompany the model code of choice. This provides the best overall fire protection and code management. Some jurisdictions do not have good relationships with their building departments and have adopted other codes in lieu of the companion documents referenced by the model building codes. While this can work, it gives rise to many conflicts and interpretational issues that confound designers and regulators alike, not to mention the plan reviewers and inspectors. We strongly recommend using companion documents for ease and consistency in the implementation and enforcement of fire prevention codes. As you go through this text and learn of the importance of building coalitions, you will see that no relationship is more critical than that between the fire official and the building official. We find that in many jurisdictions, fire marshals and building officials do not relate well. This is not a productive environment. These officials should make every effort to develop good relations. This will become even more apparent as you read Chapter 8, "Construction Document Review."

Today more than fifty percent of building codes usually refer in some way to fire protection.

Model Code Organizations

Most communities or fire protection districts adopt a code from one of two major model fire code organizations:

- *National Fire Codes* — published by the National Fire Protection Association (NFPA®)
- *International Fire Code®* — published by the International Code Council (ICC)

These two principal codes are the only two American ones available. Only outdated versions of some older model codes may remain. While many of the requirements from each of these organizations are nearly identical, their parent codes or building codes have slight differences.

How NFPA® Codes Are Developed

Fire codes are typically developed through input from special interest groups or through consensus. Usually, proposals are sought from various individuals or organizations, reviewed by technical committees, and then discussed in open meetings. Revisions or new drafts are provided for additional votes and comments. Once sufficient discussion has taken place, votes are cast and decisions rendered.

Most codes are developed and created in a similar fashion. The specific rules and procedures used by all code groups are far too detailed to print here. However, in a very general sense, codes are written as follows:

1. Technical committees or groups formulate drafts of new codes as requested.

2. A Standards Council or other similar group generally issues the approval for certain codes or standards to be drafted and or revised if it is not already on a set printing schedule. Technical committees may also hear numerous appeals throughout a code cycle.

3. Once the assigned committee or group makes drafts or revisions, the drafts are then put out for public comment. The technical committee receives the comments and then makes changes. In some cases, depending upon the type of comments, staff makes the changes rather than a technical committee.

4. The revised proposal is then republished for final public comment. Hearings are often held to discuss the proposals and to potentially receive comments from the floor. In other instances, participants submit comments again and the technical committee makes one last revision.

5. The code is then ratified, accepted, and printed.

The intent is to make sure that the code development process is fair, equitable, and representative of the people with a vested interest in them. Codes or standards can be time consuming and expensive to make. The process is technically complex and requires tremendous collaboration, technical knowledge, and strong interpersonal skills. We recommend that anyone making a career of the fire service should try to get on a code committee. There is no better way to shape your knowledge of the code and standard-making procedure than participating in the overall process. In the past, it was often very costly to travel to the code hearings in order to provide input. Today, many of the code hearings can be viewed online and code change comments easily submitted electronically.

The appropriate companion documents or codes should always accompany the model code of choice.

Model Codes and NFPA® Codes and Standards

Most code renewal cycles repeat approximately every four years. Standard renewal or revision cycles occur at about the same frequency. Regardless, codes are developed as reasonable and expeditiously as possible. The important thing to learn from this is that codes and standards can become outdated as technology, processes, and products advance. Even if you are using the most current version of a code, the discussion and committee votes for that code or standard likely occurred one to two years prior. This can be a disadvantage to industries or designers who are on the cutting edge of a process or development because there can be gaps in available knowledge to verify the safety of a particular design or product.

The Reality of Fire Codes

Anyone who is making a career of the fire service should try to get on a code committee. There is no better way to shape your knowledge of codes and the overall process of how they are made.

Although fire code development has progressed significantly in the last decade, problems still remain. Limited research, as was pointed out in *America Burning*, still delay these documents' advancement and, unfortunately, other technical issues still leave much of the progress to result from experiential data derived from case studies of actual fires. These experiential data make it much easier to address arguments against code provisions since past history already shows the issues are relevant and dangerous, but "averaging" requirements into a "one-size-fits-all" approach makes them inequitable for many.

Performance-Based Codes

Performance-Based Code — A code that allows designers to determine how best to meet an individual building's unique fire protection needs.

A new breed of design requirements has come into play called **performance-based codes**. Performance-based codes are used in various countries, such as New Zealand and Australia. They are also integrated in a variety of ways in the United Sates. These codes, although labor intensive, provide a designer an opportunity to design extraordinary projects, as long as they meet appropriate goals and objectives for fire and life safety. This is much like what the National Aeronautic and Space Administration (NASA) has done and continues to do today. They are embarking on new and different technology daily. They have no model codes or standards for new concepts to reference and, therefore, must create adequate engineering designs based on the desired outcomes and potential risks. Such is the objective of performance-based design. The labor-intensive nature of performance-based codes comes from the initial goal and objective design process and the amount of time required to validate and test for final inspection and certificate of occupancy. Future follow-up for the life of the building is also difficult, as any changes or modifications as the building ages will require significant evaluation and monitoring to make sure the initial assumptions and design parameters are still appropriate.

Prescriptive Code — A code that lists specific design requirements, such as the number of exits, fire separation, construction type, and fire suppression systems.

For the most part, current building and fire codes are prescriptive. As we have already discussed, the codes define exact requirements based on the occupancy classification of the building. A **prescriptive code** specifies the number of exits, fire separation requirements, construction type (what the building is made of), and fire suppression (automatic sprinklers) requirements. The authority having jurisdiction applies the code requirements during the construction document review process.

This prescriptive approach works well to meet the needs of the owners and architects for most structures. However, there are situations where applying these types of codes may not be practical, An example of this may be a large, irregularly shaped facility with complex industrial processes, or a building with highly sensitive operational requirements. In these situations, prescriptive codes will not always address all of the concerns of the building's owner, the architectural design team, or the AHJ. For example, the occupancy-related or prescriptive codes may not address all of the fire-resistive requirements needed for a given fire scenario that could occur. A larger than normal heat release rate may be generated, or longer travel distances to the exits may be necessary because of the structure's shape and associated processes. These difficulties lend themselves to a unique opportunity to use an alternate method—performance-based design.

Along with such opportunities, however, come additional responsibilities to ensure that the scope of the fire safety problems being addressed are well understood, that the tools and methods being used are applied properly, and that the resulting designs and levels of safety afforded are tolerable to the community and society.[5] American performance-based codes, which have yet to hit the market in full force, will not be nearly as textually complex or comprehensive as the prescriptive codes we use today. The reason is that most of the process is left to the determination and design of objectives, outcomes, and testing methods, which will not appear in a code document, but in the design criteria for a project on a case-by-case basis.

The concept of performance-based codes is to design fire protection measures to a level of safety that the owner, the designers, and the AHJ can all agree upon. This concept has also been referred to as a "quantitative assessment" or an "engineering approach." Even though this type of design references a performance-based code, it is actually a performance-based design process. This "code" simply provides the guidance and direction that must be followed in developing the criteria and measurements. The latest editions of the model building codes include provisions to allow for performance-based designs. The difference in application is that the prescriptive requirements in the code are not followed under a performance-based design approach unless specifically referenced.

Performance-based designs use scientific theories, carefully weighted assumptions, empirical formulas, and mathematical calculations to determine the probable outcomes of a fire scenario. This can be accomplished through the use of complex computerized fire modeling performed by a fire protection engineer or other design professional. The performance-based design approach identifies specific hazards and risks and the appropriate methods or strategies to protect them. It considers the entire building's features in relation to the assumed occupants' behavior and addresses the unique building features that have triggered the need for performance-based design. When done correctly, the performance-based design provides a comprehensive fire protection plan that addresses the uniqueness of the structure based on its anticipated conditions.

When done correctly, the performance-based design will provide a comprehensive fire protection plan that addresses the uniqueness of the structure based on its anticipated conditions.

A performance-based design is all but certain to fail unless everyone involved understands, communicates, and decides upon all aspects of the project.

The use of performance-based codes necessitates the diligent and committed use of all forms of human communication among fire officials and the design team. Failure is all but certain unless everyone involved understands, communicates, and agrees upon all aspects of the project. Interpersonal skills play an important part in the overall process since participants are forced to meld engineering science with practical firefighting tactics. The accuracy and commitment to this blend is all that stands between success and failure of the system.

To provide good performance-based design reviews, a fire prevention division needs to have its own technical experts in fire protection engineering or fire protection engineers on contract to evaluate proposed designs and models. Computer modeling is an essential part of this practice and, as with any models, is not necessarily as important as the assumptions plugged into it. The fire service needs to be very careful in its approach to this process. It is a far cry from the old days of rubber-stamping plans with the comment, "Hydraulic calculations are the responsibility of the designer." The responsibility of a good fire safety design rests not only with the design or project engineer but also with the fire department plan examination staff. All must engage in dialogue about risk, expectations, and outcomes. Otherwise, disaster is a significant possibility.

Performance-based codes require the fire service's participation. While these codes can be beneficial and flexible, a significant pitfall awaits. If a particular building is designed for a given occupancy and all of the assumptions, modeling, and design are done to accommodate that specific occupancy, what happens twenty years later when the occupancy changes? Remember that the fire service is responsible for buildings during the entire period that they remain standing. Questions that beg to be answered are:

- Will we know if the occupancy changes?
- What impact will various changes have on the original performance-based design?
- How will the occupancy changes impact the performance of the life safety and fire protection systems already installed?
- Can a given building, designed under a different premise, be modified to accommodate a new occupancy at a reasonable cost?
- What role do we have in accommodating all of these changes?
- How do we inspect a building that changes from one performance-based design to another and what skill level will an inspector need to accomplish this?

Many disconcerting issues surround performance-based designs. We must keep a proactive approach and focus our thought processes on this change and make certain we keep abreast of changes as they occur. Change is generally good; however, it still remains to be seen when the fire service will adjust fast enough to this dynamic design process, particularly in light of our most recent down-turn in the economy and the impact of the last recession. These effects and future recessions will have a significant impact on the staffing and resources of fire departments, which may create significant impairments to proper and effective performance-based designs.

What does this mean to the fire protection professional? Using performance-based designs requires the local fire marshal or his or her staff to either obtain the technical training and education to review performance-based designs or solicit the expertise of someone who already has it. It also means the fire official must have obtained sufficient data during the design for fire inspectors who conduct inspections later to verify that conditions used in the performance-based design have not changed to alter the outcome of a fire event. Ongoing fire safety management must be carefully monitored in facilities designed and constructed with a performance-based design. The local AHJ must understand the complexity of performance-based designs and establish a method to review them before one lands in his or her in-basket. Secondly, the AHJ must have a method to retain the supporting documentation of the performance-based design for future fire safety management.

Ongoing fire safety management must be carefully monitored in facilities designed and constructed with a performance-based design.

Summary

Building and fire codes have existed for some time. The codes themselves provide minimum requirements. The use of codes in the United States was developed from pressure of the insurance industry to provide a method for quality construction. Codes are rewritten through a process that permits input from the public, business, and industry. Because of rapid technological changes and the time it takes for the code revision cycle to be completed, codes are typically not as current as they should be.

Today, model codes are better consolidated, resulting in two major codes related to fire protection. They include the *National Fire Code®* and the *International Fire Code®*. Most codes are considered prescription-based codes. They outline exactly what is required. However, the codes also allow a performance-based approach that enables engineering analysis to determine what is required and how it will be measured.

Codes differ from standards. Codes typically cover the who, what, and when concerning requirements. Essentially, codes tell you what you have to do. A standard tells you how to do it. The enforcement levels of the adopted codes range from the local level to the state and the federal level. In most cases, the local authority cannot enforce local codes in state or federal facilities. In recent years, however, state and federal government agencies have begun working with local municipalities in order to meet their requests for compliance.

Case Study Discussion

The situation is touchy as the building has been in existence for some time. Likely, it is not a change of occupancy classification, and no remodeling is taking place. Therefore, does not need to comply with today's building code. The first thing you should do is try to research the fire and building codes that were or may have been in place in the 1930s, or when you think the occupancy became what it is today. This would be the guidance you should use to enforce major code issues such as exits, sprinkler systems, and the like. In some jurisdictions, if they abide by the *Life Safety Code®* or they have adopted the existing building chapter in the *International Fire Code®*, those problems are spelled out more clearly and specifically about what can and cannot be done.

A second issue is the hood and duct extinguishing system. It is outdated and cannot be serviced anymore, so that should be replaced with a current U.L. 300 compliant system. This is necessary because the system can no longer be serviced or maintained with original equipment, which forces the system to become current in manufacture and code requirements. However, no other requirements should be necessary unless local amendments call for it or some other major work is driven by this change.

The last, but most important, issue to be concerned about is the occupant load. As the fire code official, if you feel the facility is dangerous given the number of people capable of visiting the location, you may wish to change what is permissible or work toward other protective solutions. You must keep in mind though that this will require collaborative selling and partnership with the owners and managers. While you may have the authority, remember that they are paying your salary and you work for them. Safety is critical, but be careful and respectful with how you enact "new" requirements and work toward solutions that are reasonably achieved.

Chapter 3 Review Exercises

3.1   What are some of the historical issues or events that caused codes to be developed? _____

_____

_____

_____

3.2   What is a *code*? _____

_____

3.3   What is the function of the ISO? _____

_____

_____

_____

3.4   What is the best ISO rating that a community can receive?

_____

_____

3.5   What total percentage can Fire Alarm System grading provide to your overall rating? _____

_____

3.6   What is a *standard*? _____

_____

3.7   Explain the difference between a *code* and a *standard* and give an
      example of each. _____

_____

_____

_____

_____

3.8   What is significant about appendices to *codes* or *standards*?

_____

_____

_____

_____

3.9   What are the three categories of fire safety regulations?

_____

_____

_____

3.10 What do building codes govern? _____

_____

_____

_____

_____

3.11 What does the term *maintenance code* mean? _____

_____

_____

_____

_____

3.12 Name three factors that are driving information integration into the
      current fire codes? _____

_____

_____

_____

_____

3.13   What is a *performance-based code*? _____

_____

_____

_____

3.14   When are *performance-based codes* used?

_____

_____

_____

_____

3.15   What are the advantages of *performance-based codes*?

_____

_____

_____

_____

3.16   What are the disadvantages of *performance-based codes*?

_____

_____

_____

_____

3.17   Explain the *code-development* process._____

_____

_____

_____

3.18   What determines when codes are changed or modified?

_____

_____

_____

_____

3.19   Can a local jurisdiction mandate a federal institution to comply with local code? If so, how? _____

_____

_____

_____

3.20 Which legal authority has more clout: state or local? Explain your answer._____

_____

_____

_____

3.21 Develop and explain a general board of appeals process.

_____

_____

_____

_____

_____

_____

_____

3.22 Who has more authority: a building official or a fire marshal? Explain your answer. _____

_____

_____

_____

3.23 Explain why codes are not necessarily current.

_____

_____

_____

_____

3.24 What is the AHJ?_____

_____

_____

_____

Notes

1. Arthur E. Cote and Jim L. Linville, Fire Protection Handbook, 16th ed., (Quincy, Massachusetts: National Fire Protection Association, 1991), p. 6-140.

2. Cote and Linville, pp. 15-95–15-97.

3. Cote and Linville, p. 6-145.

4. Cote and Linville, p. 6-142.

5. Brian J. Meacham, "Addressing Risk and Uncertainty in Performance-Based Fire Protection Engineering," Fire Protection Engineering (Spring 2001), p. 16.

Development and Implementation of Fire Prevention Bureaus

Table of Contents

Key Terms

Learning Objectives

1. Identify laws, codes, ordinances, and regulations as they relate to fire prevention.*

2. Understand code enforcement as it impacts life and property loss.*

3. Define laws, rules, regulations, and codes and identify those relevant to fire prevention of the authority having jurisdiction.*

4. Define the national fire problem and role of fire prevention.*

5. Define the functions of a fire prevention bureau.*

6. List opportunities in professional development for fire prevention personnel.*

7. Understand today's changing work environment and the impact on fire prevention organizations.

8. Identify potential methods to retain and motivate fire prevention personnel.

Development and Implementation of Fire Prevention Bureaus

Chapter 4

Case Study

During an economic recession, the fire department elected to eliminate the fire prevention bureau for the City of Nowhere. The city has a population of 55,000 people and two stations serve this city. A recent increase in fires involving two child fatalities has resulted in elected officials directing the fire chief to create a fire prevention bureau. Identify the steps the fire chief needs to take to develop his new fire prevention bureau.

Fire Prevention as a Public Business

No other function within the fire service is more like a business than the fire prevention bureau. Fire prevention bureaus have as much contact with the customers they serve as does the Emergency Operations Division. As an entire Branch of the fire department, it typically performs more high profile problem-solving activities than any other division. The fire prevention bureau deals with irate customers, assists with complex designs of processes and facilities, and settle complaints among neighbors, landlords, and a multitude of other people. In addition, the Bureau generates revenue from permits and other activities. This division, more than any others in the fire department, can directly impact economic development through the business environment, regulatory complications, or ease any other political aspects.

Fire prevention bureaus must be empathetic to their customers **(Figure 4.1)**. They must enforce codes and standards to protect lives and property while performing in a professional and diplomatic fashion. As fire protection professionals, we must remember that although doing our job in fire prevention is an essential service that protects lives, it may not be perceived the same as the heroes who ride the big red trucks. We sometimes aggravate people because we cost them money by pointing our deficiencies that require fixing. Justifiably then, this increases the importance of establishing a relationship with the customer and sometimes providing service in a consulting or advisory role.

Figure 4.1 Fire prevention personnel, just as any business owners, must understand the importance of customer service.

It is important to reflect the fact that we work for them and are looking out for their best interest as well as their clients or our public. In remodel or new construction projects, we will be here to serve them and their customers long after their contractor leaves the job site. In tough economic climates, it is more important than ever to ensure that our service is portrayed as a problem-solving service and not a hindrance to the economic environment and development process.

Fire prevention is the most important job that the fire service can perform. When our efforts prevent a fire from happening or keep it as small as possible, we protect not only our community but also the firefighters who respond. Our job is sometimes the toughest and most ubiquitous, but it should be held one of the most important nonemergency functions a fire department performs. However, the importance of fire prevention services and the associated staffing are easy to overlook when trying to balance a fire department's budget. One of the greatest challenges the manager of the fire prevention bureau faces is having sufficient staff to carry out fire prevention services.

In times of economic growth, as well as during economic downturns, the ability to effectively address staffing issues plays a critical role in the organization's operational success (**Figure 4.2**). This is true whether the organization is private or public. Today, both the private and public sectors are struggling to do more with less. To address the dynamics of a changing work environment, all types of staffing options must be considered, even outsourcing the service. In the future, fire departments will continue to face the need to do more with less. The reduction of funding for fire suppression activities will likely become common in many areas of the country. Well-staffed fire prevention bureaus may play a more significant role in providing fire protection services along with encouraging more built-in fire suppression systems. The escalating costs of fire suppression activities combined with the direct and indirect loss from fire will continue to force fire departments to place a greater emphasis on active and passive fire protection systems.

Though we stress prevention's importance, a large portion of many fire departments' budgets are allocated to suppression activities. A common distribution is around 3–5% for prevention and 95–97% for the rest of the department. Staffing and equipment costs continue to rise, in turn increasing the expense of mitigating a variety of situations that fire departments face daily. Salaries and benefits are the most expensive element in a fire department's budget (excluding all-volunteer departments). Watching the news over the last couple years, pensions are being hunted, which is also adding to the "cost" of salary and benefits. In fact, studies have shown that the costs of wage and benefit packages are outpacing the consumer price index, which could lead to a future when departments simply cannot afford to continue staffing as they do today. The recent cuts faced by fire departments during the last recession may take many years to rebound from, if it is possible at all. Now more than ever, we must always consider our cost effectiveness, benefit to the citizens, and make sure we are demonstrating our importance and value to the communities we serve.

Figure 4.2 When a fire station is forced to close, there may not be funding for fire prevention efforts.

Government entities at all levels have been portrayed frequently as overstaffed and underworked. The need for exploring a variety of staffing options was not always the case with government organizations in years past. Many elected officials have taken staffing and cost criticism seriously, seeking ways to perform services at a maximum level with minimum staffing and at a reduced cost.

In the future, this may force the fire service to be very creative in its way of doing business. Privatization of fire prevention bureaus was not seen as a common occurrence. Today, privatization along with outsourcing is a positively perceived option to provide additional staffing to support the existing fire prevention bureau. If a community outsources its staffing, it benefits by not having to provide wage and benefit packages. Health care has another huge impact on all industries and businesses. Government is no exception. The use of private companies and outsourcing to perform municipal fire prevention bureau services is explored later in this chapter during our discussion of staffing fire prevention bureaus.

A variety of fire prevention staffing options is available to meet the needs of the community. Identifying the best staffing option and developing the organizational structure of a fire prevention bureau is not easy. Doing so requires managers to consider a number of factors. Even though fire departments throughout the United States provide similar services, how they provide those services and the level at which they do so varies throughout the country. The fire prevention bureau manager, fire marshal, or fire chief must consider several critical elements in order to lead the fire prevention bureau in the right direction. These elements serve as guidelines in creating organizational structure and developing staffing options. The following steps do not include all of the tasks needed to develop the framework for a fire prevention bureau, but are merely a road map for getting there:

- Ensure that the organization's mission statement includes the fire prevention bureau's primary function and purpose.
- Ensure that fire prevention is part of the fire department's strategic planning process.
- Adjust the organization as needed while monitoring the environment for internal and external changes and opportunities.

Ensure the Organization's Mission Statement Includes the Fire Prevention Bureau's Primary Function

The first step in staffing a fire prevention bureau or similar division of an organization is to evaluate where its function fits into the entire organization's goals and mission statement. As we discuss in chapter five, the function of fire prevention may not be in the same department, division, or section as the suppression service. However, fire prevention is still a function of the fire department, even if it does not have a fire prevention bureau. The fire department or organization must create or reevaluate its mission statement. The mission statement should be on track, but if it is not, involve the team in recreating it to spell out the things you should be doing. After that, do the same thing with vision and values statements.

Vision Statement — A brief description of how the fire department or more specifically, the fire prevention bureau will operate.

Value Statement — A summary of the ethical priorities for everyone's behavior when working on the mission.

Mission Statement — A description of the purpose of the organization.

The **vision statement** should be a brief description of where the fire department, or more specifically, the fire prevention bureau wants to be. A **value statement** can and should reflect ethical priorities for everyone's behavior when working on the mission. Some departments may build on this and create a code of conduct. In any case, it is important for people to be able to see how they will do what they need to do.

A **Mission statement** describes the purpose of the organization and varies from being exceptionally brief to very comprehensive. Regardless of length, a major component should specifically spell out fire prevention functions. The fire service fights fire in two ways: the first is prevention and mitigation, and the second is emergency response. Why would we want to serve our public only after a bad thing happens rather than stopping it from happening in the first place? Prevention is a critical function of the fire service and must be included in the department's overall mission.

When the fire department's mission statement clearly establishes the role of the fire prevention bureau, it informs the entire community (the customers) as well as the fire department of the fire prevention bureau's significance. The fire prevention bureau must not be structured to function independently within the fire department, but to function with the other divisions of the organization. The fire prevention bureau must understand the goals of all other divisions, as well. Conversely, other divisions within the organization must be capable of identifying the organization's fire prevention focus. Each division must share the mission of the fire department and understand the roles of all of the divisions to carry out the mission. You can still have fire prevention as part of your mission statement even if you don't have a full-time dedicated staff to perform that service. Think of this as similar to a fire department, including EMS as part of their mission, but not providing an ambulance service. They may still provide EMS!

Fire department mission statements may be common in many areas of the country, but this was not always the case. Like many other management issues, looking to the private sector for management models has not always been widely accepted as a useful tool.

The mission statement should answer the following questions[1]:

- **Who are we?** This simply identifies the organization and its service. For example, we are the fire department, fire and rescue, public safety, or fire and emergency medical services.

- **In general, what basic social or political needs do we exist to meet? What basic social or political problems do we exist to address?** The answers to this question frequently address each of our community's risks and how the organization will address them. A good response might be that we are reducing the number of children and senior citizen fire deaths through extensive fire safety education while focusing on the community's needs and the fire department's capabilities to address those needs.

- **In general, what do we do to recognize, anticipate, and respond to these needs or problems?** That is, how do we identify problems and what actions will we take to address them? In other words, what justifies our organization's existence?

- **How do we respond to our key stakeholders?** The question simply asks, "What are we going to do?" The answer is simple. We provide fire, rescue, special teams, and fire prevention services. The stakeholders are the citizens, or our customers.

- **What is our philosophy? What are our values, and culture?** This sweeping question determines our attitudes, beliefs, and/or what type of people make up our organization.

- **What makes us distinctive or unique?** The business world or private sector may refer to this as a *niche*. Why are we better than the rest? What do we do that makes us stand out compared to similar organizations? Possible answers might include that we are all volunteer, we are cost-effective, and that we provide a quality service.

These are the guidelines for developing a mission statement for the entire fire department. The challenge is to answer the questions to include the efforts and needs of the fire prevention bureau. This is the framework for why the fire department exists—the "why are we here?" question. The individuals learning and applying the mission statement are not just those assigned to the fire prevention bureau. They include every person employed by the fire department. The desired outcome of incorporating fire prevention in the mission statement is to make it everyone's responsibility. With that in mind, one person or one division cannot develop the mission statement **(Figure 4.3)**.

Figure 4.3
Fire department organizational meetings are a great tool to help communicate the department's vision and mission.

If a fire department does not consider input from the local government, it may mistakenly create a mission statement to address a problem its governing body does not consider significant. Keep in mind, the mission statement should not be so complicated that the average citizen does not understand it. The citizens are your customers or stakeholders, and your mission statement should have the power to become one of your fire department's best public information tools.

The following is an example of a mission statement from a fire department:

- Colorado Springs Fire Department: Provide the highest quality problem solving and emergency service to our community since 1894.
- Colorado Springs Fire Prevention Division: To promote a safer community through hazard mitigation, fire prevention, fire code development and enforcement, fire investigation, public education and injury prevention, hazardous materials regulation, and wildland fire risk management.

Ensure the Fire Prevention Bureau Is Part of the Fire Department's Strategic Planning Process

Fire departments are proud of their local tradition and heritage. If you ask why a department does something in a certain way, a common response is, "We have always done it that way." However, one thing is certain. The fire service today is not the same as it was yesterday. The entire world is constantly changing. Consider the evolution of the computer, the rapidly changing diversity of our culture, demographic changes, state and federally mandated compliance, and a sometimes volatile economy that can literally change overnight. Did we ever think we would see the day where layoffs of police and fire department personnel would take place in so many locations throughout the country and in such high numbers?

If we go from an anecdotal explanation to a more scientific, statistical analysis, the call distribution for most of the U.S. fire service is composed mostly of medical incidents (assuming medical calls are handled to some degree). The frequency and volume of fires are gradually declining or remaining mostly steady. In fact, the typical urban fire department is seeing a decline of roughly two percent annually in population-adjusted fire incident rates.[2] Examining various fire departments throughout the country in light of this trend, we see that the decrease is commensurate with an increased involvement in fire prevention activities, such as engineering materials or devices, plans review, inspections, and risk management. This, too, is substantiated in a survey conducted by the Colorado Springs Fire Department. Looking at what urban fire departments are doing to reduce their overall increasing workload, fairly recent programs aimed at fire prevention appear to be having a dramatic impact:

- Proactive education programs at nursing homes
- Low-hazard inspections reduced from annually to every five years
- Fewer vehicles responding to selected types of incidents
- Public education programs for EMS
- Smoke alarm blitzes in key areas like housing projects
- Aggressive inspection programs
- Public education and awareness programs
- Ambulance service providers taking nonemergency medical calls[3]

Five of the eight functions in the previous list are related directly to fire prevention. In fact, one of them showed an adjustment to the historically significant fire prevention function of low-hazard fire inspections. This adjustment is possible because we are better targeting our efforts and able to quantify the net results of those efforts. For this reason alone, a fire chief would be remiss in excluding fire prevention activities as a major part, if not the most important part of strategic, long-term planning.

The best way organization members, either public or private, can cope with constant changes is to think strategically. Strategic planning is the best tool to accomplish this. A number of books and courses available at the National Fire Academy address strategic planning. Fire prevention bureau managers should learn as much as they can about strategic planning and take an active role in the fire department's strategic planning process. This strategic planning should involve a comprehensive risk analysis, which is summarized later in this book.

It is beyond the scope of this text to go into great detail about the strategic planning process. However, since it is so important, a rough template is provided that could serve as a starting point. It is not easy, nor should it be taken lightly, but it is possible and can be accomplished at any level.

Strategic planning produces a number of benefits to the organization. The most obvious is the ability to think and act strategically. A benefit of taking part in the strategic-planning process is soliciting and providing input into the plan and its implementation. It is here that fire prevention becomes crucial to the organization. When the fire prevention bureau is involved in the planning process, it has the opportunity to communicate the challenges it expects to face. One outcome of the strategic planning process should include a cohesive plan addressing the challenges of the entire fire department while producing a number of other benefits. For example, the strategic plan should:

- Define the purpose and objectives for specific hazards based on your organizational mission and values.
- Create an environment where your entire department has participating ownership.
- Create synergy by pushing everyone to be his or her best, all at the same time.
- Communicate goals and strategies to your clients and the policymakers.
- Help make sure that you are using your department's resources to their fullest.
- Provide a benchmark or baseline from which to measure progress.

The results of a strategic planning process will most likely last into the future and benefit the fire department and the community for many years to come. For example, a fire department in the late sixties or early seventies may have included new approaches involving a paradigm shift, such as beginning the delivery of emergency medical services or starting a fire safety education program. Today we look at these as obvious functions of most fire departments. However, fire departments at that time did not commonly provide these services.

Conducting this type of planning effectively is not easy or quick. The process should not be undertaken if the developed strategic plan is not likely to be implemented. Planning takes significant time and resources. To develop a plan for addressing identified problems, it is imperative that the organization have the ability to collect and interpret data to identify the problems it faces. You cannot begin developing a strategic plan if you cannot identify your problems or challenges. With the proper information, the strategic planning team will be better equipped to arm the fire chief with the data to justify and allocate the necessary resources and to implement the solution to the problem. At some point, the fire chief will need to bring the plan forward for implementation. This will most likely require resources in the form of personnel or equipment, particularly technological support. One step to help ensure the governing body endorses this endeavor is to get its support before you begin the planning process. In fact, the governing body may have challenged the organization and be anxious for a plan to be implemented. It may also be beneficial to have some of the governing members involved in the process, particularly in the form of focus groups, subject matter experts (SME), or the like.

A number of strategic planning models are available from which to choose. Almost every strategic planning model includes the steps of developing the mission statement, creating the vision, and developing strategic goals. Developing strategic goals can generate in-depth discussion. Keep strategic goals specific enough that there is no question about what you are trying to accomplish. However, keeping them vague enough that you do not get bogged down in the strategies for accomplishing the goals.

Document your plan. This is the last step, but it is very important. Put it in a format that is easy to read and that can be distributed to everyone. Some organizations have communicated their plans in the form of a contract. Do whatever you feel is comfortable, but then have the team sign the document, much like a charter or declaration, affirming their commitment and understanding, as they helped create it. Then all you have to do is implement, evaluate, and correct. If you have completed these steps, you have just completed your first strategic plan.

Determine the Level of Fire Prevention Services

Given that the mission of the fire department must include fire prevention, the question remaining to be addressed is what level of fire prevention service to provide. The mission of the fire department commonly includes an element of fire prevention, such as reducing the number of fire deaths or the amount of fire loss. Achieving this is not an easy task, especially after the last economic recession. The commitment and ability to meet these goals must begin with the governing body of the fire department. Meeting these goals may be the most critical element in the establishment and management of a fire prevention bureau. It creates the foundation for staffing levels and for the fire prevention bureau's organizational framework.

Fire prevention services vary considerably across departments. In some jurisdictions, fire departments or fire districts may provide services for several communities or municipalities. This may require the department to perform different types of plan reviews or inspections, depending upon each locale's preference. As an example, one town's building department may do all of the new construction plan reviews and inspections while leaving the routine fire inspections to the fire protection district. Others may want the jurisdiction to handle all plan review activities and inspections. These differences can be challenging as staffing and resources can be difficult to manage and juggle. Service delivery should be determined and agreed upon through an intergovernmental agreement (IGA), a contract, or at least a memorandum of understanding (MOU). We strongly recommend that these agreements lock in time frames for service delivery and also include service performance criteria. The agreements should contain a clause or phrase that spells out periodic, regular discussions among the various clients and the jurisdiction to facilitate proper planning and prevent any surprises.

Strategic planning is another method of determining what level of fire prevention services to provide. If your elected official, board or governing authority approves the strategic plan, they are committed to supporting and continuing your mission and direction. This is an excellent mechanism that can provide many resources without the continual inconveniences and distractions of having to come back time after time to ask for approval. You may just have to periodically reference *their* approved plan!

Retain and Motivate Staff

Family demands tend to spur the employee's compensation needs and desires for more time away from work. Studies have shown salary alone is not a key factor to motivate or retain employees. Smaller fire departments probably have a greater need for high quality work, as they do not typically have numerous local experts in every field, such as fire protection engineering. Ironically, these smaller departments cannot generally afford the salaries needed to attract highly trained staff, let alone retain them. Upward mobility in the organization is another problem.

Whether the fire department uses civilians, sworn personnel, or outsourced staff, it definitely needs to attract, motivate, and retain fire prevention bureau personnel. Nothing is more frustrating to a fire prevention bureau manager than having a good employee spend two years in fire prevention and then go back to suppression or, in the case of a civilian, leave for another department. The bureau has spent considerable time training the individual and now must start the process all over again **(Figure 4.4)**. This raises costs and lowers productivity.

According to the Employment Policy Foundation, the workforce is changing. In the last fifty years, the traditional workforce of the "stay-at-home mom" and "working husband" has declined. Married couple families, in which both spouses work, represent 70 percent of married families and one-half of all families. With job demands forcing more parents to be away from their children, employees focus more on quality of life than on pay. The increased demands on the working parent

Figure 4.4 Rotation of suppression personnel into the fire prevention division may require the manager to continuously train staff.

Flextime — Scheduling system that allows employees to choose their work hours within limits established by the employer.

Compressed Workweek—Scheduling system that permits full-time employees to perform the equivalent of a week's work in fewer than five days.

Telecommuting — Working at home or sites other than the workplace through digital means or devices over various hours or days.

spill over into the workplace. Employees need flexible conditions to balance the needs of their families with the demands of their jobs. This is becoming more essential as millennials are moving into the workforce.

Work scheduling has changed significantly in recent years and continues to do so. Alternatives to the five-day, eight-hour work schedule come in a variety of forms. Alternative schedules, commonly known as flex schedules, **flextime**, and **compressed workweeks** may soon become more prevalent in public agencies, such as fire prevention bureaus. **Telecommuting** is also an interesting approach to fire prevention work schedules and work styles.

Flex schedules have been in the workplace since the 1970s and consist of allowing employees to choose their work hours, within limits established by the employer. The band of start and stop times may vary by as little as fifteen minutes; some situations allow employees to periodically balance shorter and longer days over a week or more. The government and the private sector offer a similar extent and variety of flextime programs. Flextime is a more dynamic scheduling which allows flexibility during the current work week. It allows employees to leave work early and come later or during mid-day to take care of their needs, as long as they work their minimum 40 hours and get their job done.

The Bureau of National Affairs indicates many employees value flex schedules because they allow employees to adjust work schedules to meet family obligations and other personal responsibilities. Flex schedules can allow employees to avoid rush-hour traffic, accommodate childcare schedules, coordinate with public transportation, and fit the work schedule of the employee's spouse. Many employers have argued that problems, such as scheduling meetings and the lack of supervision, can arise from a flex schedule policy. Flex schedules can also involve legal issues, such as meal periods, timekeeping, and state overtime/maximum-hour requirements.

Another scheduling alternative is the compressed workweek. This arrangement is increasing among employers interested in helping employees manage family and work demands. A compressed workweek schedule permits full-time employees to perform the equivalent of a week's work in fewer than five days. Compressed workweeks can achieve many work scheduling objectives, such as improving recruitment and decreasing turnover, increasing employee loyalty, extending customer service hours, improving scheduling flexibility, reducing work and personal conflicts, and increasing the opportunity for the employees to further their education. As with any policy, the compressed workweek does have some disadvantages. These include problems accounting for holidays, the impact on other employees, and legal considerations, such as the Fair Labor Standards Act and record keeping requirements. Because the compressed workweek means working fewer but longer days each week, employees may become tired and less productive at the conclusion of those longer days.

Even with the disadvantages, work scheduling alternatives can offer many benefits to a government entity, such as a fire prevention bureau. The bureau can use scheduling alternatives to attract and retain employees. The variety of staffing options available for the fire prevention bureau may include individuals used to the traditional fire department schedule of working twenty-four hours and being off forty-eight. Even if they are still sworn members of the fire department assigned to the fire prevention bureau, the change in scheduling a forty-hour workweek may affect the individual personally. One disadvantage of the forty-hour workweek usually associated with fire prevention is the lack of scheduling flexibility offered to sworn or shift personnel. A variation of the compressed workweek is also moving to four ten-hour days.

The nonprofit sector is examining retention strategies to counter the corporate world's incentives for potential and current workers **(Figure 4.5)**. Many nonprofit employers, such as fire departments, are realizing they cannot match for-profit organizations' cash. Still, it is important that fire prevention bureaus are staffed with the best-qualified individuals. Those individuals may or may not be sworn members of the fire department. Having something that may entice them to join the fire prevention bureau can benefit recruitment and retention of that individual for an extended time. Work-life benefits and flexibility are two ways for nonprofit organizations to retain employees.

The fire department should not focus only on benefits as a method of retention but also on programs within the department that address the needs of the work force. The fire department can start by focusing on the same human resource programs as the private sector. For example, for a number of years the private sector has been successful using mentoring programs for employees. Mentoring programs provide the opportunity for the new employee to learn from one of the more seasoned or veteran employees. The concept is to provide an atmosphere conducive to continuous learning. Also, the new employee has a person he or she can trust and ask questions of when the need arises. This benefits the company as well because an individual who is winding down his or her career can find a sense of self-worth by being needed to train the newer employee. Hopefully, the veteran employee's guidance can keep the new employee from getting frustrated and possibly leaving the organization.

Figure 4.5 Nonprofit organizations can offer quality-of-life benefits to offset monetary compensation offered by the private sector.

Adjust the Organization as Needed While Monitoring the Environment for Internal and External Changes and Opportunities

One thing is certain, nothing remains the same. As fire protection professionals, we must be able to change the way we do business to meet the needs of our customers while mitigating risks to the community. For many of us, this challenge is even more complex because we are being asked to justify our existence and perform our fire prevention service with less staff and money. The private sector has placed stringent demands on the fire prevention service. Therefore, the fire prevention services is forced to carefully examine its efficiencies and the worthiness of various programs and task effors.

In addition, just like many businesses, customers can change. In fire prevention, the community may undergo demographic changes that create a new target audience for fire and life safety education. For example, if a large senior citizen housing complex or development was constructed in a small community, the community would have a significantly increased senior population that required a different approach to fire and life safety education. The number of citizens in the high-risk group for fire deaths also has increased.

Regardless of the cause for the change, fire prevention bureaus must manage the change and the change process. Those involved in the modification will react to the change. Reaction to change is a normal part of the process. The challenge to the change manager is to make the most of the positive reactions and minimize the consequences of the negative ones. Different individuals will accept and support changes, comply in action without actually supporting the changes, or be as resistant as possible. Ownership through the process is one of the most enabling aspects of change management.

The key to the success of any change is in communicating it to all those involved while giving them the opportunity to offer feedback. This communication needs to be continuous. If change is directed from the top of the organization down, then there must be a means to communicate back to the top of the organization. Communication is the most important aspect of dealing with change. Fire protection professionals must also keep in mind that they, too, are agents of change. In many situations they are actively trying to cause change. Sometimes they may wait to take advantage of either the political climate or a pertinent event before they begin the change process. Remember our previous discussions of code modifications that resulted from tragedies? The individuals responsible for those modifications initiated them when the climate was conducive. They were monitoring their environment for appropriate opportunities.

Creating change is not always an easy task. The fire inspector may be trying to change how an industrial operation stores flammable liquids, or the fire and life safety educator may be trying to modify the behavior of adults who do not see the importance of testing smoke alarms or practicing a home escape plan. Fire protection professionals are definitely agents of change and while effecting changes, they should keep the following simple guidelines in mind:

- Help people let go of the old stuff before we expect them to grab onto the new stuff.
- Remember Einstein's quote: "Thinking as we are has brought us to where we have already been. In order to go somewhere else, we must think in a different way."
- Identify your steps (beginning, middle, and end). People like closure and want to know the status and progress of ongoing processes.
- Remember that change generally involves some type of loss or failure.
- Allow people to have feelings. They do not necessarily need to act on them, but they should have them.

Summary

Fire prevention is one of the most important functions that a fire department performs. While in America, it typically garners only 3–5 percent of a fire department's total operating budget and is probably the most important loss-control function the department can provide.

The cost of providing reactive fire protection services is constantly increasing. In fact, it could easily be argued that the cost is outpacing most communities' ability to pay for it. For this reason, fire prevention—mitigating and preventing incidents before they occur—is truly their best bang for the buck.

Mission statements are critical to fire department administration, but it is imperative that fire prevention bureaus have clear and concise mission statements. Change is inevitable and, if done right, will keep your operation on the cutting edge.

Chapter 4 Review Exercises

4.1   Identify ways in which the fire service can be more customer friendly.

_____

_____

_____

4.2   What percentage of a fire department's overall budget is typically allocated to fire prevention?_____

_____

_____

4.3   Why do fire prevention bureaus need to consider staffing options?

_____

_____

_____

4.4   Identify ways a fire prevention bureau may be alternately funded and staffed._____

_____

_____

4.5   Explain why developers may want to privatize a fire prevention bureau. Is this a good thing? Why? _____

_____

_____

_____

_____

_____

_____

4.6   Is there a best method to staff a fire prevention bureau? If so why?

_____

_____

4.7   Identify advantages of having sworn fire inspectors?_____

_____

_____

_____

4.8   Identify disadvantages of having sworn fire inspectors.

_____

_____

_____

4.9   Identify the advantages of having civilian fire inspectors.

_____

_____

_____

4.10   Identify the disadvantages of having civilian fire inspectors.

_____

_____

_____

4.11   Other than pay, what are some methods to attract and retain fire
       inspectors?_____

_____

_____

_____

4.12   Why do fire protection professionals need to be concerned with change
       in the community and work environment?_____

_____

_____

4.13   Give three examples of how fire protection professionals can cause
       changes that will benefit the community. _____

_____

_____

_____

4.14   What six questions should a fire department mission statement
       answer?_____

_____

_____

_____

_____

_____

_____

4.15 On a seperate piece of paper, create a mission statement of your own, addressing the six important points previously identified.

4.16 Why is it important for the fire department to include the fire prevention bureau in its strategic planning process? _____

_____

_____

4.17 Identify five ways any fire department can utilize fire prevention programs to reduce its workload. _____

_____

_____

_____

_____

4.18 Name four reasons why strategic planning is important to fire prevention activities. _____

_____

_____

_____

4.19 Are promotional opportunities important to fire prevention staff? Why or why not? _____

_____

_____

_____

_____

_____

_____

_____

_____

Notes

1. Adopted from John Bryson, Strategic Planning for Public and Non-Profit Organizations: A Guide to Strengthening and Sustaining Organizational Achievement, rev. ed. (Hoboken, N.J.: Jossey-Bass, 1995). This is a must read for fire service personnel!

2. William H. Wallace, Colorado Springs Fire Department, Summary of Survey Responses, page 25, January 3, 2003.

3. Wallace, p. 11.

Fire Prevention Bureau Organizational Structure and Function

Table of Contents

Key Terms

Key Points

1. The head of the fire prevention bureau can function best when the position reports directly to the fire chief.

2. The head of the fire prevention bureau is responsible for planning and implementing the three Es: education, engineering, and enforcement.

3. Progressive fire prevention bureaus must nurture strategic partnerships with designers, builders, regulatory agencies, manufacturers, and ordinary citizens.

4. The basic goal of designing and implementing an organizational framework is to provide the number of resources necessary to provide the best possible service.

5. The greatest asset a fire department has is its personnel.

6. Filling fire prevention positions with educated, qualified, and technically proficient individuals is a must.

7. Performance-based designs require the authority having jurisdiction not only to grasp the design method but to ensure that the design parameters are maintained.

8. The key to choosing the best staffing option for your organization is to identify the strengths and weaknesses of each in relation to your needs.

9. If careful management and thoughtful processes are used to match the right people and resources to the right job, success will reign.

10. Identifying problems or verifying code compliance issues early helps ensure increased protection, not only for occupants or businesses but for fire crews when they respond to an incident.

11. It is imperative that all fire prevention bureau personnel have the interests of fire suppression personnel in mind when they are performing their fire prevention duties.

12. It is necessary that you keep track of decisions, changes, and modifications throughout the design and construction process.

13. Inspections of existing buildings are intended to prevent hostile fires when possible, mitigate the effects of a fire should one occur, and minimize hazards and thereby the risk.

14. Public information should be a year-round effort given high priority and substantial support.

15. The line companies are the best people to conduct preincident planning.

16. The occupant services section can help occupants greatly by tying up all the loose ends or serving as the go-to person after an incident.

17. Wildland risk management requires emergency service agencies to be more interdependent.

18. If we begin treating structures as part of the wildland/urban interface problem, rather than just the vegetation, we will go much further in mitigating wildfire threats.

Learning Objectives

1. Identify laws, codes, ordinances, and regulations as they relate to fire prevention.*

2. Understand code enforcement as it impacts life and property loss.*

3. Define the national fire problem and role of fire prevention.*

4. Define laws, rules, regulations, and codes and identify those relevant to fire prevention of the authority having jurisdiction.*

5. Define the functions of a fire prevention bureau.*

6. Describe inspection practices and procedures.*

7. Identify and describe the standards for professional qualifications for Fire Marshal, Plans Examiner, Fire Inspector, Fire and Life Safety Educator, and Fire investigator.*

8. List opportunities in professional development for fire prevention personnel.*

9. Identify the duties and responsibilities of fire prevention.

10. Identify staffing options for fire prevention bureaus.

11. Understand the advantages and disadvantages of the various fire prevention staffing options.

FESHE Objectives (USFA)

Fire Prevention Bureau Organizational Structure and Function

Chapter 5

Case Study

The Anytown Fire Department protects a population of 75,000 people with four fire stations. The fire department staffing at the four stations includes:

Station One

- One battalion chief (shift commander)
- Two paramedics (ambulance)
- One captain, one driver/operator, two firefighters (truck)
- One lieutenant, one driver/operator, two firefighters (engine)

Station Two

- Two paramedics (ambulance)
- One lieutenant, one driver/operator, two firefighters (engine)

Station Three

- Two paramedics (ambulance)
- One lieutenant, one driver/operator, two firefighters (engine)

Station Four

- Two paramedics (ambulance)
- One lieutenant, one driver/operator, two firefighters (engine)

The fire prevention bureau is staffed by all sworn personnel consisting of one deputy chief, three fire inspectors with the rank of Lieutenant, one fire and life safety educator with rank of Lieutenant, and one Captain serving as the function of plans examiner.

The Command staff consists of one fire chief, a deputy chief, a battalion chief of training, a battalion chief responsible for EMS, and one administrative assistant.

Budget limitations have required the fire department to evaluate the necessity of the functions of the fire prevention bureau. The fire chief has been asked to present options for staffing the fire prevention bureau to provide the minimum level of service needed. Presently, shift personnel do not perform any fire prevention functions.

Administration

The most recent economic recession impacted many Fire Prevention Bureaus and fire prevention services provided by fire departments. During this economic climate, Fire Prevention Bureaus were restructured and some of them were absorbed by other departments.

It is now not uncommon to have fire prevention functions completely outside the fire department organization. What was once a rare occurrence to relocate fire prevention services out of the fire department has become a more prevalent means to address budget constraints. Budget constraints and reduction in personnel have resulted in mergers of various departments with municipal organizations, such as building departments and Fire Prevention Bureaus. Fire prevention functions, or bureaus themselves, can have many variations in configuration and organizational layout. Some bureaus report directly to the chief, while others report to other chief officers or department heads/supervisors. What follows are descriptions of functions and organizational examples that work to manage the mission, although they may be positioned differently than traditionally established.

In this chapter, organizational designs are presented that have provided the best results in accomplishing a fire prevention and mitigation mission. Additionally, an overview of fire prevention bureau functions and options to perform those functions are provided. Details of the specific fire prevention functions are examined in following chapters. We also recommend becoming familiar with National Fire Protection Association 1730, *NFPA® 1730: Standard on Organization and Deployment of Fire Prevention Inspection and Code Enforcement, Plan Review, Investigation, and Public Education Operations.* This 2016 standard provides guidance and performance criteria for how fire prevention bureaus or functions should be established and managed.

As with any loss-control function, whether for a large industrial company, municipal fire department, or fire district, the supervisor of the fire prevention bureau can function best when the position reports directly to the chief executive officer or, in the case of the fire service, the chief of the department. This direct communication is important, not only for the functions of the bureau, but also the needed communication with top administrative officials. The supervisor of the fire prevention bureau is typically an assistant chief, deputy chief, or battalion chief. However, depending upon the size of the department, the position may be that of a captain or any rank as determined by that jurisdiction. Typically, the person responsible for supervising and managing the fire prevention bureau has a title of fire marshal or fire code official, which may or may not have a specific rank because the position may be filled by a civilian and serve in title only.

The fire marshal, fire code official, or Authority Having Jurisdiction (AHJ) is basically an agent of the chief regardless of where the position is located. Typically, adopting the fire code formalizes the Chief as the Code Authority; therefore, whoever is identified as the Fire Code Official must still work under the authority of the fire chief. In order to change this, the fire code must be amended to modify this authority. We cover more specifics later, but having now established who is in charge, we will next discuss some of the organizational structure's functions and layout.

The leader of the fire prevention division, bureau, or office shoulders the responsibility for planning and implementing the "Three Es" (education, engineering, and enforcement). Consider this important note regarding the first E, Education. Traditionally, education is achieved through public educators. However, the formal public education function may not transfer through reorganization or redeployment efforts and must be included in the fire prevention organization's critical functions. Education is still a critical part of the code enforcement function for businesses, developers, or contractors.

The leader of the fire prevention division, bureau, or office may wear many hats. In fact, in smaller departments or districts, he or she might conduct all fire prevention functions or activities. While reading this chapter, consider how you might be positioned within an organization and what functions you might be charged with performing. There is no absolute or correct solution other than the one that needs to be tailored to best meet the needs of the community served. Remember, our discussion focuses on the "typical" department, one that is comparable to a larger department with an experienced staff and dedicated resources. Modifications are certainly acceptable; as our intent is to communicate concepts to you, not a rigid layout or organizational blueprint. Consider this as more of a road map of how to get there.

During tough economic conditions, fire prevention and the related functions are often subject to scrutiny and mere justification of their existence. Fire prevention and the related functions can easily be argued as being the single most important fire protection mission in any department. This runs contrary to some views, whether in the fire department itself or in other offices such as a building or other managing departments. Prepare your argument and remember, as fire service professionals we deal with hostile fires in two principal ways:

- We mitigate, control, or prevent fires from occurring (best and most economical choice).

- We respond to fires and try to put them out when prevention fails (better than none and most costly choice).

Now more than ever, fire departments must foster progressive prevention programs that nurture strategic partnerships with designers, builders, regulatory agencies, manufacturers, and ordinary citizens. Communication continues to be our downfall here. Nurturing these relationships will help us to surmount many of our current shortcomings and will contribute to successful community building (**Figure 5.1, p. 88**).

We must also acquire a much broader view of our political obstacles. The salary and service of government employees have taken a center stage in most communities. Citizens may have the opinion that parks and recreation is more important than fire apparatus. Others may think police protection is a higher priority, or maybe potholes and transportation are in worse shape

Figure 5.1 Fire departments can foster progressive fire prevention programs through the development of coalitions and strategic partnerships.

than the fire department. We live in a world of "Nothing is burning now, so what's the emergency?" How many fires do we actually have? What are the greatest demands for our service? Is it EMS? Is it special rescue? The many competing needs in any community must be funded from the same limited resources. Politics is a fact of life and the very core of the bureaucratic society in which we live. Resources cannot be detained without political support, and political support cannot be garnered without citizen support. Think back to the days immediately following 9/11/01 and the outcry for more police and fire personnel. Who would have thought of cutting services then? Now, years later, many have forgotten or are at least refocusing their priorities. The will of the people must be enabled with political capital. As paid professionals or volunteers, our services must be balanced for the good of the communities and the taxpayers — the ones for whom we work and support.

Let's look at some common fire prevention functions and methods to perform those functions. Note that not every department may perform all of these functions. Other departments may have one person perform them all, or they may be performed under a different title or name or even by a different entity altogether. Yet, other departments may have positions not even mentioned here. In any case, the field is varied and the competencies many. This chapter will explain what *tasks* fire departments should perform, not necessarily what positions should be filled. The objective is to provide guidance only and not offer answers for every situation. Different jurisdictions may have different needs that require different arrangements. Any combination is acceptable as long as it meets the objectives and, most importantly, the needs of the community being served.

Identify the Staffing Levels to Provide the Desired Level of Services

Before the staffing levels of a fire prevention bureau can be considered, some critical points must be examined. The first step is to determine the desired level of service to be provided by the fire prevention division or section. What fire prevention service level has been directed to the fire chief and how has that level of service been justified?

For example, if the elected officials or governing body have indicated it is important to the safety of the citizens that all occupancies are to be inspected once a year, begin structuring your organization based on the number of properties to inspect over that period. Assuming the inspector's only task is to conduct fire inspections, estimate how many hours he or she works a year. Deduct hours for conferences, vacations, holidays, and sick time. When you have determined the number of hours available, you can then factor in the number of inspections required. Some inspections take longer than others, and many require reinspections to achieve compliance, depending on depth and complexity. It is your responsibility to provide the recommendations of your service needs to the elected officials. You must support and document the positive outcomes of our service as well as the consequences of not performing the services at the recommended levels.

Accordingly, to address overall staffing levels, examine every function or service of the Fire Prevention Bureau and correlate it to the level of service the community has determined should be provided. The next step is to decide the total number of individuals needed to provide that level of service. Keep in mind that elected or policy officials will be asked for appropriate funding and will determine our level of service. As we consider the staff needed to provide the determined level of service, also examine support functions. These can include clerical support, training, vehicle maintenance, and information technology. A fire department large enough to have a substantially sized Fire Prevention Bureau, will greatly impact the organization. When creating an organizational framework, include the Fire Prevention Bureau in the fire department's strategic planning process. This is when other divisions or sections become involved in determining additional staffing needs department wide. Examples could include:

- Information technology support (how many computers and computer fixers)
- Number of vehicles (trucks, cars, four-wheel drives)
- Training staff for inspector training
- Staffing or enhanced software for bigger payroll
- Geographical information systems support

In most cases the supervision of fire prevention personnel is dependent on the size of the Fire Prevention Bureau and the makeup of the other division's supervisory levels. For example, an engine or truck company may be composed of a crew of four. One of the crew members is the officer or supervisor, such as a lieutenant. This gives a span of control over three people. Another individual, such as a captain, may manage multiple companies or engines in a station. The next level, such as a battalion chief, may manage a group of stations in a district or in the entire community. A common practice is establishing a span of control not exceeding five to seven subordinates to one supervisor. If you double the same span of control that your fire department organization uses to staff the emergency response division, you can get a fairly accurate idea of the type of supervisory or management structure that should be in place for your part of the organization.

Determine the Optimum Organizational Framework for the Staffing Levels to Provide the Desired Level of Services

The design and implementation of an optimized organizational framework is dependent upon the tasks required. Basically, you should provide the number of resources necessary to provide the best possible economic and customer service.

For example, consider one organization's estimate of how many inspectors and plans examiners are needed to process a particular workload. This example is based on a workload study performed in the Colorado Springs, (CO) Fire Department's fire prevention division. The study provides an approximate number of inspections and plan reviews that could be expected from each staff member. Note that this number includes a typical number of reinspections and resubmittals of plan reviews. This gives a fairly accurate representation of "typical" inspection and plan review functions. An inspector could be expected to perform 1,000–1,200 generalized inspections per year, and a plans examiner could be expected to perform 900–1,050 typical plan reviews annually.

The following represents average totals of inspections and plan reviews defined as follows:

- *Type of inspections* — New construction, existing, occupancy, fire detection and alarm installations, fire suppression system installations, hazardous materials, complaints, and referrals. The number basically averages to about 2.2 to 2.4 inspections per address which is equates to 400-500 addresses annually.

- *Plan reviews* — Development, new construction, detection and alarm plans, fire protection systems, water main, hazardous materials, and miscellaneous permit-required drawings or sketches.

The numbers allow the fire prevention manager to roughly determine how many staff members are required to perform these functions. Obviously, if you are forming a small department, one person may be able to split the duties handling around 1,000 total inspections or plan reviews. Larger departments may need a larger complement of staff divided into different sections as shown in **Figure 5.2.**

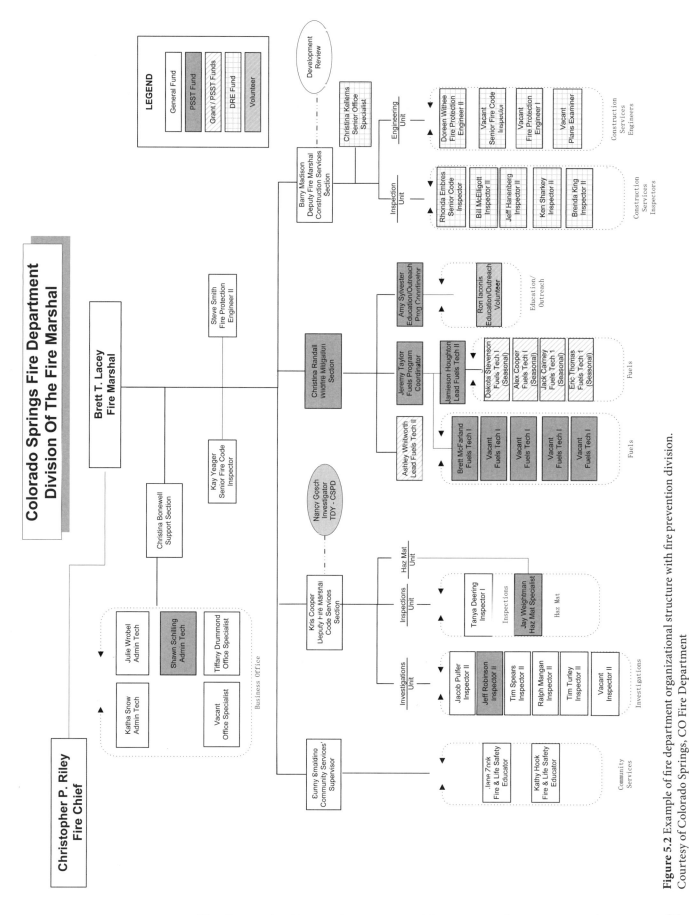

Figure 5.2 Example of fire department organizational structure with fire prevention division. Courtesy of Colorado Springs, CO Fire Department

This process may not be as straightforward when determining the other staffing levels for other Fire Prevention Bureau functions, as the tasks can be quite varied and nonspecific. For example, the division should have enough fire investigators to handle the number of fires to be investigated. This forces the division to implement policies after asking questions such as:

- Do you investigate all vehicle fires?
- Do you investigate all fires regardless of dollar loss?
- Do you work 24-hour shifts or FLSA 43-hour workweeks? (**NOTE:** The Fair Labor Standards Act requires that firefighters who are sworn police officers acting in that capacity as fire investigators must be considered as police officers and must receive an overtime rate after a maximum 43-hour workweek.)
- Do you have one vehicle per investigator?
- Do the investigators take vehicles home?
- Are investigators on standby or on-call? Do you have a defined policy for this?
- Can your budget handle on-call and standby pay?
- How many fires per year require investigation?
- Can some of the investigations be performed by on-duty company officers or firefighters?

Only after you have answered these or similar questions can you begin to address workload and staff requirements. Too many variables exist that are based on case load, cooperation with the district attorney, etc., that we have no rule of thumb for the number of investigators vs. the number of cases. This is a very specific issue based on the local jurisdiction and the enacted policies and procedures. However, the example should help to determine your own department's optimum number of fire investigators.

Life Safety Educator. The life safety educator position is very hard to quantify, particularly if you follow the recommendations in the following chapters. Specific work tasks and objectives are required, and you have to complete specific time studies to make any reasonable estimate. Conferring with business consultants and professional educators may be appropriate, as these positions generally conform to a typical professional business and educator functions more than the other technical positions.

Fire Protection Engineers. Fire protection engineers' positions can vary. If their functions are similar to those of plans examiners, the numbers previously discussed will work. However, a wise fire marshal will use these positions to perform higher-level applications such as fire modeling, risk analysis, technical research, and higher level technical studies, process management, and support to fire department operations at emergency incidents. Again, these positions can be closely aligned with a standard engineering business model that is easily time tracked. However, doing this requires a specific set of performance expectations in order to construct an accurate evaluation model. If you define exactly what you expect fire protection engineers to do, evaluating the model is easy.

Staffing Options for a Bureau

Many successful managers have stated time and time again that the greatest asset a company has is its personnel. This is especially true for a service organization like a fire department. Fire departments do not mass-produce products for distribution. The product they provide is service to the tax paying citizen. The citizens and business community are, in fact, the customers of the fire department.

Fire departments are really no different than the business community when it comes to selecting the people best suited to do the job. Both want to select the most qualified people. What can differ between fire departments and the business community is the process of selecting the individual for the position.

In years past, some fire departments would use the Fire Prevention Bureaus as dumping grounds for poor performing or injured personnel. Personnel were not always assigned to the Fire Prevention Bureau by choice. In some fire departments, people who were hired but then did not "fit the mold" as a firefighter were sent to the Fire Prevention Bureau. An old phrase labeled fire prevention as the depository for the sick, lame, or lazy. Instead of receiving a disability pension, injured firefighters have been sent to the fire prevention bureau to finish out their careers. Other departments have utilized these along with other staff positions as required tasks for promotional eligibility. This in itself is not bad, but typically the bureau is the last place these individuals want to work. Most personnel are hired on to be firefighters and not to conduct fire inspections or to perform other fire prevention functions. Significant problems can result from forced transfers or promotions into the Fire Prevention Bureau. These problems might include increased turnover, job dissatisfaction, poor work attitudes, apathy, and (most obvious) poor performance. Nothing is worse than having someone in a position just biding their time, waiting to move along.

Poorly motivated or ill-prepared individuals face considerable challenges if they are suddenly thrust into a Fire Prevention Bureau. There have been considerable advances in construction materials, engineering, building techniques, and in the codes themselves. New buildings, combined with the rapidly changing technology, make it difficult for just anyone to work productively in a Fire Prevention Bureau. Technically complex fire suppression and alarm systems are becoming more and more prevalent (**Figure 5.3, p. 94**). Most fire departments adopt ordinances making significant local amendments to the nationally recognized codes. The many activities required of the fire prevention professional in the past decade have placed fire prevention positions among the most professional and technically challenging assignments in most fire departments. Fire prevention personnel in the 21st century will be some of the most technically trained people in the fire department. To provide the best, most technically proficient service available, filling these positions with educated, qualified, and technically proficient individuals is a must. We have long since left the age of just "doing" inspections.

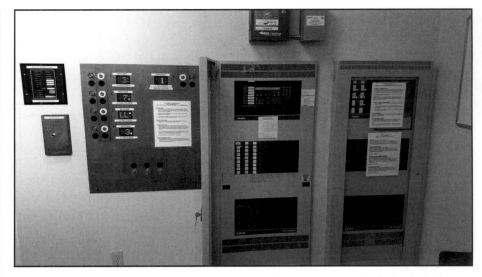

Figure 5.3 Fire alarm systems are more complex and require a greater level of technical expertise.

As an example, fire departments amend code requirements for sprinklers and fire alarms. It is not unusual to require automatic sprinkler protection in all new construction regardless of square footage or occupancy classification. Whether in a home or a convenience store, sprinkler protection may be required. The complexity of sprinkler and fire alarm systems has increased so much that newer and more effective sprinkler or fire detection devices are routinely available. From 1955 to 1981, the choice of which sprinkler to install was simple; there were only three types, an old style, a pendant, and an upright.

In 1981 sprinklers were manufactured and tested by listing agencies to be used in specific occupancies (hazards) rather than in a one-sprinkler-fits-all-situations approach. Until 1991, automatic sprinkler systems were permitted to be installed based on what is referred to as the "pipe schedule method." The pipe schedule design based the size of the sprinkler piping on the occupancy hazard of the building and a predetermined number of sprinklers supplied by the size of the piping. To evaluate the design, the construction document reviewer would count the number of sprinklers on the sprinkler piping and then refer to a table in a code book to see that the system was designed properly. Today, pipe schedule sprinkler designs have little use and are a permitted design method in only limited applications, such as pre-existing installations. Sprinkler systems today are hydraulically designed. This type of design was developed as a means to reduce the cost of sprinkler systems. The size of the sprinkler piping is not based on the number of sprinklers the piping supplies but on the amount of pressure and gallons-per-minute available for the system to operate at the most remote sprinkler head. The size of the pipe is limited basically to a mathematical proof of how much water will be available at how much pressure. This is much more complex than the pipe schedule method and requires significantly more training and expertise to verify.

Similarly, fire alarm and detection devices are changing as fast as the electronics and computer industries. Fire alarm panels as well as the detectors themselves change faster than many experts can keep up.

Fire codes have not changed just in the area of sprinkler and fire alarm design and installation. Until recently, most fire and building codes used what are referred to as prescriptive codes. Prescriptive codes prescribe, or dictate, specific requirements or combinations of requirements, much like a "cookbook" approach. As an example, a common prescriptive code requires a typical stairway be constructed with a minimum width of 44 inches (1.1 m). Using this type of approach, prescriptive codes permit very little flexibility in the design and construction of a building.

Today, most all building codes have begun including an alternative to the prescriptive approach. Alternate means and methods, or performance-based design approaches, are now permitted in building codes used throughout most of the world. The performance-based design option was included for the first time in the 2000 edition of the *Life Safety Code®*.

Performance codes detail objectives and establish criteria for determining if they have been met; thus the designer and builder are free to select construction methods and materials as long as they can be shown to meet the performance criteria.[2] The methods to achieve the objectives or goals are proven by a sequence of complex calculations and computerized fire modeling based on assumptions agreed to by the designer and the AHJ. The computer fire modeling calculates the effects of a fire and associated products of combustion in a particular room or space. Additional modeling indicates the potential effects on the occupants. The overall outcome of the fire modeling will determine the construction requirements for the building.

The design engineer and architect may reduce the cost of the prescriptive code requirements by using the performance-based design option. If the designer can show the authority having jurisdiction that the design can meet the goals for the level of safety, the prescriptive code requirements do not have to be met. This forces the AHJ to ensure that the engineer and architect are accurately representing a computerized fire model and sequence of calculations. The major stumbling block of this process is that the parameters of the buildings design must remain constant for the life of the building. The conflict that surfaces is that buildings never stay the same. After a building that incorporates a performance-based design is constructed and commissioned, the long-term maintenance of the design becomes a significant factor during the inspection of the building.

Performance-based designs require that the authority having jurisdiction not only grasp the design method but to ensure that the design parameters are maintained. This has propelled the authority having jurisdiction into a realm of higher technical expertise that requires extensive education, training, and experience. If the AHJ does not have this training, it likely will or certainly should have staff members that do. The AHJ can also use thrd-party engineering firms to address those issues.

Many smaller fire departments throughout the country do not have an individual on staff capable of reviewing and inspecting performance-based designs. Some fire departments are not convinced that performance-based designs are a reality or a probability, and subsequently they are not prepared to handle them.

Staffing with Sworn Personnel

Many fire departments staff the fire prevention personnel with assigned firefighters, known in many jurisdictions as sworn personnel. Typically, sworn personnel are civil service employees of the fire department who have taken an oath to perform their duties. Fire departments that assign firefighters to the Fire Prevention Bureau do so in a variety of ways. Some fire departments assign them to the fire prevention bureau. Others ask for volunteers to transfer into the Fire Prevention Bureau or require them to work in the Fire Prevention Bureau for a predetermined amount of time as part of the promotional process as already mentioned. Others have fought hard to keep firefighters in the bureau; however, due to a significant lack of interest, bureaus have started offering incentives by creating promoted positions to generate interest. Typically though, these are short-lived because the increased cost of personnel forces departments to pursue alternative, less expensive staffing, such as civilian or nonsworn personnel.

Most firefighters were attracted to the job of fire fighting because of the ability to work on shift, fight fires, and be on the trucks. Generally, if people are forced to do something they do not like or want to do, motivating them can be difficult. Forcing an individual into a position is never a good idea, unless it is absolutely necessary. Certain skills for fire prevention are far different than those for suppression. On the other hand, there are those sworn personnel who eagerly volunteer for the opportunity to work in fire prevention. Putting one of these individuals in fire prevention can be an asset to the community as well as the department.

Fire departments in areas of the country where firefighters carry police officer powers may prefer to use sworn personnel to conduct fire investigations. Historically, however, many firefighters placed into fire prevention generally have not had the necessary qualifications to perform the assigned duties. Most of their training was done on the job (OJT), and just when they got good at what they were doing (after two years or so) they would be transferred back to the line. Promotions of people within the Fire Prevention Bureau can be another downfall. In the fire service, how can employees better demonstrate their skills, determination, and leadership than by taking promotional exams and being successfully promoted? While this stands well for these people, the bureau loses high-quality performers. Of course, in many cases, successful leaders in the fire service typically held many staff positions, which helped them round out their careers by increasing and expanding their overall knowledge for the job. Historically, promotions within Fire Prevention Bureaus are limited. Those sworn fire prevention bureau personnel successful in the promotional process are typically promoted "out" of the fire prevention bureau.

Staffing with Civilian Personnel

Civilian personnel are individuals who are not typically sworn into their position through an oath of office. In some fire departments, civilian personnel (sworn or not) may wear a fire department uniform and also hold a rank. Civilian personnel do not serve under a board of fire and police commission or other governing body such as civil service. Most civilian positions do not require an employee entrance exam and are subsequently not protected under civil service laws. These individuals are typically hired on an open-competitive basis, this means that individuals must interview and show their best qualifications for the job while competing among a field of many candidates, which is similar to the private sector.

Occasionally you may find departments that debate the issues of civilian vs. sworn and if a sworn person can take direction from a nonsworn person. In some cases, sworn ranks may develop an elitist attitude over civilian rank or status, thinking they should never work under or for a "civilian." Many of the chiefs in fire departments are not within the civil service rank and file, and although they likely swore an oath, they are very similar in many ways to civilians in a fire department. They have been appointed by an elected official or board of civilians. The city manager, mayor, or other public official who is the chief's authority figure is likely a civilian. Most certainly, the public we serve are civilian, and they are not sworn but are undoubtedly employers. The fact is, all sworn personnel are employed and directed by civilians. In either case, it is the manager's job to make sure everyone works to the best of their abilities and accomplishes the mission, regardless of their status. Some advantages of staffing with civilian fire prevention personnel are:

- Cost savings
- Consistency
- Quality
- Decreased turnover
- Specialized technical education

Our discussion of sworn personnel emphasized logistical problems, such as having a sharp individual become well trained just in time to rotate back to the field. Civilian staffs typically allow fire prevention bureaus to hire someone for a specific job who can provide productive and consistent longevity. Individuals with specialized training may be more productive in the long run and have lower turnover rates than sworn personnel. An example of an employee with specialized skills and training whom the fire prevention bureau can recruit is an individual with a degree in fire protection engineering. Most fire departments have personnel with degrees, but few have sworn firefighters with degrees in fire protection engineering who are willing to work in the Fire Prevention Bureau. As previously discussed, the escalating responsibilities and duties of positions in the Fire Prevention Bureau have created a need for highly skilled, technically qualified individuals.

Disadvantages of using civilians for fire prevention work can include:

- Lack of fire service experience and fire-related education
- Limited promotional opportunities
- High turnover due to salary issues compared to sworn personnel in similar positions and responsibilities

The issue of fire service experience can be overcome by posting minimum fire service requirements for the fire prevention job. The position may require experience or fire service training. For example, the City of Colorado Springs has utilized this type of requirement for some time and has been very successful at recruiting highly trained personnel who came from other fire departments, retired from within their own organization, attended a fire recruit academy, or was an active volunteer firefighter in another department.

Limited promotional opportunities can be frustrating for ambitious employees. In some cases, advancement is limited in a fire prevention bureau unless the community experiences substantial growth and warrants a larger bureau. This can become more of a policy issue within the department, which might be modified to retain these individuals. Refer to the various organizational charts to see how different departments address this issue. Individuals may realize the only way to advance in their careers is to move to a larger department, particularly if the management-level positions in its Fire Prevention Bureau are civilian or it has different job classifications and levels of inspectors, life safety educators, or investigators.

The turnover rate is more often than not attributed to a lower pay scale for the civilians. Many departments have strong labor/management relations tied to civil service. These departments typically fight for and receive higher pay for sworn ranks. Civilians within the same department may tend to feel that management or the city administration views their jobs as less important than that of the firefighter. In many fire departments, recent economic conditions dictated budget cuts resulting in replacement of sworn fire prevention personnel with a lower salaried civilian.

In some organizations, civilians may not have the equipment, skills, or expertise to take an active role at emergency incidents. In others, they may have a significant role: they may coordinate tasks such as emergency operation or coordination center functions and evacuation implementation. Civilians may have an active role in Incident Management Systems (IMS) such as planning, logistics or finance, damage assessments, and flood monitoring. Civilians can contribute to the organization if they do not have any fire service background. Each staffing option has advantages and disadvantages. The key to successfully choosing the one for your organization is to identify the strengths and weaknesses of each option in relation to your needs. Select solutions that address the weaknesses of the individual, which will determine the best staffing and training options available. What is necessary is the establishment of good policy, proper procedures, and sufficient training to accomplish the required functions and tasks — regardless of whom you employ.

Outsourcing Fire Prevention Services

As previously mentioned, the country's economic conditions placed many fire departments in a position to seek somewhat nontraditional ways of providing a cost-effective service to their citizens. Historically, when the demand for an increase of quality services at lower prices evolved, outsourcing of services became an option to consider. Now outsourcing has become prevalent in all forms of government and even in the private sector. Outsourcing may be used for specific services such as construction, document review, or as general as inspections. Due to other circumstances, such as an onset of high workload and service demands, it may be beneficial for a bureau to outsource or contract a third-party plan review or inspection service. Outsourcing can reduce some of the bureau's work while still holding contractors and designers accountable. In many instances, the contractor or developer pays for the third party plan review service, and the report is sent directly to the fire department for data recording and permit issuance. Essentially, the contractor is paying the fees for the service while the work is being performed for the fire department. It is critical that the obligation and service performance of the third-party service is clearly defined in the agreement or contract. Their work is clearly for code enforcement for the fire department or agency doing fire prevention work, not for the design professionals.

A common method of using third party plan review services is to have the contractor submit plans directly to the third-party reviewer, who will in turn do the review and bill the client directly. However, in our opinion, it is more appropriate to have the construction documents submitted to the fire department who, in turn, forwards the documents to the third-party company for review and back. This eliminates invoicing issues for the department and allows the client to deal directly with the third-party contractor.

The contracting of fire department services is also not limited to fire prevention functions. In the scheme of bureau outsourcing, the contracting of services is a form of privatization. **Privatization** is the transfer of functions or duties previously performed by a government entity to a private organization. This strategy is sometimes used to address either personnel issues or funding in a fiscally challenged organization. Many municipal administrators throughout the country have used outsourcing to solve their budget woes. By outsourcing, municipalities or agencies limit legacy costs such as vacation, sick leave, retirement, workers compensation, etc. Most administrators initially examine outsourcing as a means to save on these costs or address concerns regarding the quality of service. A major reason that outsourcing might reduce costs is that private organizations have incentives to keep costs low. They need to operate effectively and efficiently in a competitive environment. Most government entities do not have the same incentives as private business to keep down the cost of their service. One of the superior benefits of privatizing services is the competition. Often, government agencies, since they are the sole source of their services, become complacent and unmotivated to change. Businesses on the other hand, must change in order to remain competitive with one another. One method that private companies use to stay competitive is to reduce costs and still provide quality innovative services. Management in private industry can achieve this by establishing incentives that encourage cost savings. The difficulty for government is that it typically enforces its existence by demanding compliance under threat of penalty rather than trying to do the "right" thing.[4]

Rural Metro is an example of a successful privately-owned company that had been specializing in providing both fire and emergency medical services in various parts of the country, since 1948. This department serves several metropolitan areas in California, Arizona, and Alabama. When they were in Scottsdale, Arizona, its operation relied heavily on prevention and mitigation, such as mandatory fire sprinkler installations throughout all new construction.

Privatization — The transfer of functions or duties previously performed by a government entity to a private organization.

American Emergency Services provides fire, EMS, and fire prevention services under a contract with a fire district. The fire district is surrounded by municipalities that operate traditional municipal fire departments. Gary Jensen, the company's owner, has served as fire chief for the department since the company began providing fire protection for the fire district. This privately-owned fire department functions in the same manner as the neighboring municipal fire departments. It is a full participant in the suburban mutual aid program and has automatic aid with neighboring communities.

Another solution, known as **enterprising**, may actually become more prevalent for Fire Prevention Bureaus than for any other division of the fire department. An enterprise organization is run like a business, and the services it provides are avenues for the entire organization or a division of the organization to operate fully on a cost-recovery basis. These departments generate revenue and in turn use it to cover their expenses. As we already stated, the demand for qualified, technically competent individuals with specialized skills has been identified within fire prevention divisions. The foreseeable increase in performance-based designs, complex hydraulic sprinkler calculations, and computerized fire alarm systems are just some of the issues facing the person responsible for construction document reviews. These issues become overwhelming for an inexperienced review and inspection staff. It is highly likely the skills needed to address these technical demands may not be available from the pool of fire department members.

Like other businesses and divisions within fire departments, Fire Prevention Bureau Personnel may periodically see an increase in the demand for their services. Today, staffing levels have been reduced in many fire prevention bureaus. They cannot handle an increase in demands for service, and hiring additional staff is not feasible. The demand for services may increase seasonally or temporarily, as during construction seasons or development booms. It can be difficult to justify hiring a person to assist with these short-term work demands. In situations such as these, it may be necessary to identify services that must be reduced, scaled back, or eliminated.

Staffing a Fire Prevention Bureau in Volunteer Departments

Often, Fire Prevention Bureaus in volunteer departments are staffed by part-time paid staff, by paid staff, or sometimes by the volunteers themselves. Obviously, there are different requirements and expectations for each of these. Some volunteers have more zeal and dedication than "professional" staff in other departments. In all cases, if people are properly trained and adequate numbers of staff are provided, the results should be the same. The aspect of paid versus volunteer should only matter in the realm of remuneration and corresponding laws, policies, and procedures, and not in the way our clients are treated or our tasks are performed.

Tasks Within the Bureau

Fire Protection Engineering or Plans Examination Section

Fire protection engineers are typically responsible for conducting high-level plan reviews and design work on a multitude of design and construction aspects such as development plans, plats, zone changes, new building design and construction, fire detection/suppression systems, and process design or methods. Plan reviews provide the first opportunity to critique a design before it is permitted for construction or actually installed. It has increased in popularity over the past decade or so because of the realization that fire service input and review provides long-term benefits to managing risk. Since many departments realize its importance, this is a position often outsourced to a third-party entity. Identifying problems or verifying code compliance issues early in a design process helps ensure not only increased protection for occupants or businesses but protection of fire crews when they are responding to an incident. Fire protection engineers seek not only to protect the building occupants but the firefighters as well. It is imperative that all Fire Prevention Bureau personnel have the interests of fire suppression personnel in mind when they are performing their fire prevention duties.

Fire protection engineers are becoming more commonplace in the design industry and consequently, more prevalent in the fire service. Since most fire departments of any significant size review construction plans for new buildings, processes, or systems, they need a competent staff to perform that function. In years past, those departments that hired engineers or architects rather than taking sworn firefighters off the line to handle this task were generally large municipalities or large departments, and they were the exception by far, not the rule. Because of the changes occurring in the model codes and other standards, departments should seriously consider individuals with engineering degrees or science backgrounds to act as liaisons as well as resident experts.

Not all departments can afford qualified engineering staff. However, if a smaller community or organization desires plan reviews, it is not precluded from having them; it only means that their level of competency will be less than that of a qualified engineer. Therefore, any individual who receives training and certification in such activities can review plans. The National Fire Protection Association and other model code groups still offer certifications for plans examiners based on NFPA® 1031, *Professional Qualifications for Fire Inspector and Plan Examiner.*

A larger engineering section might be divided as follows:

- *Life Safety Group* — Reviews new or remodeled construction plans and hazardous materials, related operations, or processes. The bigger the jurisdiction, the larger the volume of construction plans there will be. Hazardous materials processes or systems are generally not that large in volume, but frequent enough to require dedicated positions based on the complexity of processes and newer chemicals. By combining both of these functions, the work group can verify that the construction methods and requirements are commensurate with the needs of both the building code and the fire code. New construction projects that involve hazardous materials require extensive communication. Combining both in the same group of assigned responsibility should achieve the best of both worlds.

- *Development Review Group* — Responsible for all conceptual and proposed projects including annexation, development plans, plats, zone changes, and wildland/urban interface issues if appropriate. This group is also attractive to the local economic development organizations, as it is a potent resource for clarifying hard-to-answer questions for businesses being recruited into the community. The review group needs to communicate carefully and diligently with the life safety group because commitments or decisions will likely be made here as the first of many steps in the construction phase of a project. The client will want the process to flow well, and if these two groups misstep, they can cause major problems and delays.

- *Systems Group* — Responsible for handling water plans; fire detection, alarm, and suppression plans; special hazard systems; and anything else that falls in this category. The systems group will also have to interface with the life safety group as theirs will be the last set of plans reviewed in the long line of plan reviews to come through the office.

There will undoubtedly be debate about how or why to break these groups up. Why not keep from specializing and make everyone generalists? Depending upon the size of your department and the workload you process, that may be feasible. In some fire departments, the size of the municipality may not dictate the need for separate groups, but the functions will be the same. However, the larger the demands, the higher the workload, the harder it is to maintain the expertise, quality, and control over all these areas. For that reason, it is best to segregate them when possible.

We recommend cross-training groups so that a baseline of support can be expected in the event of a required absence such as schooling, illness, or vacations. This is also beneficial to staff so they do not get stuck in one area too long and stagnate. The reality, though, is that no one person should be expected to be completely knowledgeable in every area. As we have said, small departments may have only one person doing these reviews. While one person can do them all, there will become a break point on capability and expertise and the quality may suffer.

Document all plan review activities as they not only become a permanent record of your work on given projects, but they also provide a technical reference for the designers, architects, and inspectors who will be following up on the project. Document meetings and phone conversations and file all correspondence. It is astounding how much communication can take place, particularly on large projects (See Chapter 14 for data management).

Fire and Life Safety Education

Fire and life safety education is one of the most valuable services of public fire protection. In most common practices, the fire service utilizes education for two purposes:

- Fire prevention and injury education
- Behavior modification

We propose a much wider usage. Assuming a fire department provides varied services such as emergency medical response, upwards of 70 percent of its call volume may be for medical emergencies. This being the case, why would public education venues only be limited to fire topics?

As in the engineering section, any number and type of individuals may staff the public education or fire and life safety education section of a department. An increasing number of fire departments are utilizing professional educators for performing the directed tasks. These individuals are highly skilled and talented at utilizing the best educational methods to transmit our message to their selected audience. The standard referenced and used for training and maintaining qualified staff in this section is NFPA® 1035, *Standard on Fire and Life Safety Educator, Public Information Officer, Youth Firesetter Intervention Specialist, and Youth Firesetter Program Manager Professional Qualifications, 2015 edition.*

Chapter 7 discusses fire and life safety education in detail as well as who should be targeted as an audience. Typically, the very young and the very old have been targeted. While this makes sense, as those two age groups are most at risk, consider targeting the other age groups who are responsible for the young and the elderly. Think about the ramifications of educating businesses in the proper use and maintenance of fire alarm systems. In Colorado Springs, Colorado, if educating this group on the topic reduced the incidence of false alarms and alarm malfunctions by just 10 percent, it would directly reduce overall annual workload responses by over 300 calls. In this particular community, that is nearly the same number of calls that determines the need for the construction of a new fire station at a cost of nearly $1.5 million a year.

You must consider all the incidents which impact your organization and think how to best educate your public. It may be important, too, to distinguish education from awareness. Most of the adults in the middle target group we spoke of do not want to be educated, but they will accept being "made aware" of various issues. It is a simple semantics issue, but sometimes can be critical in dealing with the masses. Topics need to be presented in a "teaching not preaching" approach. Even though these people may perceive themselves as simply becoming aware of issues, they are actually learning. Trained education professionals will help keep this issue in perspective.

For those departments that may not have the resources to hire specific staff for this purpose or that cannot afford to dedicate sworn positions, we strongly recommend forming collaborative relationships with local educators. Many groups and associations can provide input, volunteer support, guidance, and other tools to help you get the job done (see Chapter 7). Remember to pay special attention to cultural groups. Demographics play a part in our overall fire problem. Enlist help from a diverse group of people who represent the community you serve. These groups will help to make sure your message is right and will be heard in the manner it is intended.

The National Fire Protection Association uses two principal programs. The older and more traditional is Learn Not to Burn. The newer and more proactive program is Risk Watch. Both programs target specific elements of fire safety with Risk Watch being a more global approach to overall injury prevention. Many departments have implemented a more holistic approach to community risk reduction (**Figure 5.4**). An excellent resource is the International Fire Service Training Association's (IFSTA) *Fire and Life Safety Educator* manual.

Public education should be viewed globally. The fire service has done an exceptional job of marketing "Big Red" the fire truck and firefighters. We should take a more active marketing approach to all the services we provide, keeping our important safety messages and efforts intertwined so as to keep our entire community up-to-speed with our mission and relevant topics to keep them safe.

Figure 5.4 Fire departments have a holistic community risk-reduction effort that includes other educational elements besides fire. Courtesy of LaGrange Park (IL) Fire Department.

Fire Inspection and Code Enforcement

Fire inspections historically have been and still are primary functions of a Fire Prevention Bureau (see Chapter 11). Fire inspections are based on local or state laws that may require or recommend specific types of inspections for various operations or occupancies. Examples could include state licensing of daycare facilities, special-hazard occupancies, or high-rises. Many fire departments perform different types of inspections not relegated to licensing, such as new construction, existing buildings, fire protection systems, or hazardous materials. Various voluntary inspection programs may also be established, such as home and business inspections. All inspectors should be trained to meet the appropriate level of inspector qualifications based on NFPA® 1031, *Professional Qualifications for Fire Inspector and Plan Examiner.*

The details of these inspections vary depending upon the types of inspections to be conducted. Many departments perform voluntary inspections through suppression companies assigned to specific fire stations, while the more technically complex inspections are handled through the fire prevention bureau or division. Consider the quality and complexity of an inspection. As the fire code and its related standards become increasingly complex and detailed, properly training line officers and firefighters in these duties is becoming harder. Today it is very common to require company officers to be proficient in the basics of fire prevention activities. Some promotional processes at the company officer level include fire prevention topics. A great resource is the book titled, *Fire Prevention Applications for the Company Officer,* 1st Ed. Smaller departments may have a better ability to train their responders due to their smaller call volume, proximity to centralized training facilities, and overall cost efficiencies. Larger departments may find this more difficult because of numerous factors.

Example: Let's say you have 18 stations, seven of which are double-company stations (more than one principal piece of staffed apparatus). Your department runs three platoons or shifts. You want to put on a two-hour course for all officers, which means a total of 25 people each shift. You have a minimum staffing policy, which means you may have only three over hires per shift. To accomplish your goals for only a two-hour class, you may be able to pull only five people per shift for a bare minimum of 16 days, assuming only one makeup day for a 24/48 schedule. Other departments may be able to accommodate more training if they have audio/visual support, but this example shows that providing detailed, technical training is difficult at best. For this reason, most technical inspections are assigned at the fire prevention level, where personnel are in a better position to handle and schedule them.

Fire inspections help ensure a reasonable degree of fire safety based on fire code compliance. You can divide inspections within a prevention bureau or division into three basic types:

- Existing buildings (annual inspections)
- New installations or new construction
- Other (for example, complaints, inspections for fire prevention permits, or business licenses)

The following discussion will focus on existing buildings and new installations, which are the two types that a Fire Prevention Bureau most commonly encounters.

Inspections of existing buildings cover any and all elements, behaviors, or conditions that increase the hazard of a given building or process. These inspections are intended to prevent hostile fires when possible, mitigate the effects of a fire should one occur, and minimize hazards and thereby the risk.

Figure 5.5 Unsafe storage practices may also hinder fire department operations.

Housekeeping issues are common problems for correction in these inspections. The unsafe storage of combustible materials, debris, and access to attractive nuisances are just a few of the conditions an inspector should watch for **(Figure 5.5)**. Existing buildings typically have many mechanical and other features that should be checked periodically. These include mechanical and elevator equipment rooms, smoke control systems, fire doors, detection, and alarm and suppression systems, all of which should be checked or tested periodically to make sure they are in proper order. This may be as simple as verifying a private contractor has completed the required regular maintenance and systems check, or it may be as complicated as requiring that the test be done and witnessed by the inspector.

Occasionally light-duty personnel may be assigned to the fire prevention bureau. These are the firefighters or civilian staff that have been injured and can do some work but have not been medically cleared to return to their regular work duties. These individuals can be very helpful in handling lower level or basic inspections even re-inspections; however, it is strongly recommend to place them through some basic fire-inspection-training program prior to sending them out. Keep in mind our previous discussion of the "old way" of doing business and how all it did was frustrate people, get others in trouble, and make enemies of our clients and customers.

Public Information

The public information function has been named many things, such as community relations, public relations, media relations, and so forth. Public information should be a year-round effort given high priority and substantial support. This function is more important now than ever before (Chapter 8 explains this function in greater detail). The department's effectiveness in providing public information directly impacts other efforts such as fire and life safety education.

We live in the media age. CNN and web-based news are in everyone's home, or even more frequently, tweets and Facebook updates are on their pocket phone. Instant reports from around the globe keep us apprised of regional and international events. Because we are an "information starved" society, we desperately need to feed the machine by keeping people informed about what we are doing, what we can do, and what we plan to do. The media inevitably will portray an image of the fire department. It is the fire department's responsibility to use the media to its advantage.

The fire service typically is a public governmental agency. It is supported and funded by public dollars, and therefore needs public support. In today's tough political environment, the chief needs all the proverbial ammunition he or she can get to keep important issues in front of the department and in the open for everyone to see. The media can be either a godsend or a curse. The only way to maintain your media as a godsend is to be proactive and form partnerships with them (**Figure 5.6**). They can help you in ways you could never imagine.

What better way to assist your fire prevention mission than by using the media? Why wait for a catastrophe to hit before talking to them? Let them carry your public education messages. Let them carry your message of engineering services. Let them support your inspection program. Think of the contacts you can generate.

Figure 5.6 A chief meets with media. Courtesy of Ron Jeffers, Union City, NJ

The public information functions are critical to a department of any size. Smaller departments may defer the job to the chief. Larger departments may delegate these duties to officers or other individuals filling different full-time positions. Still others may dedicate staff and resources to the job. In any case, it is important to spend time on this function. This position not only can keep your customers (citizens) informed of periods of fire danger, help communicate when evacuations are necessary, or warn of restricted areas due to fire or a hazardous materials release, but it can also help your department maintain a positive image year-round.

Remember, the heroes of the job are the emergency responders. They are the ones who show up at your customers' door when they need help on the worst day of their life. Prevention personnel are the ones who show up when someone least expects or needs them, finds nothing but problems, and forces the client to spend hundreds or possibly thousands of dollars to fix them, just to have them come back again next year and find something else in need of fixing. Who needs the image help here, firefighters or fire prevention?

Today's media includes the internet. Taking advantage of social media and the internet is an excellent way to stay in front of the public and constantly present a message. Tweeting and using Facebook are the most prevalent social media available today in which they have the ability to touch every aspect of the world.

Preincident Planning

Preincident planning is essential to fire crews who respond to emergency incidents. Unfortunately, many departments do not spend nearly enough time preplanning. More attention should be given to preincident planning as it is now included as part of dangerous building placarding in the International Fire Code®.

We will start by explaining what preincident planning is or at least what we think it should be. Our view may be contentious, as it differs some from other people. Again, the amount of detail in preincident planning depends on the size of your department; however, we propose keeping it simple, understandable, and easy to use.

Preincident planning is the process of identifying tactical operations or procedures for specific occupancies, buildings, or locations should an emergency occur. It is a common practice to first identify target hazards. **Target hazards** are locations or buildings that are unique and differ from the "typical" buildings or structures throughout the response area of your fire department.

What does typical mean? We define "typical" as those locations that are bread-and-butter from an operational standpoint. For example, a typical bread-and-butter location would be a detached single-family dwelling. Anything "different" may require an automatic second or greater alarm, such as a potential for a significant number of lives lost or high life hazard (hotel, nursing home, high-rise assisted living etc.), a hazardous material, or an occupancy with unique construction features that may hinder or danger fire department operations (**Figure 5.7**).

Figure 5.7 Target hazards may include occupancies that need additional resources for evacuation or potential for large life loss.

Once you have a list of target hazards or dangerous buildings, you should develop a form, or emergency "incident cheat sheet," that provides critical information at a quick glance to assist the Incident Commander or company officer in making quick tactical decisions. Develop a customized form to assist you with policy or guidance decisions based on your department's operational tactics. Another way to look at preplanning is through comparing it to a simple risk analysis. Evaluating the following items can assist you in this task:

- Potential life risks (firefighter and occupant)
- Building or structure contents
- Construction features

- Built-in fire protection
- Suppression resources needed for mitigation

Limit the information on each location to no more than one sheet of paper. It is hard to fit three-drawer file cabinets in the cabs of most apparatus. Driving to a fire at 0300 hours in blinding snow, talking on the radio, and trying to open a three-inch, three-ring binder while people are yelling at you can be stressful. You need good information immediately available in small bites. Keep it simple. Some fire departments have elected to keep it simple by placing the information in an electronic format. Mobile data terminals and good Geographic Information System (GIS) support is invaluable (see Chapter 14).

The first due suppression company is the best to conduct preincident planning. The line is far better served if the members survey the premises and then record those findings they view as important. Firefighters in Colorado Springs (CO) perform this duty regularly and are encouraged to complete an inspection form if they see violations while doing the survey. The principal interest is for the crews to complete the survey with a secondary emphasis on completing the inspection form. In some fire departments, the Fire Prevention Bureau works hand-in-hand with shift personnel in maintaining the preincident planning data. The information obtained by the fire prevention staff during fire inspections is used to update the preplan, which was most likely developed or drawn by line personnel. This works well in some communities that do not have enough line personnel to visit the occupancies frequently and need another means to keep the preincident plan current.

Fire Investigations

Fire departments throughout the country deal with fire investigations in extremely varied ways. Some do nothing because fire investigations are the responsibility of the local law enforcement agency. Others will perform origin and cause determination and leave the investigation of arson to the law enforcement agencies. Still others perform all of these functions and have staff members who are certified peace officers with the authority to investigate and arrest.

Many jurisdictions delegate the responsibility of fire investigation to the fire chief. In these cases, it is common for the fire department to have dedicated members perform the investigative duties (**Figure 5.8**). Sometimes these individuals also double as fire inspectors. Fire investigations are a critical and integral part of the overall fire prevention mission. The information gleaned from fire investigation and post-incident analysis is important to feed back into code development and public education or public information. This information is critical for prioritizing fire safety actions. Without this link, a department can run on the same or similar incidents time after time without ever eliminating their cause. Specific information from a good fire investigation program is key to determining the specifics of the local fire problem. Fire cause trends can be established, fire prevention issues identified, and then measurement of changes evaluated.

Figure 5.8 Fire department personnel investigate the origin and cause of fire.

Deliberate communication with local or state law enforcement agencies is also a must. Collaborative and cooperative relationships must be nurtured for different agencies to communicate developing commonalities or criminal trends. Often, arson is not these individuals' only crime. They commit other crimes, which if unnoticed cannot be connected, thereby allowing them to get away. The fire investigation unit's organization within a department is just like the other groups we have discussed previously. Remember too that juveniles are one of our largest arson perpetrator groups. A good juvenile firesetter program coupled with quality investigations goes a long way toward eliminating or greatly reducing this problem.

The standard of good practice to follow is NFPA® 1033, *Professional Qualifications for Fire Investigator.* The success of your program will depend on resources and commitment. However, we offer the following guidelines:

- Develop a good process or system for fire origin and cause determination. This is the most critical part of any investigation. You cannot prosecute criminals if you cannot determine they caused the fire. You also cannot correct a fire safety issue if you are never certain how or why a fire occurred.

- Develop good policies or guidelines for conducting the operations. Typically, you will find that the individuals who gravitate to this job are independent thinkers who will need a solid structure to keep them on track.

- Establish very detailed, yet flexible data recording systems. This will be invaluable to keeping track of all the data you may want to review later.

- Establish formal relationships with your jurisdiction's legal authority (district attorney, etc.), your law enforcement agencies, BATF, FBI, state fire marshal's office, and all other local, state, and federal agencies with which you may come in contact. Due to the events of 9/11/01, there are more mandates for communication and cooperation among the various agencies than ever before. Take advantage of these channels and you may find opportunities and resources you never knew existed.

Wildland Risk Management

Wildland risk management is a relatively new commitment for most fire departments. Although it affects only those departments that have a wildland/urban interface, it is very involved and complex, requiring real commitment (see Chapter 13). Many parts of the country have wildland fire exposure that has never been specifically identified in the past. If you don't believe us, look in Florida, Oklahoma, or Texas.

Relationships are hugely important to wildland risk management. All through this chapter we have emphasized the importance of maintaining relationships, but wildland risk management is more interdependent than any other task. Wildland areas present many challenges that are in proximity to a city or location within a county or state. You must consider private lands as well as public, local, state, federal, and maybe even tribal lands. Each landowner must meet specific responsibilities and expectations and abide by specific restrictions and rules, let alone potentially having to enforce them. Managing wildland areas is a political love-fest like no other. There are environmentalists to deal with, users, clients, adjoining property owners, and people who you never even knew existed, each having concern and care for land that may burn.

The job of managing the wildland/urban interface is tough. The first thing you must determine is whether you are the right authority to deal with the problem. Generally, the fire department is the first agency called to fight a fire, but if the fire is in a national forest next to your jurisdiction, how much authority do you have? These are the types of questions that must be addressed.

Once the areas of authority have been established, you must determine your real risk. Again, Colorado Springs, Colorado, is an excellent example of a progressive, interactive wildland risk management program. Its web site not only illustrates how involved and complex the issues are but also explains how the program determines its risk. The most basic issue is determining a boundary between your forested areas and your urban core. Draw this line on a map, and you have defined your wildland/urban interface (**Figure 5.9**).

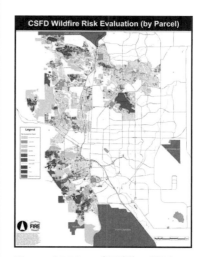

Figure 5.9 Map of Wildland Urban Risk.

Once the risk has been identified and valued, you can begin to manage it. This will be a long and ongoing process, as it is dependent upon the will of your policymakers to deal with the risk. Include the following considerations in the course of framing your program:

- Is this a new and upcoming problem or an existing one?
- Will an ordinance or regulation settle the issue completely?
- Can we use public education to influence behavior?
- What resources will we need?
- Prioritize the available mitigation methods from the most effective to the least as they relate to your community. Your choices are:
  — Fuels management
  — Construction methods
  — Public awareness

Planners have to realize that structures play a more significant role in the wildland/urban interface than previously thought. A review of the Los Alamos, New Mexico fire revealed that the most important role in the destruction of the community's structures was played by the structures themselves, not the forest. That does not mean the forest had nothing to do with the firestorm; however, it does mean that if we begin treating structures as part of the problem rather than just the vegetation, we will go much further in mitigating wildfire threats.

Careful consideration needs to be given to this task. More and more American communities are facing drought conditions. This shortage of water increases the threat of fire to dry vegetation. Major fires in forested watersheds cause tremendous pollution that costs millions of dollars to correct. Damage to watersheds in an already drought-stricken region only reduces the available water that much more. People are moving away from urban cores to more rural settings. Insurance companies have to cover the cost of unprotected homes destroyed by wildfires, and ultimately we all pay for insurance losses. This is a community-wide problem that must be addressed differently than in the past. Remember Albert Einstein's words that we quoted at the end of Chapter 4: "Thinking as we are has brought us to where we have already been. In order to go somewhere else, we must think in a different way."

With all these functions and potential responsibilities, the number and types of staffing required for wildland risk management can be significant.

Training

Given the importance of maintaining a technically strong, competent, and professional staff, how do we recruit them? How do we maintain them? Training is a critical part of this objective, and it requires money and commitment.

Many training courses are available: short courses, seminars, college courses, on-line classes, and even guest lecturers. The more qualified your staff, the more expensive it is to keep them schooled. For example, if you have engineers on staff, they typically have annual or two-year re-registration or re-certification requirements. Meeting them requires time in courses that generally are not given locally. Other staff members who are certified also must receive contact hours on coursework related to their certification.

Unfortunately, the fire service is ill-prepared to support the prevention staff. We do a good job of training our recruits and of keeping our line personnel up to speed on evolutions and tactical operations, but we do a poor job of training our engineers, inspectors, educators, and others. Chiefs must be encouraged to address this need. Typically it is the prevention staff that provides classes for the line personnel on systems, water supply, investigations, and the like (**Figure 5.10**). Without a doubt, it is in the best interest of the fire service to begin addressing more specifically the required training for the experts in the fire prevention divisions upon whom the rest of the department so heavily relies.

Figure 5.10 Fire Prevention personnel can be a resource to assist with providing training to suppression crews.

Summary

Fire departments deal with hostile fires in two ways, prevention and suppression. The structure of the division responsible for fire prevention will vary depending on the size of the fire department was well as the organizational structure of the municipality or fire district.

Typically, Fire Prevention Bureaus or offices charged with these functions should perform three core functions: education, engineering, and enforcement. Other duties commonly performed by Fire Prevention Bureau personnel include fire investigations, public information, and assistance to the Incident Commander during emergency operations.

A Fire Prevention Bureau staffing can utilize sworn fire personnel, civilians, outsourcing, or a combination of all three. Each staffing option has advantages and disadvantages. It is up to the fire department to develop a staffing plan that will enable its Fire Prevention Bureau to provide the level of service needed by the organization. Today's Fire Prevention Bureau staffs require a diverse technical background. The level of professionalism in the Fire Prevention Bureaus will continue to increase.

Chapter 5 Review Exercises

5.1   Identify methods to staff fire prevention. _____

_____

_____

_____

5.2   Compare and contrast the use of civilian personnel and sworn personnel to staff a Fire Prevention Bureau. _____

_____

_____

5.3   Explain outsourcing. _____

_____

_____

5.4   What are the advantages of outsourcing a service?_____

_____

_____

5.5   Identify and explain three functions performed by Fire Prevention Bureaus._____

_____

_____

5.6 What sworn fire department rank is common for a fire marshal?

_____

_____

_____

5.7 Identify two common ways fire departments deal with hostile fires?

_____

_____

_____

5.8 List five items to consider when starting a new Fire Prevention Bureau. _____

_____

_____

_____

_____

5.9 Define *span of control.*_____

_____

_____

5.10 Name the types of reviews a plans examiner is expected to perform.

_____

_____

_____

5.11 Explain the historical fire department practice from which staffing of fire prevention suffered and why it was overcome._____

_____

_____

5.12 What are performance-based designs? _____

_____

_____

5.13 Discuss how turnover impacts both firefighter and civilian-staffed fire prevention bureaus. _____

_____

_____

5.14 Discuss options for outsourcing fire prevention services.

_____
_____
_____

5.15 Contrast both private and public Fire Prevention Bureaus.

_____
_____
_____

5.16 Briefly explain what the following sections do: engineering, fire and life safety education, inspection, investigation, public information.

_____
_____
_____
_____
_____

5.17 What is NFPA® 1031? _____

_____
_____

5.18 What general type of inspections can the inspection section do?

_____
_____
_____

5.19 What is preincident planning? _____

_____
_____

5.20 What is NFPA® 1033? _____

_____
_____

5.21　List four key areas of focus a good Fire Prevention Bureau investigation section should address.

_____

_____

_____

_____

5.22　What is wildland risk management? _____

_____

_____

Notes

1.　Oren Harari, "A Leadership Primer from Colin Powell," Management
Review (December 1996).

2.　Arthur E. Cote and Jim L. Linville, Fire Protection Handbook, 17th ed.,
Quincy, Mass.: National Fire Protection Association, 1991.

3.　Ronny J. Coleman, "It's the Fire Service, Not the Fire Business," Chief's
Clipboard, Fire Chief Magazine (April 1997).

4.　Ibid.

5.　Cote and Linville, Fire Protection Handbook.

Risk Assessment

Table of Contents

Key Terms

Key Points

1. The risk analysis should be most concerned with those hazards that will be unusual and likely will need to be treated differently.

2. A credible risk impact analysis requires looking at response history and loss experience and determining how many similar events or incidents to expect at the target hazard.

3. As a fire protection professional, your role in decision making should be only to inform and advise your customers of the facts.

4. Remember that a good risk assessment is necessary to properly identify the target.

5. Fire prevention professionals must address the young adults who are responsible for taking care of the very young and the very old.

6. Although public process is complex, time consuming, and labor intensive, it is the only reasonable method to engage and understand the public will and your mission.

7. If the public knows the facts, it will be reasonable and prudent in its choices, decisions, and opinions.

8. Risk assessment is a community-based decision process that must involve everyone.

9. The one common issue in risk assessment is money, which flows through the community like ripples in a pond.

Learning Objectives

1. Identify laws, codes, ordinances, and regulations as they relate to fire prevention.*
2. Understand code enforcement as it impacts life and property loss.*
3. Define the national fire problem and role of fire prevention.*
4. Define the functions of a fire prevention bureau.*
5. Understand risk.
6. Identify the benefits of a risk assessment.
7. Understand mitigation.
8. Understand prevention.

FESHE Objectives (USFA)

Risk Assessment

Case Study

The community you work for is annexing 50-square miles of property that is planned for aggressive development, including industrial occupancies, light commercial occupancies, and residential developments. There are currently six fire stations in your jurisdiction. In order to properly protect the additional planned area, an estimated four additional stations will be necessary to provide roughly equal protection as what you have currently in the rest of the community. Your department provides fire and first tier EMS response. Principal funding for your community is from sales and property taxes. The problem is that the city has insufficient funding to construct four new stations, let alone staff them. As a result, you have proposed drafting a mandatory fire sprinkler ordinance for this entire area before construction starts. The reasoning is the department should be able to delay and potentially modify additional fire station construction, potentially needing only two of the four fire stations planned if the bulk of the fire problem is engineered out.

If each station costs about $1.5 million to build and equip, and it costs roughly $1.5+ million to operate each station on an annual basis, this proposal could save the city 3 million dollars in one-time capital costs and over 3+ million a year in salaries and benefits from here on out. Consequently, this is a good tax saving move for the community, but there are many objections to the proposal that must be identified, understood, and resolved. To start, here are a few the Chief has already heard:

1. Affordable housing will be unattainable.

2. Why should new home builders pay for what the rest of the city will benefit from?

3. This proposal eliminates future firefighter jobs.

4. How will you maintain your emergency response time standard by reducing the number of fire stations?

Given some of these questions and concerns, how would a community risk analysis help you to accomplish your goal? What possible answers and other problems might come out of a risk analysis? What would you consider to be a list of stakeholders that would need to be involved in this process? Should a risk analysis be conducted? Is a risk analysis even the right approach to use in this situation?

Defining Risk

There are many publications available that discuss and describe risk. Some explain in great detail the specific science involved in a good risk-assessment process. However, Webster's simple definition of **risk** is the exposure to possible loss or injury. While this is pretty clear, let's see how it applies to fire service applications. We can group all of our fire service activities into two separate umbrella categories:

- Preventing, mitigating, or controlling hazards to minimize risk
- Responding to emergency incidents to minimize injury, suffering, and loss

It is also important to note that throughout this chapter, the term **injury** includes monetary impact, job loss, aesthetic impact, and personal injury. It basically refers to any aspect of community impact, hence injury.

In many respects, our job in the fire service is basically that of risk managers. In his book, *Seven Fires*, Peter Charles Hoffer cites a definition of risk management for the fire service that was applicable in 1904. "...officers need to balance the danger to the firefighters against the desire to do the job." This definition is still relevant today; however, we must expand this beyond just the firefighter to the citizens we protect *[Seven Fires, the Urban Infernos that Reshaped America, Peter Charles Hoffer, Public Affairs, Perseus Books Group, 11 Cambridge Center, Cambridge, MA 02142, 2006]*.

Since we are trying to become skilled and motivated fire protection and suppression experts, we need to discuss a paradox. While we traditionally call our work "fire prevention," people in the business can easily debate the reality of what we actually prevent. Some argue that we do little in the way of actually preventing incidents and, at the very least, have a difficult, if not nearly impossible, time proving or articulating what we actually prevent. Preventing deaths, however, is something we do have documented evidence of self-evacuation and rescue as a result of early detection and/or suppression from the use of smoke alarms or fire sprinkler systems. Most departments can likely say their efforts of advocating and installing smoke alarms and fire sprinklers have resulted in someone getting out of a structure fire alive because they heard the alarms or benefitted from fire suppression system activation before the situation became untenable. While there are many anecdotal relationships to this potential success, the actual statistical corroboration remains elusive in many cases. Because of this, we see some departments marketing our functions differently using terms like *fire and life safety services* or *hazard mitigation division*. We do prevent fires and incidents, but more frequently we mitigate their effects. We define **mitigation** holistically as the prevention and reduction of the severity of an undesired event. In the quest to mitigate community injury, our objective is to reduce, or limit as much as possible, serious loss of life and property incidents.

Risk — The exposure to possible loss or injury.

Injury — Personal injury, monetary impact, job loss, or aesthetic impact.

Mitigation — The prevention or reduction of severity of an undesired event.

Let's think about how we address risk and the hazards we are expected to deal with. First, we must recognize the assets at a community's disposal. What emergency response equipment or methods are we able to respond with? Answering this question provides much information about the type of emergency your community can handle. Doing so requires extensive interaction with those fire departments and other emergency personnel responsible for emergency response. Seeking their input in the process is essential.

For example, let's say you are a Chief and manage a department with two stations, and you have two engines and one ladder truck. Then assume your response tactics are to respond with the first-due engine and begin a "quick attack" with that crew. The second-due engine responds and secures a water supply and backup lines for the first-due company. Now let's assume the ladder truck responds to provide forcible entry and ventilation. We'll also say that one of these companies will also be responsible for primary and secondary searches and any other functions your policies and procedures call for.

In this example then, we could deduce that for a "typical" single-family dwelling fire, the response would likely be adequate for containing and extinguishing a fire in an area or room of origin. This adequate response could then be determined sufficient to handle your community's "typical" fire.

Now let's consider a two-story office building that is roughly 20,000-square feet. Depending upon the situation, could your typical response handle a fire in one of the floors of this building? If the entire floor is open office cubicles, the total area or room of origin may be as much as 10,000-square feet. Maybe the office is a mix of closed offices and open cubicles. If the fire starts in one of the enclosed offices, it likely would be no different than a room-and-contents fire in the single-family dwelling. In either case, only you can make the appropriate determination, based on your department's operational capabilities. Let's say we predict the fire in an open-cube arrangement would likely grow beyond our typical response capabilities of containment. This being the case, we could further presume the fire will progress, likely consuming the entire structure, or at least creating damage sufficient to render the structure unusable.

In summation, it would validate our example that a single-family dwelling should be considered your "typical" response. Larger buildings that are not subdivided into similar compartments (compartmentalized) and are of a similar lightweight wood frame construction are not typical but should be considered a higher hazard risk. This allows us to simplify our model: a low-hazard risk on its own would not burn as much, so it likely would not damage adjoining properties or exposures or create significant risk to your fire crews. Moderate-risk occupancies pose a "moderate threat" to adjoining structures and, without your department's intervention, could cause a conflagration and potentially minor injuries to your crews. A high-hazard risk would be considered an unusual event that could likely extend beyond your existing capabilities of control, assuming desired expectations or outcomes are the same as those of a low or moderate hazard risk.

The point here is that the overall risk analysis effort need not give much consideration to typical hazards or risks, at least not yet. The hazards you should be most concerned with are those that will be unusual (not typical) and likely will need to be treated differently. So, if the office building in our example is mostly divided offices, it likely presents a typical scenario because a hostile fire can be physically confined to smaller areas. However, if one or both floors are open cubes, you may need to treat the building as an unusual or higher-risk occupancy, thereby developing alternate protection or mitigation strategies **(Figure 6.1)**.

Figure 6.1 Depending upon the size of cubicle spaces on office floors, such as this photo of a burned office building, the risk may be typical or unusual.

What we are discussing is the definition or delineation of risk. Any risk that is moderate or lower does not require much more than our usual thought process; we can likely handle that risk because it is what we are trained, supported, and expected to do. Any risk greater than that (an upper-moderate to high risk) needs to be identified and targeted for more consideration and thought, particularly by your policymakers or your governing body. This high risk may need to be treated differently or consciously debated as being an acceptable loss as long as everyone has sufficient factual information and shares realistic expectations. Let's dig a little deeper.

What Is a Risk Assessment?

Risk assessment can be very complex and studied and subdivided into many parts. A basic **risk assessment** can consist of analyzing the risk impact and the risk perception and then combining the results.

Risk Assessment — Process of analyzing the risk impact and the risk perception and then combining the results.

Risk Impact

Risk impact is simply a measure of the probability that something will occur and the severity of its results. This risk impact then can be termed the actual or factual loss. This can be done any number of ways. How many buildings will we lose? What is their valuation? How many lives could be lost? How many serious injuries could occur? How much total money will this event cost? (Do not forget to include both the direct and indirect costs.) Statistical information is collected for nearly all of the events we respond to in the fire service. Some departments obviously do a better job of collecting information than others. Those that do well with risk assessments typically do an excellent job of information gathering before and after an event. Good risk-impact evaluations start by tracking the amount and type of loss incurred based on what is valuable to your community. Chapter 9 covers data tracking in more detail. Understanding and defining community value is explained later in this chapter. This concept of risk impact can be difficult to grasp. Some may measure the severity in square footage and dollar amounts. Others may measure the severity in tenant spaces and dollar amounts. Still others may measure it in dollar amounts and injuries or deaths. To perform a credible risk impact analysis, you need to look at your response history and loss experience and determine how many of those events or incidents you should expect on or in your identified target hazard. If your community is small, look to larger cities or towns with similar demographics and economics and use their figures.

Risk Impact — A measure of the probability that something will occur and the severity of its results.

Risk Perception

A **risk perception** determines the "perceived" value of the risk. Basically, this is what we will discuss as public opinion fostered through a stakeholder process. The risk perception study for a municipal department must involve stakeholders (the public it serves and the political body or policymakers who govern it). This determination of risk avoidance or acceptance is in essence a business decision borne by the community.

Risk Perception — The actual "value" of a risk.

If you operate a private fire department, for example at a petrochemical plant, your defined risk perception would be much easier to establish as it is basically left to the plant manager's opinion or knowledge. For example, let's say the plant fire chief has convinced plant management to follow a policy that requires all buildings on the site to be protected by automatic fire sprinklers. An outbuilding storing only steel piping and flanges with limited or no ignition sources is currently unprotected by automatic fire sprinklers. If the pipe and flanges are worth $2,500.00 and the building is an old metal garage that is worth more torn down than it is standing, why would the plant manager consider installing a $28,000 sprinkler system just because your policy says to (**Figures 6.2a and 6.2b, p. 126**). His or her determination whether to install automatic sprinklers will be a business decision. After thinking about it for about 15 seconds, the plant manager will have conducted a risk-perception evaluation and determined nothing in or around that building was worth $28,000.00, so the best decision is to "let it burn!" This is one example of a good risk-perception, decision-making process. The decision was based on exposure, risk perception, cost, and benefit.

Figure 6.2a Risk analysis is different depending upon the type, size and location of a structure, not to mention how it needs to be dealt with in an emergency.

Figure 6.2b Power Plants are critical infrastructure locations that require more consideration than the small shop shown in Figure 6.2a. *Courtesy of Colorado Springs, CO Fire Department*

A policymaker's view will nearly always gravitate to "What is the best bang for the buck?" What are citizens of your community willing to pay for certain levels of fire protection? Is it more economical to lose a 20,000-square-foot office building that is insured and owned by someone outside the state than to pay a million and a half dollars more in taxes each year for an additional fire station? These are the business decisions that your public and your policymakers must make. As a fire protection professional, your role should be only to inform and advise your customers of the facts. These facts include capabilities, expectations given specific resources, and cost estimates. As a fire protection professiona,l you will generally be working for someone else. It will be very rare, indeed, for you to have autonomous authority and discretion to issue community mandates and implement whatever "rules" you chose or believe are right. The rules and requirements will typically be political decisions (perceptions) based on your input and the input and opinions of others. We want to emphasize the importance of this stakeholder process. Involving the stakeholders is important because, when a big loss occurs, people frequently will be upset and mad, demanding answers as to why nothing was done in advance to prevent that loss. You will need clear evidence of everyone's participation in the decision that led to that loss. Not only is that cover for you and your boss, but it helps the community and others see the results of decisions borne from informed consent.

Why Do a Risk Assessment?

A risk assessment is critical for long-range strategic planning in the fire prevention bureau, as well as the entire fire department. A risk assessment basically clarifies what your problems are now and likely will be in the future, and it provides a basis for determining how best to address these problems.

This planning is crucial not only to allow people within the department to make appropriate decisions, but also for those outside the department such as the city council or village board. A municipal city council or county commission must be kept regularly informed on the status of the community's fire problem. By identifying the risks, the fire chief is able to communicate better with the policymakers, thereby helping set priorities for dealing with the appropriate issues in the desired time frame. Nothing could be worse than a fire department trying to fix a problem that the elected officials do not believe exists. Not only will fire department members become frustrated and feel abandoned, but the policymakers may develop serious doubts about the fire administration's grasp of reality in addressing community concerns.

The important thing to remember here is that a good risk assessment is necessary to properly identify the target. You must know where you need to go, how you need to get there, and what you need to support the venture. Without a relevant map, how would you ever find your way around an unfamiliar city or state? Yes, you can use the "Columbus" method (search and discover), but in today's competitive economy, you likely will not have that luxury, nor will your customers appreciate slow arriving and potentially unsuccessful results. We live in a society that demands instant gratification and solutions. Everyone wants his or her problems fixed today, not tomorrow, let alone five weeks or two years down the road.

Issues to Address in the Risk Assessment

A good risk assessment provides many benefits. Not only does it describe your problems, but it also allows you to forecast future demands and prioritize solutions. Perhaps the largest single benefit is the ability to communicate critical concerns and needs to your policymakers and the community. Often times the policymakers are reticent to invest in big change; however, if the community understands and is on-board, they will nudge the policymakers to make the right decision. This assessment provides an honest and reasonable picture of where you are now and where you are heading.

Any numbers of tools are available for communicating this analysis, some of which are detailed later in this book; however, the key is gathering the appropriate data for dissemination among those who control the resources to accomplish your objectives. Important considerations include the following:

- Engaging the public on risk
- Your public's opinion of their level of acceptable risk
- The truth hurts
- Paradigm shifts
- The nuts and bolts

The remainder of the chapter discusses these concerns.

How to Engage the Public with the Risk Concept

Our adult voting populace (those in their twenties to late forties) is very busy. They have children in school, parents needing care, soccer practice, little league, and tons of other obligations and activities. They may have new jobs (sometimes two) and many other distractions while trying to live a "normal" life. How can you possibly tell them everything you think they should know about the fire service in order for them to make the "right" decision? You can't.

In their view, that's why they hired you. They expect you to know. However, they are typically interested in knowing what the real issues are, particularly when it comes to spending their money or raising their taxes. So, in reality, you are in a pickle. They want you to do the work and make the tough decisions but not without their approval and, by the way, they don't want to take time to listen to the issues in order to give their approval. Now, while this is a dramatic example of the situation, it is not far off. They want to know what the potential issues are that can affect their family and what the potential outcomes could be. They will likely want to know what you can do currently to mitigate or solve the problems and how much it will cost them to achieve a different outcome. This is especially true if property tax (mill levies) or other funding increases are being requested for increasing or maintaining services. The public is not stupid, but they are likely ill informed. They, like us, are not looking for issues to discuss, as they are too busy. However, they do want to know about issues that can impact them. Therefore, it is important to keep the issue focused on what is important to them. Emphasizing it is not about you or your department as much as it is about them. The following steps will help you to work toward a solution:

1. Succinctly identify the real (actual) problems—not the perceived problems.

2. Prioritize the issues from most hazardous and greatest risk to least.

3. Recruit a good cross section of engaged stakeholders from the community (businesses, citizens of diverse incomes and locations, regulatory agencies, and anyone and everyone you can think of who has a vested interest in the outcome). Do not forget the people from economic development and your policymakers.

4. Break these people into focus groups if the number is too large. Remember, the more you involve, the more accurate your outcome. Don't expect huge amounts of participation, (sometimes 2 or 3 people is all you get), but be ready as the people involved will typically want to be very engaged. A word of caution is to be careful in designing meaningful possible outcome objectives and be clear about them. An engaged group will work hard. But this same group will quickly abandon you if they feel they have no purpose or are wasting their time on a fruitless or unorganized task.

5. Discuss with these people the problems' understandable (simple) outcomes or threats. This is where you want them to discuss and list their "perceived" problems. Honor this input, as it may seem trivial or silly at times. Remember, it is very important to them and, therefore, very important. These then get factored and listed with the real problems. Anecdotal references go a long way here.

6. Accurately define for them your capabilities for solving or mitigating problems and what they should reasonably expect. It is critical in this step not to exaggerate or dramatize your wants as opposed to your needs and capabilities. If you find they expect more than you can deliver as an organization, don't argue or shut them down, just make sure they

realize the additional costs. This can also be where more of their previously identified perceptions get boiled out, leaving a good solid list of actual problems. This process makes them aware of factual issues and corrects misconceptions or poor perceptions.

7. Engage these people or groups to provide at least two (and no more than five) possible solutions with their resultant outcomes. Limiting the list is not to discard the remainder, but simply provide focus and accountability for solution. The remainder from the list can be placed on the strategic long-range plan to address later. Consider this the menu of choices you will use to propose recommendations to your policymakers.

8. Sort these recommendation options in descending order from most to least costly while identifying the overall risk-reduction benefit, most to least.

9. Present this final list to the same groups for confirmation or readjustment.

10. Publish the document of findings and use them as the key component of your risk analysis as you have just defined their risk perception.

The Colorado Springs, Colorado Fire Department has used this process successfully several times to accomplish different objectives, all tied to defining the acceptable risk and achieving community goals, objectives, and appropriate expectations. The major community impacts were:

- A resolution establishing single company response times of 8 minutes for 90 percent of the time and effective force response times of 12 minutes for 90 percent of the time. This facilitates future station planning, expenses, and program development with little explanatory effort.

- Adoption of a community-wide comprehensive Community Wildfire Protection Plan. (C.W.P.P.). This has been a resounding community success impacting over 36,000 addresses.

- Passage of a wildfire mitigation and structural hardening ordinance.

Although this process is extremely time-consuming and labor intensive, it is the only way to gauge the public's degree of acceptable risk and generate the trust that is so critically important. This process will help you gain supporters and help you understand those things you may not be appropriately focusing on based on your community's desires. Remember, the answers you discover may not be the ones you would choose or even thought possible. However, the community will support and defend you all the way to the bank. Let us be clear. We are not saying the entire public will completely agree with you because they won't. We are simply working to have the naysayers grudgingly go along with your analysis and, ultimately, the plan. This is the greatest victory you can reasonably expect to achieve, and when this occurs, you should celebrate the win.

What Does Your Public View as Their Level of Acceptable Risk?

Ask us for the single, hardest question to answer in the fire service, and this is probably the one. What does the general public view as their established comfort level with their acceptable risk? (Keep in mind that your public includes your elected officials.) The truth is that they probably do not even know what "acceptable risk" means. If they do, their perceptions likely will be all over the board because we in the fire service have failed to make tax-paying adults aware of what this really means to them.

Based on our nation's fire history, we have emphasized fire and life safety education for the two principal target groups with the greatest risk of being injured or dying in a fire: the very young (stop, drop, and roll; call 9-1-1; etc.) and the very old (take a pot holder with you when cooking, no smoking in bed, etc.). However, we are finding we have frequently failed to communicate our issues to those people responsible for taking care of the people within these target groups: the working class adults in their twenties to late fifties. This is the same vote-wielding group that determines our destiny with regard to municipal bonding, tax increases or rollbacks, and other key financial issues. We must target them, inform them better, and gather their feedback to help us map our planning.

For example, consider the following analogy. A thirty-year-old father of three decides to go for a ride on his bicycle with his oldest child. The father cares very much for his child's safety and therefore makes sure the child wears a helmet during their ride. However, without consciously thinking about it, he makes a decision about his acceptable risk by choosing not to wear a helmet himself for whatever reason: only a short ride; I've got 20 years' experience riding bikes, etc.

Riding along, they round a corner and hit a layer of sand deposited by a recent rainstorm. They both lose traction and crash their bikes. The child ends up with multiple abrasions while the father strikes his head against the curb. The father's injury results in a depressed skull fracture causing permanent brain injury, leaving him debilitated and nursing-home-bound for the rest of his life.

This injury has results in loss of current principal earnings for the family, long-term health care bills that are now the wife's responsibility, at least a short-term loss of a father-figure role model, and unknown future problems compounded by the loss of half the family nucleus.

Did the father think about all this before the incident? Likely, he did not. As an average American, he probably thought he and the child were only going out for a few minutes. Thus, his level of acceptable risk was to gamble on a very infrequent event of very great consequences.

How can we engage the public on issues of fire and life safety, getting them to consciously make informed decisions on how to best protect themselves and when to rely on a public or private service to step in and provide that extra level of protection? The answer lies in public process.

What is public process? Our definition of **public process** is the act of engaging the public with factual information concerning issues that are important to them and learning from them "what they want and can't live without" **(Figure 6.3)**. These include:

- Services the fire department provides today (fire protection, EMS, hazmat response, etc.).

- Factual benefits and outcomes of that level of service (response times, basic life support outcomes, advanced life support outcomes, efficiencies or benefits of those services, availability, etc.).

- Cost of current services (personnel, stations, capital, ongoing; generally shown in a cost per capita).

- Public desires and expectations (finding out what services they approve of and expect and those they want but are not currently receiving).

- What additional costs are they willing to pay for (again, typically given in a cost per capita figure)?

- What services you are providing that they do not think are necessary?

Figure 6.3 Public input meetings are difficult and trying but necessary for proper risk-analysis input.

You should also offer the public your factual explanation of what you view as risks. In response, you should obtain from them:

- Their agreement or disagreement on what they view as risks

- Consensus on what community risks are and are not

Again, public process is complex, time-consuming, and labor intensive. However, it is the only reasonable method to engage and understand the public's will and your mission objectives.

Determining your public's acceptable risk is crucial. Otherwise, the fire service will continue to be in the position of "telling everyone what is good for them" based on our opinions and biases, rather than performing our job, which is to do what the public wants and, in fact, pays us to do.

The Truth Hurts

Granted, the solutions and answers we get from our public may be contrary to our opinions. The reality is that if our boss understands the facts of an issue and tells us to do something, we most likely should do it. This is not to say that in all cases the public's or a boss's view is the answer of choice; however, the reality is that if they know the facts and trust us, they will defer the decision to us and usually allow us to act upon our will, or make a decisions that is at least informed. We propose the public should be viewed as reasonable and prudent in their choices, decisions, and opinions if we have done a good job of communicating. After all, our nation was founded based on a government whose decisions were intended to represent the will of the people.

Consider this example. As a fire protection professional, we believe that the more fire safety inspections are conducted, the lower the fire incident rate will be, and the magnitude of fires that do occur will be equally low. So, you propose to hire ten more inspectors for your fire prevention division. However, doing this will require an increase in taxes, which requires a vote of the people. The vote is cast, and your public says not only no, but *heck no!* Are you still going to hire the ten inspectors? Not likely. Is the public's decision wrong or bad? No. That is, not if you have provided them with all of the relevant information necessary to make an informed decision. Through the vote, they made a decision based on their perception of their acceptable risk. They likely believe that the potential increase in fires does not offset the cost of adding ten additional employees to prevent them.

Our point is that the public's decision is legitimate. You as a fire protection professional need to accept it. Do not confuse the aspects of informed decision making. Your job as a professional is to inform your citizens and public officials—your stakeholders—of the relevant factual issues surrounding the decision. What is relevant information? Therein lies the true test of your ability and professionalism, not to mention leadership. As long as the public understands the issues, their decision—defining their acceptable risk—will be reasonable.

Paradigm Shifts

What is a **paradigm shift**? Stephen R. Covey refers to a paradigm as a way of understanding and explaining certain aspects of reality.[1] Shifting your paradigm means you get rid of the old ways of thinking and open the door for new insight, new methods, and different views. Basically, you start looking at things through different goggles.

For many years, the public has relied upon the fire chief to tell them how best to protect themselves and the community from fire and, sometimes, other disasters. As a single administrative position or department head in most communities, the chief has had tremendous leeway, respect, and autonomy. It was the chief, however, who was left holding the bag when disaster struck, and it was the chief who most easily walked in front of a governmental body to tell them what needed to happen to fix a dangerous issue. For years, the chief would be the one to give his or her professional opinion on what needed to be done, and the public would often blindly agree because "the chief said so." This led to power and authority potentially being misplaced or misused, particularly if the chief really did not know the true issues or if inappropriate political will was being exercised. That is not to say such things happened a lot, but the possibility was always there. The evolution of our country's political system has led to mistrust of government and constant questioning of nearly everything government does. The net result in many communities is a public who questions every cent spent and offers opinions of what is right or wrong, regardless of their knowledge, just because they do not trust the government or the people who run it. Since 2007, through the last great recession, there was an increase in public frustration directed at nearly all government employees because of their high pension, benefit costs, and salary increases.

Our challenge is with a community that over and over again has a very difficult time perceiving what disasters or hazardous situations are prevalent until they actually happen. Then, of course, they are quick to blame someone for not telling them or taking action sooner to prevent such an occurrence. Americans are very apathetic. Thus, our current communities tend not to listen as well as they think. "It can never happen to them"—until it does. This is especially true if you analyze the number of wildland urban interface fires we are suffering over the last ten years. We know what can be done to minimize these disastrous occurrences but most don't believe it will ever happen to them. People who live through these events suddenly understand the risk is relevant and real. Another example is the changes we have undergone since 9/11/01. Would you say every American is on their toes with caution and as aware of our threats from terrorism as we were on 9/12/01? We propose not.

We, as fire service, must rethink the way we communicate to the public, evaluate risk, and propose solutions. Our vocation needs to shift paradigms. Rather than telling the public what is good for them, we must begin crafting our facts and our anecdotal stories to educate our public, particularly our middle-aged population, as they impact both the young and old. This education must be constant and relevant, changing the messaging as necessary based on the changing environment. Most importantly, our voting populous should become more aware of what "real" risk means and what the "real" hazards are. When we cannot define those real risks and hazards, we must let this group define them for us. We have the professional expertise available to propose good solutions. However, unless the public understands the risk, the expectations, and the outcomes, how can we expect them to help us solve the problems? Remember, these problems are not ours (the fire service). They are the community's. We all suffer or we all benefit. It is a community-based decision process that must involve everyone. No longer can we tell the public what to do or what is good for them. As a society, we have become far too educated and inquisitive for that. We must change our way of approaching our communities and help them to help us find the right answers.

The risk of doing this is that we may hear answers that we do not agree with. Remember our example at the beginning of this chapter? As fire service members, we have been trying to win the zero loss game. No deaths, no property damage. This objective is hardly attainable. People will die and property will be lost. Our job is to provide reasonable protection from reasonable incidents. Good risk management will not address every single event unless someone provides the money to support that objective.

The National Aeronautics and Space Administration (NASA) is a good example. They spend billions of dollars to prevent fires in the space station because they cannot call the Houston Fire Department if someone up there smells smoke. Therefore, they do support the financial commitment to the elimination of fire potential through engineering, education, and enforcement with a multitude of design, policies, and procedures. In a large community here on earth that has from hundreds of thousands to millions of people in various jurisdictions, this solution is not reasonably attainable. So, we use a cadre of methods, means, and processes. We need to implement these differently than we have typically done in the past. Proof is evident in some of the more progressive departments throughout the country. They have had to learn how to look at problems and think them through differently, if for no other reason than sheer survival. This is particularly true given our most recent economic challenges, which are worldwide.

The Nuts and Bolts

Now that we have laid all the groundwork on risk analysis, we will close this chapter with a nuts and bolts discussion of how to do one. This process is relatively generic and can be used for many types of specific applications. Don't forget and don't discount your use of stakeholders—as many as you can garner participation from. The higher your numbers of participation are, the stronger that your position will be.

As we have discussed, start by defining the relevant and credible threats and outcomes. Remember, you cannot protect everyone from everything. Make sure the threats you include are among those issues that your mission statement addresses **(Figure 6.4)**.

Once you have defined the threats, ask yourself these questions. What are the possible outcomes? What are the potential losses? Explore both the direct and indirect effects. What are the monetary impacts to individuals, the community, and other areas of concern? What is the potential duration of these affects? Identify the costs of recovery and reclamation, not just the actual damage. What are the life threats **(Figure 6.5)**? Despite death's tragedy, our experience has been that people tend quickly to become apathetic after a fire death occurs. For reasons that only credible psychologists can dare explain, people acknowledge the horror and tragedy of a fire death but soon afterward take the approach, "Sad it happened to them. Good thing it will never

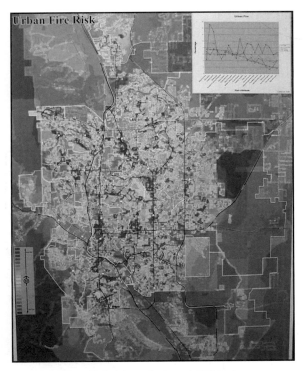

Figure 6.4 Detailed risk analysis and GIS tools can assist in not only identifying risks but communicating them pictorially. *Courtesy of Colorado Springs, CO Fire Department.*

happen to me." You will find that the one common issue is money. Money speaks. It ripples through a community like ripples in a pond, particularly given our current world economic climate. Even considering wildfire risk, timber (logging) has a price and aesthetic value (someone will always pay for pretty land, including the Bureau of Land Management or the United States Forest Service). Historical structures have a value (intrinsic, restoration, and maintenance). Manufacturing plants have a value (number of employees and lost income, tax income, structure, business interruption costs, lost customer base, etc.) (**Figure 6.6**).

high likelihood-low impact Educational E	*high likelihood-moderate impact* Small residential care R3 - R4	*high likelihood-high impact* Nightclubs-Dormitories Apartments-Hotel-Motels A2 - R2 - R1
moderate likelihood-low impact Low hazard storage Church S2-A3	*moderate likelihood-moderate impact* Moderate hazard storage Factory-Fabrication S1- F1- H3- H5	*moderate likelihood-high impact* Hazardous manufacturing H1- H2
low likelihood-low impact Low hazard manufacturing Arenas-Jails- Agricultural U - F2- A4- A5- I3	*low likelihood-moderate impact* Theatres-Office-Day Care- Retail A1- B- I4- H4- M	*low likelihood-High impact* Assisted Living Nursing homes-Hospitals I1- I2

Figure 6.5 Once risks are analyzed, they can be prioritized and classified based on low to high rankings such as the occupancy types listed here in this risk matrix. *Courtesy of Colorado Springs, CO Fire Department.*

Figure 6.6 Monetary cost is a significant factor in evaluating risk, whether it involves the value of timber (a) or the cost of restoring a historical structure (b).

Conduct a comprehensive analysis of your local incidents by area of study. Geographic information system (GIS) support will be an invaluable tool in accomplishing this. Other computer-generated data will also be beneficial in this process. These data not only provide you with the number and types of incidents but give you a graphical representation of their location and magnitude. The adage that "a picture is worth a thousand words" is never truer than in this type of analysis communication (see Chapter 14 for utilization of technology to assist in this endeavor).

Next, determine the risk impact. Study the frequency and history of events by parcel, occupancy type, or whatever other data field best suits your objective. At the same time, look at the impact of your losses as a result of specific consequences for these specific data fields. For example, how many apartment fires are you having per year and in what parts of town? Next, ask what are the life and monetary losses for each of these fires in each of these areas. Overlay and compare the data to get a perspective on which locations are in the worst shape. This can also be used for emergency medical response issues. If your department runs on medical calls, your distribution is likely somewhere around 70% medical to 30% other calls. That reveals a huge target for examination and strategic planning for reducing that high medical call volume.

Rank your risk impact results. Again, the order of priority will depend on what you are trying to accomplish. We recommend classifying risks by occupancy types to start with. Next, determine your department's level of awareness; explore your management capabilities to deal with the identified areas and your currently available resources to address the risk. These data will be critical for the risk-perception phase.

Conduct your risk perception following the steps outlined previously in this chapter. This is a critical but slow process. Do not try to shortcut or dilute the process. It will anchor everything that follows. At the same time, develop a good understanding of national standards of good practice. What are other people doing about these problems? What other codes, standards, or ordinances can address it? What other various ways (think out of the box) can we use to mitigate or solve this problem? Once you have completed this step and can readily evaluate and communicate your risk impact in combination with your risk perception, you have completed your analysis. Remember, your analysis only provides a tool to communicate your risk issues. It does not solve the problems.

Once your risk analysis is complete, combine those results with your standards of practice information. Plug this combined information into your previously determined risk classifications and forecast through modeling the possible effects of these solutions on your overall identified risk. Assuming the proposed solutions will change the outcomes for the better; determine what resources you will need to accomplish your objective. These should include all three elements of fire prevention work: engineering, education, and enforcement.

The next step is to try to implement your recommendations (**Figures 6.7a and 6.7b**). Obviously, this will require support from many sources. Develop a comprehensive short- and long-range plan that strategically outlines how you can achieve your goals. Establish benchmarks and dates for completion. Have

Figure 6.7a Dead Gambol Oak was identified as a high risk to this neighborhood. *Courtesy of Colorado Springs, CO Fire Department.*

Figure 6.7b Once identified, the Dead Gambol Oak was removed, thereby reducing the risk to the community that surrounded it. *Courtesy of Colorado Springs, CO Fire Department.*

some of your stakeholders involved in evaluating and monitoring your progress. They will not only continue their engagement but develop ongoing ownership in the process and the mission.

Once you have implemented your recommendations, regularly evaluate your results. Providing detailed analysis of your run profiles, statistics, and overall emergency response performance will provide excellent feedback as to how your strategies are working or not working. Determine what effects both mitigation and response have had on the outcomes and plug these results back into your risk-perception process and incident analysis by area of study. This information needs to be given back to your core groups that helped you in your process. Be prepared to make adjustments and take advice for course corrections on your implementation or celebrate the affirmation that your plans were on track. Continue to make changes as necessary for the program's overall success and effectiveness.

Summary

Risk assessment should be a main staple of fire prevention functions. To deal properly with community injury problems, the finite target and actual risk must be understood. Risk is often misunderstood or attempts to deal with it are misdirected. Risk is commonly defined as the chance of injury, damage, or loss. However, in a broader sense, we in the fire service may define it a little more practically, based on whether or not we can deal with an issue. If we have enough resources and equipment to handle an unwanted event, such as a particularly hostile fire, we may consider it a low risk. Having insufficient resources or equipment to handle a potential event at a given location may identify an area to be considered a high risk.

To identify methods of handling risk, we must conduct a risk assessment. From a fire service perspective, we should keep our focus fairly basic and keep things a little more simple; in this light, a risk assessment involves identifying the risk impact and the risk perception and then combining those results in order to properly identify our target.

Risk impact is a measure of the probability and severity of a particular event, series of events, or calamities. It speaks to what we can term the actual risk. A risk perception is the community weighted value of a risk or the perceived risk. It is what the tax-paying public believes is the actual value of protection protects. This assessment is essential for effective long-range strategic planning and provides a clear road map for future endeavors.

By identifying your public's acceptable risk, you will in turn identify what you need to do to satisfy their desires. Maybe we as fire safety professionals worry more about things than our "bosses" (the public) think we should. It is better to know risks up front rather than spending a ton of time and resources trying to achieve a goal that will not be politically viable or, worse, will not be funded and supported. We as a fire service must also look at shifting some of our own paradigms. This is essential if we plan on hitting the right target and going after the right problem. We must also help our public understand their risk and our capabilities. We must then use these findings, decisions, and outcomes to mitigate those risks that can be dealt with. Remember, prevention deals with actually preventing unwanted events from occurring. Mitigation is simply the modification of an unwanted outcome to a level that is sustainable, survivable, or nearly nonexistent due to active protection measures.

Discussion for Case Study

Taking some of the discussion points in the case study at the beginning of this chapter, a risk analysis is the BEST process to use to resolve this issue. The dilemma is your department needs to be prepared to hear the answer, as it may not be what seems intuitive or desirable, particularly by the administration or labor. A good risk analysis would be your best, most stable solution to this community-wide decision. First, you will provide tremendous community awareness of the issues and they will have an opportunity for great ownership in the solutions that are borne out of the process. Second, your fire department membership will become much more aware of the community's sentiments of your services. The accomplishments that are being done will be highlighted and the things that need improvement or re-tooling will be readily identified. The benefit of all this is an increased awareness by all community members, and a tremendous opportunity for collaboration, communication, and cohesiveness. Given the right opportunity, support for your department will either increase or increase A-LOT! Look at the two possibilities here. It's a win-win for both outcomes.

Answers that may come out of this process are:

1. Elimination of EMS services by the fire department, or additional support and positive changes to the EMS services your department provides. Maybe the option of privatizing EMS services is more financially beneficial. Maybe your department should get into transport or enhance your capabilities from basic life support (BLS) to advanced life support (ALS). You may have an opportunity to provide dynamic deployment of medical units or provide additional performance conditions in an existing EMS transport contract.

2. You may discover different types of apparatus and station designs being proposed, particularly if you are changing the way you fight fires through these engineering solutions. For example, you may choose to combine truck and engine functions by purchasing Quints or Squirts.

3. There may be animosity among various portions of the community, as some seem to be paying more for some service or systems than others for similar protection. For example if the residents in the newer development must shoulder the cost of installing fire sprinkler systems, the existing residents benefit from the reduced operational costs (taxes).

4. Relationships with the insurance industry will likely be enhanced due to the need for their participation in providing information on trade-offs in premium costs… or not. If fire sprinkler systems are going to be provided, it only makes sense that insurance companies may benefit from lower risk; therefore, they should be engaged in discussions to consider reducing property protection premiums.

5. Political energy may be developed to provide some type of retroactive ordinance for the existing parts of the community over time. This can further help reduce costs and strain on the city budget. However, the cost shifting will be borne by the same constituents for a time.

6. The stakeholders can be nearly limitless. We have identified some, but don't let the following list constrain your thoughts:

- Contractors
- Home builders association
- Businesses
- Insurance carriers
- Utilities companies
- Public works department or personnel

- Employee associations
- EMS providers
- Neighborhood associations
- Property management companies
- Apartment associations
- Facilities for the aged

Chapter 6 Review Exercises

6.1   What is risk? _____
_____
_____

6.2   What are the purposes of a risk assessment?_____
_____
_____

6.3   What are the benefits from a risk assessment? _____
_____
_____

6.4   Contact the local fire department and discuss if they have conducted a risk assessment and then summarize what they have determined to be the most significant risk in the community they protect.

6.5   What is mitigation?_____
_____
_____

6.6   Explain the difference between mitigation and prevention.

_____

_____

_____

6.7   Provide an explanation of what "typical" risk could be and give some examples. _____

_____

_____

6.8   Define risk impact. _____

_____

_____

6.9   Define risk perception. _____

_____

_____

6.10   Why is public process important? _____

_____

_____

6.11   What is public process? _____

_____

_____

6.12   What is the barrier that many fire departments typically struggle with in conducting proper public process? _____

_____

_____

NOTES

1.   Stephen R. Covey, *Principle Centered Leadership* (Summit Books, 1991), pg. 173.

Fire and Life Safety Education

Table of Contents

Key Points

1. The main objective of fire and life safety education is to motivate the public to act in a fire-safe manner.

2. The real role of fire and life safety education is to modify behaviors or instill basic, appropriate behaviors at an early age.

3. Preventable injuries and the local cry to prevent them were driving forces for fire departments to include educational topics in addition to fire prevention.

4. Fire departments must seek outside assistance in developing and implementing their fire and life safety education programs.

5. Fire and life safety education is a core element of all fire prevention programs.

6. An effective fire and life safety education program will enable the audience to understand and apply the behaviors being taught.

7. The bottom line of an effective fire and life safety education program is the reduction or elimination of the community's fire incident, injury, and death rates.

8. Fire and life safety education programs must change to reflect the community's changing needs.

9. Before developing a fire and life safety education program, you must consider your target audience and its level of learning.

10. Every fire event contains a fire and life safety education message.

11. Fire and life safety education, public relations, and public information are all inseparable.

Learning Objectives

1. Define the national fire problem and role of fire prevention.*

2. Identify and describe fire prevention organizations and associations.*

3. Define the functions of a fire prevention bureau.*

4. Identify and describe the standards for professional qualifications for Fire Marshal, Plans Examiner, Fire Inspector, Fire and Life Safety Educator, and Fire Investigator.*

5. List opportunities in professional development for fire prevention personnel.*

6. Identify programs for fire and life safety education.

7. Identify the different types of media that can be used in fire and life safety education.

8. Identify various presentation methods.

9. Identify sources of potential funding of fire and life safety education.

FESHE Objectives (USFA)

Fire and Life Safety Education

Case Study

The full-time fire and life safety educator position has been eliminated in an effort to cut cost. The fire chief fully understands the benefits of fire and life safety education and is seeking alternative methods to provide this service. Provide an example of how fire and life safety education can take place without a fulltime fire and life safety educator.

What is Fire and Life Safety Education?

Fire and life safety education is by far one of the most effective methods used to prevent injuries from fire or other types of emergencies from both a cost and resource perspective. However, this is one of the first programs that is frequently eliminated during budget reductions.

In order for a fire to occur, a fire needs heat, fuel, oxygen, and "something" to bring them together. The "something" that brings them together is usually a person, or what is sometimes referred to as the human factor. In fire prevention, we can address the prevention and mitigation of fires by eliminating one of the three elements listed above. The elimination of the heat and fuel is accomplished most commonly through fire engineering and enforcement. Through fire and life safety education, we can focus on behavior modification. The main objective of fire and life safety education is to motivate individuals to act in a fire-safe manner, which may require modifying some of their ideas or perceptions about fire. Through fire and life safety education, individuals can be informed about the threat of fire and how or where that threat exists in their environment. A fire and life safety educator can change people's beliefs or attitudes toward a fire problem. Later in this chapter, we will explain how the fire and life safety educator's role has expanded beyond just preventing and reducing injuries from fire. This is similar to how our services have expanded beyond just fire-suppression activities.

Preventing fires is an essential element of fire and life safety education; however, fires and disasters do occur, and people do need to be prepared to react correctly should one happen in their home, school, office, or elsewhere. For instance, we routinely used to teach children how to stop, drop, and roll when their clothes were on fire. Changes in laws regarding flame-retardant clothing criteria has allowed us to change our focus now from the traditional stop, drop, and roll message to more relevant instruction, such as youngsters telling adults about where lighters and matches are when they are found and calling 9-1-1 when appropriate.

Recently, the need is to educate citizens on how to be prepared for other types of emergencies, such as terrorist attacks or natural disasters, which may require a family to relocate and have a designated meeting place for the entire family. This need has required us to partner with other organizations such as school districts, universities, or private entities with large conference centers that accommodate large numbers of citizens. Fire and life safety education is a proven method to teach people how to react and be prepared when fires and emergencies occur. The documented cases of families alerted by their smoke alarms and subsequently followed their escape plans to meet safely at their meeting place are found in all parts of the country. In just a few years, the same scenario will occur with the difference that the homes will have had little fire damage and the occupants will have been able to evacuate easily because of their residential fire sprinkler system. In addition, during disasters we are seeing more citizens taking a preparedness approach and having a plan in place to evacuate their family to safe location until the event, such as in the case of wildfire, has passed or being able to defend in place. Lessons related to this were prevalent during the 2012 Waldo Canyon Fire in Colorado. Due to education, community awareness and various drills, over 34,000 people were evacuated from the fire in less than four hours. We can see how fire and life safety education is evolving to other important areas which define the need for a community risk-reduction program, not just fire and life safety.

Why Is Fire and Life Safety Education Important?

Fire and life safety education is very important for the following reasons:

- The public must understand that the fire department cannot instantaneously be everywhere a fire or disaster (natural or man-made) may occur.
- People are a main cause of fire. If they maintain good fire-safe behavior, fewer fires will occur.
- People's basic instincts are often very different from person to person. Therefore, each individual may have different fire-safe behavior. (For example, some may try to run for help if their clothes catch fire rather than dropping to the ground to smother the flames; others may be complacent that they don't need to develop a family evacuation plan in case of a large-scale emergency).
- Changing behaviors, such as smoking while being on full-time oxygen, calls for education and awareness, not enforcement.
- As a community, citizens should be able to rely on one another in times of crisis. Nothing is more evident of this than the growing popularity of Citizen or Fire Corps organizations.

- Many families are two-income families with latchkey (home-alone) children.

- As the number of retirees and seniors increase, awareness and education must be provided to prevent significant increases in medical and fire emergency calls.

The main role of fire and life safety education is to modify behaviors or instill basic appropriate behaviors, such as developing a home escape plan, to those at an early age, **(Figure 7.1)**. As a country, we are just now seeing older adults appearing from the generation of children to whom Sparky the Fire Dog and Smokey Bear were common and familiar educational props. Only now will we be able to better measure the benefit of our efforts to reach a larger population. Our goal as fire service professionals has been to instill these proper behaviors into the entire population. Unfortunately, many immigrants who come to this country have not received this type of education. We still reap rewards, however, because their children are compensating and acting on these fire-safe behaviors as they are being taught fire and life safety education in their schools. Frequently, we see children who are recognized as the heroes in emergencies because their parents were either not home or did not speak English and did not know what to do. Some people ask, "Are these stories just public relations events?" The answer is unequivocally no. It is living proof that education and behavior modification work.

Figure 7.1 Children can learn at a young age simple lifesaving tips such as a home escape plan.

History of Fire and Life Safety Education

Fire and life safety education has had several names. In many parts of the United States, it was and continues to be referred to as fire prevention education and public fire education, or pub ed. The need for fire safety education began in the late 1800s but was well documented in 1909 when the National Fire

Time Line of Events Affecting Public Fire and Life Safety Education

1909 NFPA's Franklin Wentworth begins sending fire prevention bulletins to correspondents in 70 cities, with the hope that local newspapers will publish the bulletins as news articles.

1911 Fire Marshals Association of North America proposes the October 9 anniversary of the Great Chicago Fire as a day to observe fire prevention.

1912 *Syllabus for Public Instruction in Fire Prevention*—a collection of fire safety topics for teachers to use in the classroom—published by NFPA.

1916 NFPA and the National Safety Council establish a Committee on Fire and Accident Prevention. Communities nationwide organize Fire Prevention Day activities.

1920 President Woodrow Wilson signs first presidential proclamation for Fire Prevention Day.

1922 President Warren G. Harding signs first Fire Prevention Week proclamation.

1923 Twenty-three states have legislation requiring fire safety education in schools.

1927 NFPA begins sponsoring national Fire Prevention Contest.

1942 New York University publishes *Fire Prevention Education.*

1946 U.S. government publishes *Curriculum Guide to Fire Safety.*

1947 Hartford Insurance Group begins the Junior Fire Marshal Program, perhaps the first nationally distributed fire safety program for children.

1948 American Mutual Insurance Alliance publishes first edition of *Tested Activities for Fire Prevention Committees,* based on Fire Prevention Contest entries.

1950 In October, 7,000 newspapers receive the ad, "Don't Gamble with Fire—The Odds are Against You," developed by the Advertising Council and NFPA.

1954 Sparky® the Fire Dog is created.

1965 *Fire Journal* begins a regular column on "Reaching the Public."

1966 "Wingspread Conference" highlights the need for public education.

1967 Three Apollo astronauts die in a fire in their spacecraft, drawing national attention to the need to be prepared for fire emergencies. Later, attacks on firefighters during urban riots attract more public attention.

1970 President Richard Nixon appoints the National Commission on Fire Prevention and Control.

1973 The National Commission on Fire Prevention and Control publishes its report, *America Burning.*

The Fire Department Instructors Conference offers its first presentation on fire and life safety education, delivered by Cathy Lohr of North Carolina.

1974 NFPA and Public Service Council release the first television Learn Not to Burn® public service announcements starring Dick Van Dyke.

Fire Prevention and Control Act establishes the National Fire Prevention and Control Administration.

NFPA report by Richard Strother *A Study of Motivational Psychology Related to Fire Preventive Behavior in Children and Adults* explains the effectiveness of positive educational messages.

1975 The National Fire Prevention and Control Administration holds its first national fire safety education conference.

1977 NFPA 1031, *Standard for Professional Qualifications for Fire Inspector, Fire Investigator, and Fire Prevention Education Officer,* is published.

National Fire Prevention and Control Administration releases *Public Fire Education Planning: A Five Step Process.* National Fire Academy offers its first public education course on the same subject.

National Fire Prevention and Control Administration launches national smoke detector campaign.

1979 J. C. Robertson's *Introduction to Fire Prevention* published by Glencoe Press.

Project Burn Prevention, funded by the Consumer Product Safety Commission, develops educational strategies and materials for reducing burn injuries—a significant milestone in the shift from "fire education" to "all-risk education."

The *Learn Not to Burn® Curriculum* is published by NFPA.

International Fire Service Training Association releases IFSTA 606, *Public Fire Education.*

International Society of Fire Service Instructors establishes its Public Education Section.

1981 NFPA establishes its Education Section.

TriData Corporation releases *Reaching the Hard to Reach.*

1985 The National Education Association recommends the *Learn Not to Burn® Curriculum.*

NFPA publishes *Firesafety Educator's Handbook.*

1986 Learn Not to Burn® Foundation incorporated.

1987 The first edition of NFPA 1035, *Standard for Professional Qualifications for Public Fire Educator,* encourages civilians to become public fire educators in the fire department.

TriData Corporation publishes *Overcoming Barriers to Public Fire Education.*

1990 Oklahoma State University publishes the first issue of the *Public Fire Education Digest.*

TriData Corporation releases *Proving Public Fire Education Works.*

1995 NFPA and National SAFE KIDS Campaign begin developing all-risk school curriculum called *Safety Sense.*

This time line relies in part on information from Pam Powell's "Firesafety Education: It's Older Than You Think," *Fire Journal,* May1986, pp.13+

Figure 7.2 Time line of events affecting fire and life safety education.

Protection Association's (NFPA®) Franklin Wentworth began to send fire prevention bulletins to local newspapers in hopes of getting them published. As shown in **Figure 7.2**, fire and life safety education has made significant progress since 1909.

Some of the more notable early impacts were the Junior Fire Marshal Program, the birth of NFPA's Sparky the Fire Dog, and the establishment of NFPA® standards for professional qualifications for a fire inspector, fire investigator, and fire prevention education officer.

The *America Burning* report published by the National Commission on Prevention and Control brought the nation's fire problem to the forefront (see Chapter 2). This document, while not current, is still exceptionally relevant. Every fire service professional should review it to see what we have and *have not done* to date. Not only did this document depict the significance of the fire problem in the United States, it identified educating people about fire as one of the best ways to reduce fires and fire-related deaths and injuries. This document's impact on recognizing the need for fire prevention efforts was again recognized in the summer of 1999, when the Federal Emergency Management

Agency director formally recommissioned *America Burning*. The 2000 United States Fire Administration (USFA) report, *America Burning Recommissioned*, states, "There is wide acknowledgement and acceptance that public education programs on fire prevention are effective … no prevention effort can succeed without a public education component."[1] The commission considered that fire departments now respond to more than "just fire" and are the first line of response to an array of disasters that communities face throughout the United States. The commission determined *America at Risk* to be the correct title and orientation of the report.

Today, an effort for a national strategic plan for loss prevention exists through the collaboration of many organizations and individuals.

In the spring of 2008, the Vision 20/20 National Forum gathered in Washington DC for the development and support of a national strategic agenda for fire loss prevention (The Report is available at http://www.strategicfire.org).

Objectives of the Vision 20/20 initiative include:

- Provide a forum for sustained, collaborative planning to reduce fire loss in the United States
- Involve agencies and organizations with expertise and commitment to fire loss reduction in this collaborative effort
- Focus on actions that are needed to bridge the gap between recommended solutions and the current status of fire prevention activity
- Communicate recommendations and actions clearly with all levels of the fire safety community
- Build on the success and momentum of existing efforts

In the past, fire departments focused their educational efforts only on fire prevention and fire survival topics such as stop, drop, and roll and practicing a home escape plan. Today, fire departments have expanded their roles in educating the public to preventing injuries in a number of areas. The changing world in which we live has also created the need to educate citizens about how to be prepared prior to a disaster or terrorist event, when many of the fire department resources will be strained with response needs.

With the inclusion of these additional prevention and preparedness topics, the name of the fire department's education activities evolved to fire and life safety education.

The fire department took on these additional educational topics for a number of reasons. One of the most notable is that fire departments have expanded their role in emergency response from just fire fighting to include such incidents as emergency medical response, swift water rescue, ice rescue, hazardous materials, and in more recent years, terrorism response. With the increased emergency services, fire departments were seeing citizens injured from a variety of activities. These included drowning while swimming, fractures and head injuries from skateboarding and biking without helmets and pads, deaths and major injury from not wearing seat belts or proper infant and child-safety seats. These types of preventable injuries and the local cry to prevent them were driving forces for fire departments to include educational topics in addition to fire prevention.

During natural disasters, terrorism attacks, or other types of attacks such as school shootings, it was apparent that fire departments can become overwhelmed and citizens may not be able to rely on them for evacuation or assistance with relocation efforts. Preparedness education continues to be a primary focus for many communities given continuous events, such as tornados and the massive wildfires in Colorado in 2012 and 2013. Colorado Springs, Colorado, suffered a tragic loss of 347 homes in the 2012 Waldo Canyon fire. However, because of preparedness and planning, coupled with thousands of people being routinely educated on the dangers of wildfire and becoming familiar with what their responsibilities were, over 1,900 homes that were directly threatened by that wildfire were spared. That is an 82% save rate! While tragic, only two lives were lost in that fire that impacted over 34,000 residents. Without public education and focus on the risks, success stories such as this would not be possible.

The responsibilities of today's fire departments extend well beyond the traditional fire hazard. Fire protection expertise can be shifted from reaction and response and used for prevention activities. Today many fire departments are providing injury prevention or all-risk or community risk reduction programs to include a number of topics such as:

- Bicycle safety
- Natural disasters
- Water safety
- Ice safety
- Pedestrian safety
- Babysitter training
- Cardiopulmonary resuscitation (CPR)
- First aid
- Environmental hazards
- Seat belt and car seat training
- Medical health/wellness/education outreach
- Community paramedicine programs to offset medical alarms

In April of 2001, the North American Coalition for Fire and Life Safety Education conducted a symposium to address solutions from the Solutions 2000 symposium. The report, *Solutions 2000,* published recommendations for improved safety for young children, older adults, and people with disabilities. The *Solutions 2000* symposium was held in Washington, DC, in April 1999. The symposium offered an avenue for experts in the area of fire safety to meet with those who had specific concerns about young children (under five), older adults (over sixty-five), and people with disabilities. The attendee's report, *Beyond Solutions 2000,* suggests solutions for actions that will impact fire safety for high-risk groups (This report is accessible electronically at www.usfa.fema. gov). The report built upon the recommendations in the first report with its primary focus on targeting the high-risk groups. The recommendations of the report focus on decreasing fire deaths and injuries in the targeted groups by improving fire safety in the areas of (1) egress capability (2) early warning, and (3) fire sprinkler protection. It was clear that more aggressive education is needed to heighten the awareness of the fire problem in all of the target groups. People need to better understand the threat of fire before they will be motivated to change their behavior or their environment.[2] This only reemphasized the need for further fire safety education.

The National Strategic Plan for Loss Prevention (Vision 20/20) (www.strategicfire.org) previously discussed continues to focus on a national plan for fire and community risk reduction by implementing the following five strategies:

Strategy 1: Prevention

Strategy 2: Prevention Marketing

Strategy 3: Prevention Culture

Strategy 4: Prevention Technology

Strategy 5: Prevention Codes and Standards

Each strategy is a key component in an overall community risk-reduction program that is covered in more detail in Chapter 6.

Fire and life safety education will play a significant role in fire department services even as resources to provide the services become scarce. Fire departments will need to be creative with their resources and will need to look for the role of fire and life safety education to be part of every firefighter's duty, not just the fire and life safety educator. Partnering with other community agencies and organizations will be paramount to the success of a comprehensive fire and life safety education program. Vision 20/20's partnership model is an excellent example of building a national coalition for community risk reeducation.

Not only will fire departments take on a community risk reduction approach, they will be taking it on with other agencies. Fire departments must seek outside assistance to develop and implement its fire and life safety education programs and seek assistance for fiscal resources. A principal way of gathering this support is through coalition building. Community involvement through coalition building only helps to encourage good fire and life safety behaviors and heightens community awareness. The awareness of the problems or risks reinforces the need for preparedness before the event occurs. Among many other agencies, community coalition members may include:

- News affiliates (news media, such as television, radio, and newspapers)
- Schools
- Neighborhood associations
- Housing authorities
- Apartment associations
- Grant-providing entities
- Fast food restaurants
- Grocery stores (particularly chains)
- Public mass transient entities (bus, rail, etc.)
- Local lodging and convention centers

- Chambers of commerce
- Economic development
- Libraries
- Civic groups, such as Kiwanis, Rotary, Lions Club, etc.
- Private companies
- Mental health facilities
- Hospitals and wellness/care facilities
- Law enforcement agencies

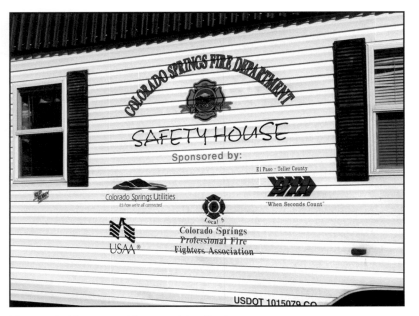

Figure 7.3 Always provide recognition for donors.

There are perhaps more private and sponsored funding options than ever before. In fact, you likely will experience more problems trying to keep up with writing grant applications because the possibilities are so numerous. There are likely local organizations in your community that provide endowments, grants, loans, and other assistance. Make sure that all donors get some sort of recognition for their generosity (**Figure 7.3**). Remember that a partner in community risk reduction may not necessarily be in the form of funding, but in a form of service. The Colorado Springs Fire Department has established a large fire and life safety coalition with numerous pillars of the community that make up the program of which some include: schools, apartments, mental health facilities, parks and recreation, public information, etc.

These pillars work to solve their issues which involve code enforcement, education, wellness, and the like, but also partner among each pillar to interact and share resources. An outstanding example of interaction and sharing resources is the issue of school shootings, which is the crosswalk between all these groups. The coalition meets regularly to strengthen and share loss control strategies for protecting our young and the at-risk population who attend these occupancies. Events such as these tragic school shootings in our country have resulted in a shift from evacuation from just fire to evacuation from other types of emergencies. Fire departments and law enforcement agencies have been asked to assist in the development of school crisis plans which include lockdown, defend in place, security, and various types of controlled evacuation components. These types of crises and evacuation plans are also becoming more prevalent in the business community, which is a critical driving factor for collaborative and interactive solutions among various groups.

To provide funding streams and support, it may also be possible to establish trust accounts specifically set up for individual programs to which anonymous donors can contribute. You can also establish not-for-profit mechanisms such as 501c(3)s offering the department opportunities to earn and collect money that it could not through normal bureaucratic channels. Be careful to check with your department's attorney and fiscally adept advisors. Trust management and accounting involve some serious issues. While trusts offer more financial flexibility, you must exercise great care to keep yourself and your chief out of trouble.

Fire and life safety education programs will not just take place in the school and local fire stations. They will also become a part of the workplace as more companies and businesses see the need to educate their workers on good fire and life safety behaviors. Nationally recognized fire codes now place a greater emphasis on employee evacuation drills. The code requirement verbiage intentionally does not use the wording "fire drills." It is recognized that employee evacuation planning is more than leaving the building when the fire alarm sounds. There may be the need to leave the building for other types of emergencies. The need to educate employees of when and where to evacuation as well as provide a means for accountability is continuing to become more prevalent in the workplace. As businesses develop their emergency evacuation plans, they will seek the guidance and input from the fire department. This not only serves as an excellent coalition-building opportunity, it ensures the evacuation plan does not conflict with fire department operations and is consistent with other community plans recognized by the fire department. NFPA® 72, *National Fire Alarm and Signaling Code,* 2016 edition, has inserted requirements for mass notification systems used for weather emergencies; terrorist events; biological, chemical, and nuclear emergencies; and other threats which are falling to fire departments to review and enforce. These types of notification systems are an integral part to the emergency evacuation planning process, as there may be limitless ways to notify or provide instruction to occupants based on the threat and risk.

Does Fire and Life Safety Education Really Work?

Fire and life safety education is a core element of all fire prevention programs. Essentially, in every aspect of fire prevention, members of the community are educated either about the fire problem or ways to reduce the impact of a fire if one should occur in their home or business. Conveying an effective fire and life safety education message is not limited to a formal presentation in front of an audience. Operations personnel may deliver the message during their interaction with a homeowner for a non-life-threatening call such as a residential carbon monoxide detector activation or during a preconstruction meeting with a developer.

For example, the owners of a new industrial facility may learn about the benefits of automatic sprinklers after they are required to install them in their new facility. At first, the owners may have been reluctant to spend the funds on something "they will never use." However, we can hope that through our educational efforts and after those owners discuss this issue with a member of the fire prevention bureau, they will understand how sprinklers operate and how effective they are. Routine fire inspections are an example of fire and life safety education constantly taking place but not always being identified. Far too often, owners or occupants are reluctant to address deficiencies identified by fire inspectors. This may be due to either the financial impact or their belief that the deficiency is not significant or important. The fire inspector's job, whether he or she is a full-time inspector or part of an engine company inspection, is not only to obtain compliance but also to educate and sell the occupant or owner on the importance of addressing the issue (see Chapter 11 for further discussion). More importantly, if the fire department's chief officer, leadership, or governing elected body does not feel that fire and life safety education works, the program may become a target for reduction or elimination, as was very evident during recent poor economic conditions.

To evaluate a fire and life safety education program, it is paramount that those individuals conducting fire and life safety education programs know their audiences and that their audiences understand and can apply the behaviors taught. For the purpose of this discussion, keep in mind we are evaluating the effectiveness of the fire and life safety education program, not the instructor.

The text, *Fire and Life Safety Educator,* further examines in detail many evaluation methods that include process evaluation, impact studies, and outcome evaluation.[3]

Process Evaluation

This is commonly referred to as program monitoring. It is used to track the evolution and outcome of a program. Accurate record keeping is required and should include the anecdotal information that documents public feedback, attendance numbers, and relevance to the issues that are being presented. Is the public gaining anything from the program in their opinion?

Impact Evaluation

This compares baseline data with post-program results and identifies what changes have been made to the targeted population. Impact studies show a stronger proof over process evaluation because impact studies measure knowledge gain, behavioral change, and modification to the lifestyle.

Outcome Evaluation

This is the highest stage of program evaluation. It tracks statistical and anecdotal evidence over a period of time. The strongest evidence that a program is working is the reduction of occurrences.

All fire and life safety educators need to learn and understand these evaluation methods. The *Fire and Life Safety Educator* manual is a must for any fire and life safety educator because it was written and validated by experienced life safety educators.

In this text, we have previously compared the fire department's operation to the private sector's. Measuring the results of a fire department's program is subject to the same comparison. The effectiveness of a fire and life safety education program should be measured by its impact on the "bottom line." In this case, the bottom line would be the reduction or elimination of the community's fire incident, injury, and death rates.

Proving the Worth of Your Fire and Life Safety Education Program

In fire and life safety education, we can compare our communities' fire-injury and death rates to those of similar communities throughout the country. If we have a successful fire and life safety education program, our profit will be shown in fewer fire injuries and deaths than the norm. Understand that fire and life safety education alone will not achieve this. We have discussed how every component of our fire prevention effort includes a part of fire and life safety education. A number of fire and life safety education programs have been implemented and evaluated. When establishing a fire and life safety education program, you may be wise to see how the program was effective in other communities.

Fire departments can compare their fire and loss records to three national fire databases: the annual surveys of fire departments by the National Fire Protection Association®, the National Fire Incident Reporting System (NFIRS) of FEMA/USFA, and the Fire Incident Data Organization operated by NFPA. This works well if your community uses NFIRS. However, not all communities use this system, and some decision makers may want to see more local statistics than national. Making your case always requires you to have some local statistics.

What Do the National Statistics Show?

Fire prevention officers can compare local statistics to national norms for similar-size communities to show the need for further fire and life safety education programs in their communities. The data can be used to compare how low the community's loss is compared to similar communities or how high its reoccurred losses are.

Today's technology and affordable databases are so user friendly that reports alone become worthless data collection instruments. For example, a database tabulates a list of 100 call types, which can then be queried. In this example, let's say we are targeting fires in schools. We tap into our trusty desktop computer and query for fires in schools. If you see only two or three, you might assume you are doing a good job. However, because you read this book, you know to look deeper. The data reveals a number of activated fire alarms at many schools. Upon further query, you are able to determine the cause of the fire alarm activation to be fire in a wastebasket.

Our point is to be careful with tracking data. Be aware of how the data is collected and who is doing it. Then develop a clear understanding of what you want. We strongly recommend obtaining a free copy of the *Fire and Data Analysis Handbook* from FEMA/USFA. This handbook simplifies using data and interpreting and presenting the results. The ability to gain insight into the community's fire problem allows fire and life safety educators to address the root of the problem with a specific program designed to meet the needs of the citizens in the area concerned.

Planning a Fire and Life Safety Education Program

In 1977, the National Fire Prevention and Control Administration published a public educator's document that incorporated a planning model for fire safety educators that is still in use today. In 2001, FEMA and the USFA published *Public Fire Education Planning: A Five-Step Process*, an excellent, free resource. This material is still mostly relevant, but you must always consider your audience. The basic steps will remain the same but the content and approach to your audience will change. The five step planning process how now evolved into the Five Step Planning Process to Community Risk Reduction.[4]

The five-step planning process consists of the following steps that can be used for a variety of fire and life safety education programs:

Step 1: Conduct a Community Analysis

Step 2: Develop Community Partnerships

Step 3: Create an Intervention Strategy

Step 4: Implement the Strategy

Step 5: Evaluate the Results

The five-step process contains a number of activities and decisions that will present program planners with a clear direction (**Figure 7.4**).

The five-step planning process must be a continual program. Communities are not stagnant. They are constantly changing in size, culture, demographics, and in injuries sustained. Fire and life safety education programs must change to reflect the community's needs, and the best way to accomplish this is through the five-step process.

Designing Presentations

In the business world, you can have the greatest idea for a product or a service, but unless you have the ability to deliver and market that product or service, your idea is worthless. The same holds true for a successful fire and life safety education program. We just identified the importance of the program; now we will examine how to deliver the program for our targeted customer.

Targeting Your Audience

Before we can discuss types of fire and life safety education programs, we must consider who our target audience is and their level of learning. It is important to determine the learning characteristics of the audience to ensure the program design will be conducive to their learning. For example, the way a preschooler learns behavior is different than the way a teenager does. Even though the message to each group may be similar, how it is presented is very important to achieving the program's goals. As another example, the way you present an educational message about lighters and matches will differ from one age group to the next. Preschoolers will learn that hot things hurt them and if they find a match or lighter they should tell an adult. Elementary students will learn that matches and lighters are tools, not toys. Finally, adults will learn to keep matches and

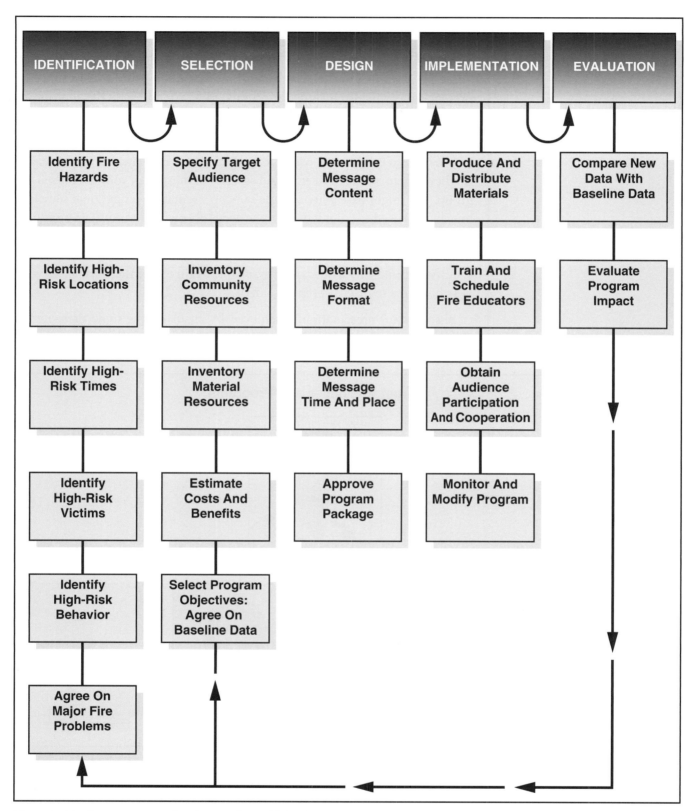

Figure 7.4 The five-step process for planning fire and life safety education.

lighters out of reach of children. The format will be different as well because of each group's level of learning. **Tables 7.1** and **7.2** indicate the cognitive development of each age group as well as each group's motor and personal development. **Table 7.3** identifies attitudes and implications for educating adolescents. The presentations will also include educational ideas conducive to the appropriate learning behavior of the age group.

The message of your presentation should be applicable to the audience. One way to help do this is to choose data to which they will relate. For instance, you might use data on the frequency and severity of injuries in your audience's age group. You might base a fire and life safety presentation on national statistics, but you will need to tailor it to meet the needs of your community. The national statistics and types of injuries per age group will not necessarily be the same as in your community. Another important point to consider is the season of the year. As we enter the winter months, people begin to use their fireplaces. This is a good time to discuss fireplace safety.

A successful presentation is achieved by ensuring the topic presented is pertinent to the targeted audience **(Figure 7.5)**.

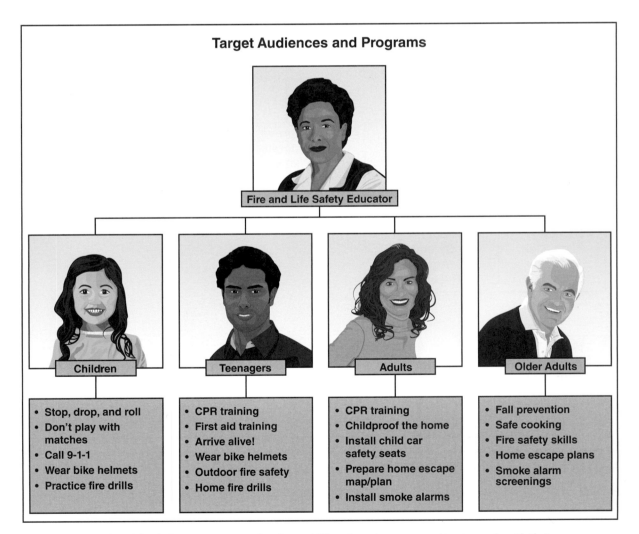

Figure 7.5 Examples of the different age groups that fire and life safety educators must try to reach with their presentations.

Table 7.1
How Elementary Age Children Grow

Age	Cognitive Development	Motor Development	Social And Personal Development	Education Ideas
5-6 years	Attention span increases, learns through adult instruction			Fire and burn safety activities can be longer and explain more.
	Can match pictures, colors, shapes, and words			Enjoys safety materials that involve matching pictures, etc.
	Can stay with one activity for about 20 minutes			Fire and burn safety activities can be longer and explain more.
	Likes to talk (and does!)			Expect children to talk about their real or pretend experiences with fire and burns.
		Colors within lines		Enjoys fire and burn safety coloring books.
		Likes to climb, jump, throw, and march		Enjoys activities that use large muscles, such as "Stop, drop and roll," "Crawl low under smoke," and escape drills.
			Likes to anticipate events, such as holidays and birthdays	Fire prevention week is an "event" for these children.
6-7 years	Begins to develop sense of time and space			Can begin to understand how fast fire is.
		Very active, but clumsy		Expect children to run, shove, and push during escape drills.
			Loves and respects (and often quotes) parents and teachers	Can be very influenced by firefighters and other authority figures.
			Wants to know what is good and what is bad	Responds well to fire and burn safety rules.

Table 7.2
How Elementary Age Children Grow

Age	Cognitive Development	Motor Development	Social And Personal Development	Education Ideas
7-8 years	Will stay with learning activities for long periods			Fire and burn safety activities can be longer and explain more.
	Understands time			Needs to know how much longer a fire and burn safety lesson will last
	Is ready for simple map work			Can draw a home escape plan
	Understands space and can find way around the community in proper order			Would enjoy walking with a parent to see the fire station
	Listens to long stories, but often misinterprets the facts		Begins to think teachers and parents are more unfair; often afraid of new situations	May still regard firefighters as heroes
	Talks about cause and effect by using "because" and "so"		Is impatient in large groups	
			Plays dress-up	
			Fears monsters or other fantasies	

Adapted from "Developmentally Appropriate Learning Activities," presented by Faye Ann Presnal, Early Childhood Specialist, at the Oklahoma Public Fire Education Conference, August 7, 1991.

Table 7.3
Helping Adolescents Cope

Attitude	What It Means To Adolescents	Implications For Fire And Life Safety Education
A sense of self-worth	Feeling respected and valued for yourself, not just for what you can do	"Because you are a valuable person, we want you to be safe."
A sense of competence	Knowing that you can do several things well—and can excel at something	"Because you know how to be fire safe, you can show others how to be fire safe."
A sense of acceptance	Knowing that groups you care about care about you	"Because you care about your friends, you care about their safety."
A sense of responsibility for others	A commitment to looking out for people you care about	"Because you care about your family, you care about their fire safety."
A future vision of self	Having ideas about what your future life will be like	"Because you want your home to be safe, you'll want a smoke detector."

Fire and Life Safety and Community Risk Reduction

The topics for the fire department's fire and life safety education programs need to reflect the community's fire problems and associated risks. The following topics tend to be relevant to most communities.

- Matches are tools, not toys
- Stop, drop, and roll
- Smoke detectors: where to put them, when to test them
- Addresses: a responder's friend
- Crawl low in smoke
- Adopt a hydrant: remove snow from your hydrant
- Home fire-escape plan
- Fireworks safety
- Hot liquids burn
- Fire extinguishers
- Call 9-1-1 for emergencies

- Make the right call (don't abuse 9-1-1, save it for emergencies)
- Fireplace safety
- Safe cooking
- Carry a pot holder with you
- Cooking outdoors
- Campfire safety
- Drown your campfire
- Remember, only you can prevent forest fires
- Don't play with fire

- NFPA® Risk Watch (a comprehensive all-risks program)
- Bicycle safety
- Water safety
- Outdoor safety (camping, hiking, etc.)
- Wildfire mitigation
- Wildfire evacuation and planning
- Wildfire neighborhood chipping and recycling programs

In Chapter 6 there is considerable discussion of Community risk reduction. A community risk assessment guide is available on line at http://strategicfire.org/community-risk-reduction/community-risk-assessment/. Vision 20/20 developed resources for with Community Risk Reduction planning. CRR is a process to identify and prioritize local risks, followed by the integrated and strategic investment of resources (emergency response and prevention) to reduce their occurrence and impact. The online Community Risk Assessment Guide (also available as a PDF) will assist fire departments and other organizations conduct a basic or more complex assessment of risks within their community. In addition, a Community Risk Reduction Planning Guide is available to assist in the CRR planning process.

Fire and Life Safety Education Programs

The fire department can conduct a number of programs on topics of the audience's choice (**Figure 7.6**). Essentially these are opportunities to have a "captive audience:"

- School visits
- Use of fire safety trailer
- Fire station tour
- Citizen fire academy
- Children's safety academy
- Children's fire safety festival at a local mall
- Block party visit by the fire department
- In front of the grocery store while the fire department personnel are shopping
- Birthday party in the station
- Sleepover at the fire station
- Fire department open house
- Sprinkler demonstration
- Neighborhood Wildfire Evacuation drills

Figure 7.6 Fire department open houses are great opportunities to showcase the fire department and educate the public. *Courtesy of the Winnetka, IL Fire Department*

What Makes a Great Presentation?

As discussed earlier it is imperative that sufficient planning and preparation take place to address the specific needs of the targeted audience.

Planning for the Presentation

Once the informational topic has been chosen and the informational or promotional materials selected, there is still planning that needs to be done for the presentation. Simply put, planning illustrates a professional concern for the quality of the presentation and a respect for the audience.

Each informational presentation, while following the speaking points, should have three components: introduction, body, and conclusion. All three components have important roles in the effective presentation. The introduction must be motivational and sufficient to capture the attention of the audience. The body of the presentation, where the new information is relayed, cannot be underestimated. The conclusion wraps it all up and may be the final opportunity to have an impact on someone in the audience. The following explains each component.

Introduction of a Presentation

The introduction of a presentation contains information about the topic and why the audience should be interested. During the introduction, the educator should gain the attention of the audience and provide motivation for them to participate. The motivation may be to view the fire truck in the parking lot after the lesson. Or, the motivation may be for older adults to maintain the ability to remain independent and in their own home.

An introduction should focus the attention of the audience on the topic at hand. It sets the scene or climate for the presentation. The introduction should explain why the educator is there, the purpose of the presentation, what will take place, and a time frame involved. The introduction might include anecdotal stories, historical perspectives, topical information, or a review of information that the audience already knows.

Body of the Presentation

The body of the presentation follows the introduction, where new information is presented. The body includes clear directions, motivational practices, and reinforcement of the audience's interest and learning. It is during this time that the effective educator reminds the audience of what they know and presents the new information, knowledge, skill, and attitude. The body of the presentation should provide encouragement and guidance to the audience, as well as to reinforce and monitor audience interest and learning. The body should make use of appropriate instructional methodologies and appropriate materials.

Conclusion of the Presentation

The conclusion of the presentation is just as important as the introduction and the body. It allows for questions and an opportunity to generate additional interest in the topic.

During the conclusion, the educator consolidates the program content and assesses the level of comprehension. This may be from a posttest, survey, or simple question and answer period. The conclusion is the time when the educator needs to connect the information to real-life, review the major points, and list the resources available to the audience for assistance, including the fire agency involved.

The Learning Environment

Everything about a presentation can be perfect, but if the learning environment is inappropriate, effectiveness will be minimized if not lost. The wrong learning environment — the physical facilities and room arrangement — can ruin even the best presentation effort.

Just as with subject matter and teaching methodologies, the most effective learning environment will depend on the makeup of the audience. For the most optimum environment conducive to learning, different age groups require different settings and room arrangements.

Preschool Children

For preschool children, the appropriate setting is the one to which they are accustomed. This should be one that encourages learning and allows the children to feel safe. Most of the time younger children will be more attentive sitting on the floor, on a rug, or in a circle. Ensure that there is adequate space before gathering large groups of young children, otherwise the group size can create distractions. An educator addressing young children should sit on the floor or in a small chair to be near eye level with them. Distractions need to be kept to a minimum. Classes or presentations to young children should take place in a classroom or similar controlled area. Trying to deliver educational presentations outside and next to a fire apparatus simply will not work.

Elementary Children

When addressing elementary children in the classroom, allow the children to sit at their usual desks or tables. Once again, keeping the groups small will assist in maintaining attention and minimizing distractions. When using drama, games, or other activities, make sure that there is ample space for these activities. When addressing elementary children outside of a classroom setting, such as Boy Scouts or after-school groups, plan ahead to limit the distractions as much as possible. If using multimedia, make sure that all the children can see the entire screen clearly.

Older Adults

When working with older adults, the following guidelines can be used when planning these programs:

- Limit programs to 30 minutes or less. Like anyone else, older adults may get anxious during longer presentations.

- Include question and answer periods in discussions. Many older adults like to participate in classroom discussion.

- Dim or control the lights if PowerPoint or transparencies are being used. Sunlight or glare may present a problem for this audience.

- Keep the room at a steady, warm temperature.

- Make sure that all members can hear the speaker. Speak clearly and make sure that a microphone is available.

Tips for Presenting Educational Programs:

- Be aware of the age group's stage of intellectual development.
- Understand and honor attention span limits.
- Limit the use of lecture-based instruction. Intersperse lecture with interactive and reality-based experiences.
- Pause after key points to check for understanding and verbally test for comprehension.
- Connect each new piece of information to what has been previously presented. Clarify its relevance and importance.
- Intersperse opportunities for interactive exchanges of information between the presenter and participants.

Getting Your Message Out

The media is indispensable for getting out your fire and life safety education message. Every fire event contains a fire safety education message that you can incorporate into the press release. An apartment fire started by children playing with matches that ended in the children safely evacuating after the smoke detector activated is a good example of how we can deliver our fire and life safety education message through the media. The press conference or press release will indicate the exact details of the incident pertaining to the response of the fire department, time of day, where the incident occurred, and so forth. The fire department spokesman should also discuss the cause of the fire. Tell the media that fire started from children playing with matches and explain that matches are tools, not toys, and adults should be aware of the dangers associated with matches and lighters. Now is also the time to go to the data. Give them the local and national statistics for fires caused by children playing with matches. Also use this opportunity to show how important a working smoke detector is and how having a planned escape route saved the family. A good way to do this would be to ask inspectors or whatever personnel might be available to check all the smoke detectors in all the units in the affected building. It is not uncommon to find only 3–6 percent of battery-operated detectors in working order. This is a powerful statistic to put on the news. The media is a great tool to reach a large number of people and, if used appropriately, can be the fire and life safety educator's greatest allies.

Is this fire and life safety education, public relations, or public information? The answer is simply, yes to all! We can use the media for fire and life safety education, but they are also an avenue for building a positive image of the fire department within the community. Conversely the media can also portray a negative image. Using the media is usually called providing public information. Public information is simply the process of informing the public of our operations or actions. As you can see, fire and life safety education, public relations, and public information are all inseparable. But, what about the nonemergency use of the media? Do we have to wait for an event to occur before we seek the media? Absolutely not! In fact, using the media before the big event helps fire departments to become familiar with the various media representatives and to gain their trust. As shown in Chapter 8, this will help during the "big event." Fire prevention, it's not just one week a year!

Summary

America Burning demonstrated that fire prevention efforts cannot be successful without public education. Public education has since evolved into fire and life safety education. This term better describes the diverse topics presented by those fire prevention personnel performing the task of public education. Today, Vision 20/20 is playing a key role in the development of a National Plan for Prevention through the implementation of five key program strategies. The fire service continues to be called upon to perform a variety of emergency services. From those services, the fire service has seen the important need to provide a comprehensive safety program that encompasses more than just fire prevention. Topics presented by fire departments may range from ice safety to bicycle helmets. As a result, fire and life safety is evolving to a holistic community risk-reduction program.

The goal of fire and life safety education is to motivate individuals to behave in a safe manner by modifying their behavior. To succeed, fire and life safety educators must create programs utilizing the five-step process. However, the programs must be appropriate for the target audiences. Fire and life safety educators can improve the success of their fire and life safety education message by selecting a delivery method that is most appropriate to the age of the audience. Using the Internet, computers, and social media is a common and important method of presentation for citizens.

As budgets for fire and life safety education are scrutinized more and more, fire and life safety educators need to look for alternative funding for their programs and think outside the box for methods to deliver the fire and life safety education message.

There are many online resources that can help you improve presentations. Here are some suggestions:
- IFSTA Resource One
- Safe Kids Training Academy
- Mr. Media Training: 25 Most Essential Public Speaking Tips.

Chapter 7 Review Exercises

7.1   Identify the potential fire department impact when fire and life education is eliminated as a service._____

_____

_____

7.2   What is the role of fire and life safety education?

_____

_____

_____

7.3   Why is fire and life safety education important?

_____

_____

_____

7.4   How did the term *fire and life safety education* evolve?

_____

_____

_____

7.5   Why do fire departments conduct presentations on topics other than fire? _____

_____

_____

7.6   Identify six safety-related programs other than fire prevention that fire departments can present.

1. _____

_____

2. _____

_____

3. _____

_____

4. _____

_____

5. _____

_____

6. _____

_____

7.7 Explain the National Strategic Plan for Loss Prevention (Vision 20/20) and how it positively impacts fire departments. _____

_____

_____

7.8 Discuss how to prove fire and life safety education works.

_____

_____

_____

7.9 List 12 fire and life safety resources available for free. Indicate where they can be obtained.

1. _____
2. _____
3. _____
4. _____
5. _____
6. _____
7. _____
8. _____
9. _____
10. _____
11. _____
12. _____

7.10 Discuss the different types of media that can be used in fire safety education and who would benefit from their use.

_____

_____

_____

7.11 Discuss the various presentation methods that fire and life safety educators can use and which audience you most likely would use them for. _____

_____

_____

7.12 Identify six sources of potential funding for fire safety education in your community and explain how you would go about securing funds from them.

1. _____

2. _____

3. _____

4. _____

5. _____

6. _____

7.13 What did *America Burning* say about fire and life safety or public education? _____

_____

_____

7.14 Read the report *Beyond 2000* and write a one-page summary.

7.15 On a seperate sheet of paper, identify 10 community businesses or organizations that could become part of your fire prevention and injury reduction coalition.

7.16 What is Community Risk Reduction?_____

_____

_____

7.17 What are the six steps of a community risk reduction program according to Vision 20/20?

7.18 How can the media help you present life safety messages?

_____

_____

_____

_____

Notes

1. US Fire Administration. *America Burning*: Recommissioned (Washington DC: US Fire Administration, 2000).

2. Beyond Solutions 2000 Report, United States, January 2001.

3. *Fire and Life Safety Educator* (Stillwater, Okla.: IFSTA, 1997).

4. Public Fire Education Planning—A Five-Step Process, Federal Emergency Management Agency. Integrated Prevention Interventions, National Fire Academy Risk Reduction Curriculum.

Public Information Officer: A How-To

Table of Contents

Key Terms

Key Points

1. We live in an age when information can be disseminated almost simultaneously with actual events and in which people routinely make judgments based on what they see and hear from the media.

2. As a part of good government practice, we should always assume everyone's approach and judgment are reasonable and prudent given appropriate information.

3. People who perform PIO functions must clearly understand their community and grasp their target audience.

4. The PIO must work very hard at establishing credibility with all members of the local media.

5. Never, ever think for one second that anything is "off the record."

6. Proficiency in all forms of written communication is essential to public information functions.

7. Using a variety of media to deliver the same message to different audiences can profoundly impact the public information officer's success.

8. To communicate fire protection or prevention messages or handle on-scene media issues, familiarity with your department's operations is crucial.

9. Know what your community expects of your department and be able to deliver it.

Learning Objectives

1. Define the national fire problem and role of fire prevention.*

2. Define the functions of a fire prevention bureau.*

3. Identify and describe the standards for professional qualifications for Fire Marshal, Plan Examiner, Fire Inspector, Fire and Life Safety Educator, and Fire Investigator.*

4. Understand the importance of good community relations.

5. Understand the role of the public information officer.

6. Identify standards for the position of public information officer.

FESHE Objectives (USFA)

Public Information Officer: A How-To

Case Study

A large municipal fire department has experienced three significant structure fires over the last few days. The wildfire danger for this community has also become a hot topic in the media over the last week. Many grass and forest fires have occurred throughout the state. The community is on edge.

One major structure fire was the second to occur in less than 12 hours at the same address. Many citizens and some elected officials in the community accused the fire department of not doing an adequate job of extinguishing the first fire (rekindling the original). This caused the media outlets to badger fire department staff for updates and statements about why a second fire, at the same location, completely destroyed the building.

Local policymakers are hearing this communication and are asking the chief if there are problems within the department. The community's concerns are being voiced because of a potentially disastrous wildfire occurring during this latest dry spell, and lack of confidence in the fire department is disheartening.

Media at the scene of the last fire overheard firefighters questioning whether the previous shift actually had a rekindle. The media also overheard complaints that information had not been provided regarding the findings of the fire investigation. One of the media outlets announced unconfirmed reports from firefighters that the second destructive fire was a rekindle from the previous fire. However, the fire investigators have told your chief that the second fire was clearly incendiary, but no information should be released until more progress is made on the investigation.

Given this detailed scenario, what are at least three significant concerns that the chief has regarding what is going on surrounding these events? What elements of a strategic plan could be implemented to address these concerns? What are five ways that information transfer could be used to assist the chief with his concerns?

The Importance of Public Information

The role of a **public information** officer is critical to maintaining the image of the fire department and to disperse accurate information to the media. From the fire service perspective, public information is the communication to the general public regarding receiving data or messages about events or issues, how to become better educated or prepared for various situations, and how to stay engaged in fire- and injury-prevention behaviors. While critical in educating or making people aware, public information is not a replacement for public education roles and responsibility. Media management is a tool in marketing and informing, but it is not "Public or Community Education" in the program sense. People may use information to make decisions, act on those decisions, and to help the fire service in mitigating situations or circumstances that may harm our overall community.

To properly perform the functions of disseminating public information, a dedicated position should be considered to fill this position. Demands for social media and multimedia management are very time-consuming. While many smaller fire departments may not have full-time staff to fulfill this function, they should work to integrate these tasks into the organization's overall operational plan. Whenever possible, they should use a person or group with some training for performing the duties of the PIO. NFPA® 1035, *Standard on Fire and Life Safety Educator, Public Information Officer, Youth Firesetter Intervention Specialist and Youth Firesetter Program Manager Professional Qualifications*, provides some guidance that can be applied to public information. In some cases you may choose to use your Fire and Life Safety Educators (FLSE). However, PIO duties are typically separate and distinct. Many state fire marshal's offices, state training agencies, or the emergency management institute provides specific courses through the federal emergency management agency.

(Whether you are an experienced PIO or just learning, the IFSTA *Fire and Life Safety Educator* manual, Third Edition, has very good information and is a great resource covering a variety of related topics.) We recommend placing this position or responsibility in the fire prevention division or bureau where the bulk of routine public interaction and information dissemination takes place.

Fire departments need to encourage citizens to receive and act on information in order to modify their behavior to prevent fires or injuries and be prepared for disasters **(Figure 8.1)**. An underlying, but no less important, motive in disseminating this information is marketing the fire department and its services. Essentially, this accomplishes the role of public relations, which is important because fire departments are engaged daily in some form of public relations activity. The general public has its perception of the services and daily activities of the fire department. Most of the interaction a person has encountered with his or her fire department is in the form of emergency services. The continued presence in the media provides a means to educate and market the services provided by the fire department. This also

Figure 8.1 Discussing hazardous fire conditions with the media is one way in which a public information officer might try to modify the public's behavior.

provides a means of exhibiting leadership to the community that can be very beneficial, particularly in times of emergency. The dispersed information need not always be related to an emergency calls for service.

Not that long ago, the release of information took hours or days to disseminate. Now we live in an age where information can be disseminated almost simultaneously with events as they unfold, whether through the evening news or through electronic communication. Current popular methods of electronic communication are tweeting, using Facebook, creating blogs, texting, skyping, etc. No matter where you are in the United States, or the world for that matter, electronic communication and the news brings you real-time images of disasters, shootings, fires, and elections, not to mention the creation of local, state, and world policy. People routinely make judgments based on what they see and hear from these sources of media. Messages flood cell phones, computers, airwaves, print, and other media, reaching more people than ever before. The problem is that with limited control over what the media and people can and will communicate, particularly among different countries, countless mixed messages proliferated throughout the world generate endless controversy. Unofficial information can also lead to many rumors and exaggerations that can create lots of anxiety if not genuine hysteria. Obviously, for a public emergency response agency, this is an exceptionally unwanted event. Whether we like it or not, responsible emergency agencies must work very hard to provide accurate, informative, and timely information. Experience has shown us that smartphones are providing self-managing means of generating, modifying, and validating pretty correct information, before emergency agencies are even warmed up to gather it. The reality, like it or not, is that if we do not provide people with accurate and timely information, they come upon it one way or another, accurate or not. We are in a much better position to control the information released so that all parties involved are protected and respected as much as possible. Management of this data is also critical for criminal or other investigations. The disappearance of the Malaysian jet, flight 370, in 2014 is a very good example of media management, and lack thereof.

One example of how important it can be to have a message portrayed and interpreted favorably is that of a fire marshal presenting a sprinkler ordinance for approval by the elected officials (**Figure 8.2**). If local developers and builders oppose the ordinance, the media or building associations may portray it as costly, burdensome, unwarranted, and of no benefit to the general public. While different people or groups may have opposing ideas or thoughts, it is important to address all viewpoints, making certain to express the facts you want represented to support your position. The key is to back

Figure 8.2 Opportunities to market or advertise how sprinklers actually work can be impactful to your efforts. *Courtesy of the Colorado Springs, CO Fire Department*

your position by facts, not your opinion, unless you are specifically asked for an opinion. Remember, everyone has an opinion. The facts speak for themselves. Oftentimes politicians and media bait you for an opinion, after which to be used as the fodder to discredit your effort. Tread carefully. Remember, the facts of the situation remain constant. Consider the same ordinance being portrayed as a means to reduce property loss, to save lives, and potentially to save insurance costs. Perhaps the community could offer incentives or tax breaks for installing the life safety system. Public opinion may become more positively influenced, all depending upon how the information is released and how factually it is represented and accurately balanced. Now, what if the ordinance specifically addressed multifamily occupancies and was being presented in a community that just experienced a multiple life loss from fire in a nonsprinklered apartment building? Depending upon the information provided, the public's view of the ordinance may be totally different from the original picture of its being overly restrictive, burdensome, and too costly. How can this be when all sides are discussing the exact same issue and have the exact same facts? It is because of the way the media received and presented the information to the community.

As a part of good emergency services practice, we should always assume everyone's approach and judgment are reasonable and prudent, given appropriate information. That does not mean everyone will agree all the time, but most reasonable people look to solve problems and understand things in similar ways. When the media present consistent and factual information, public opinion and insight should be fairly similar. However, that being said, the media frequently put "spins" on stories or messages to get attention and sell their product.

Returning to our sprinkler ordinance example, the media could likely spin the ordinance story differently after a significant fire death rather than under other circumstances, such as perceived regulatory intrusion. They will cast different opinions in the interest of balanced journalism based upon what they are told or what they see and believe. Unfortunately, the media will commonly portray an opinion or view that may be contrary to what the "sender" is trying to say, simply for purposes of sensationalizing an event or situation. Again, keep in mind what their mission is—to sell, print, or gain market-share on TV or radio.

In *Feeding the Media Beast*, Mark E. Mathis explains, "Journalists are not the enemy. In spite of what you have been led to believe, news people are not dangerous vipers. To the contrary, the vast majority of folks who work in the business are passionate professionals who want to make the world a better place. They tend to be idealists, but that's a good thing."[1] Why is that a good thing? Because if we pay attention to Mathis's statement, we know the angle from which journalists typically approach the fire department PIO. The reality is that fire protection professionals are generally idealists too (**Figure 8.3**). Without even trying, we are already in a better position to communicate with the so-called "Beast" because we understand its idealism. Not only must we try to understand the media, we must become students of the media and continually focus on improving our public information skills.

Figure 8.3 The fire protection professionals and journalists shared idealism can be the basis for successful public relations.

PIO Skills

The PIO must develop a basic set of communication skills. Seven areas that good PIOs routinely practice and that require some level of proficiency are:

- Community relations
- Media relations, operations and their processes
- Writing
- Public speaking
- Audio/visual presentation
- Fire department operations and functions
- Expertise in all social media formats

Community Relations

People who perform the PIO functions must have a good understanding of the community along with a good grasp of their target audience. One key element often overlooked is understanding the demographics and diversity of a community along with the varying culture, which can play a critical role in their perceived view point of the fire department. Also identify who works, invests, and maintains a high profile within your community—individuals, businesses, corporations, philanthropic organizations, and others. Is your focus going to be baby boomers or Millennials? The PIO should also have a keen awareness of organizational relationships and partnerships. A good understanding of these networking issues can provide not only a wealth of information and resources but also a good preface for complimenting and thanking folks when good things happen.

It is also important to understand the community's views and concerns regarding emergency preparedness, their political opinions, and volunteer involvement. Do not forget to understand the different cultures and value systems throughout the community. Although the community as a whole may have a common set of values and beliefs, it may also have subgroups, each with its own set of values and beliefs. Such background information is critical if you are going to propose a new program that the community may not yet consider important. This would be particularly true if the program required money at a time when the community may have just voted down any new tax increases. It may not make your job easier, but your chief may appreciate not being led into a pond full of alligators.

Having a well-rounded view of your community (your target audience) provides tremendous insight into how you can motivate your community. The right approach can move them to support your department's requests and endorse the department's attempts to accomplish its mission.

Media Relations

Media relations are likely the most challenging and frustrating part of a PIO's job. As the PIO, you must be willing to work hard at establishing credibility with *all* members of your local media. An old adage says, "The media will always have the last word!" This is mantra, and you need to buy into it because they do have the last word. If you do not help them get a story, they will modify, mold, or (in the worst cases) potentially fabricate any information to create the story. It is imperative that you become the media's ally. With the variety of media available, you cannot focus just on building a relationship with a single medium, such as the local newspaper. You must establish good relations with all the key participants in the business. You must be timely, factual, responsive, and know their deadlines. The better your relationships, the better your flexibility and the better the support and participation you will get from them.

Treat them ethically and reasonably, and they will do the same for you... as long as it fits their desires and deadlines. Does this sound one-sided? It is. Their job is to get a story. Everything is on the record. Never, ever think for one second that anything is "off the record." Reporters do not understand those words. But do not think they are disingenuous, either. Think about what their job is and what your job is. If you have information that should not be released or is off the record, keep it to yourself. We cannot tell you how many times we have seen individuals come unglued with regret after sharing something with a reporter by prefacing it as "off the record." This is like dropping by a fire station and telling the on-duty crew "I just saw somebody get hit by a car down the street, but don't tell anybody or do anything because we shouldn't get involved." Why would we expect the media to be any different than us?

It is also important that PIOs become familiar with how their various media work. Visit their businesses and see how they create their productions, make their newspapers, or produce their broadcasts. This insight will provide you with valuable information about their need to meet deadlines and their preferences for product. In return, it will give them ownership in a joint relationship. We like to show off our fire stations; why would we think they do not like to show off their work sites? Getting to know them before a "big event" makes talking to them during even the biggest event easier because you already have a level of trust and a foundation for your professional relationship.

Develop a perspective on the journalist community. Mark Mathis explains that "They are manufactured in America's universities in a liberal arts curriculum. Journalism schools teach students that they are 'the voice of the people.' Reporters are to stick up for the poor, the downtrodden, and the disadvantaged, in our sometimes-oppressive capitalist system. It is their job—so they are taught—to 'comfort the afflicted and to afflict the comfortable.'"[2] As the PIO, do you think you will be considered afflicted or comfortable? Many fire service personnel have spoken harshly of reporters because "they don't even respect what we do" or "they don't even know the history of why things are the way they are." Many reporters are not born and raised in your community, but are merely advancing on one of many stepping stones to another promotion in another market. The process of networking and getting to know them must never stop because their movement never stops. News is a dynamic business and must be respected as such. Remember, too, that these same media folks are snooping for that "big scoop," the portfolio, and résumé builder. This is all the more reason never to reveal anything "off the record." If they do not understand why and what you do, then you should take the time to inform them. Educate or, better yet, make them aware of who, what, when, where, why, and how you do things before they show up at your door step!

To succeed in getting your message across to the media, you should follow some basic guidelines. Below are some of the more common guidelines, as adapted from Mark E. Mathis's *Feeding the Media Beast*:

- Remember that, to the media, things that are important are not necessarily important unless they are different. This means you should consider your message and emphasize what is "different" from other similar issues and stress that difference.

- Use emotional messages to the factual messages — reporters always prefer this.

- Remember that reporters and newscasters like information in small bites. Be concise but powerful with the message using key points.

- Have a group of "canned" messages or information you want to release. You never know when a similar story may break, giving you an opportunity to interject yours, which is slightly "different."

- Make sure that the information is easy to give and easy for media to get. Help them as much as possible to entice them to play.

- Treat the media with respect by answering all of their questions but make sure you direct and redirect your answers to what you want to communicate.

- Repeat your message just like in teaching. Tell them what you are going to tell, tell them, and then tell them what you told them.

- Modify slightly important event information or messages for each different delivery. You can still pass along the same general message but remember number one above…. make it different each time to grab more attention. Remember, each member of different media would like his or her own scoop.

- Never, ever make stuff up. If you do not know, admit it, research the answer, and get back to them.

- Never, ever speak for another agency or official who has not given you permission and verified what you are going to present.

- Never, ever say "No Comment." This tells them you are hiding something. Be truthful and explain why you do not have information or why you cannot release it.

- Do not give your personal opinion. Any opinion you give will be assumed to be that of the organization regardless of how you preface it.

- Assume you are always on camera or on tape.

- If you do not want information or comments on the nightly national news, don't say it.[3]

The media generally want typical, similar information about any event or disaster. They generally focus on who, what, where, when, why, and how, and so should you. Worksheets are a good way to keep your information on track (**Figure 8.4**). Below is a list of common items that you could use to create simple, customized worksheets of your own:

- The cause, if known, and the situation encountered
- Eyewitness accounts or reports by responders
- Extent of response to incident (who was there and where they were from)
- Statistics that communicate the scope (dollar loss, lives lost or injured, firefighters and or equipment on scene, etc.)
- Types of injuries
- Survivors and their stories
- Past history of similar events
- Community (civilian) actions
- Time or duration of interruption or repair
- Action taken
- Fire and life safety education message related to the event

  Things to consider for the media if you are on the scene when they show up:

- Staging location for portable "live or remote" broadcasts
- Schedule of briefings
- Identifying their contact and giving regular briefings
- Information on phones, lodging, food, and other logistics

PUBLIC INFORMATION WORKSHEET — Page 1

PUBLIC INFORMATION WORKSHEET

DATE: ____________________ Fire _____ EPS _____ Hzmt _____

TIMES: Dispatched: __________ On Scene: __________ Controlled: __________
Tapped: __________ Cleared Scene: __________
Alarms: 2nd __________ 3rd __________ 4th __________
Zone: 1 __________ 2 __________ 3 __________ 4 __________

DISPATCH: Reason Dispatched: ____________________
Address: ____________________
Occupancy: ____________________
Owner: ____________________

OCCUPANCY: Residential Structure: ____________________
Multi-family: ____________________
Commercial: ____________________
Business/Function: ____________________

APPARATUS:

Engines	_____	Aid	_____	Cmd Cars	_____
Ladders	_____	Hzmt	_____	Medic	_____
Amb	_____	PD	_____	WNG	_____
DOE	_____	EPA	_____	PUD	_____
RC	_____	Support	_____		

MUTUAL AID DEPARTMENTS/RESOURCES: ____________________

SITUATION ENCOUNTERED/ACTION TAKEN: ____________________

Figure 8.4 Examples of PIO worksheets for varying incidents.

PUBLIC INFORMATION WORKSHEET — Page 2

SPECIAL HAZARDS/ACCOMPLISHMENTS/RECOMMENDATIONS: ____________________

INJURIES/FATALITIES: Civilian (C) ____ Firefighter (FF) ____

Name	Sex	Age	Injury	Where Taken

DAMAGE: ____________________

$ ESTIMATE: ____________________

CAUSE: ____________________

PUBLIC EDUCATION MESSAGE: ____________________

SMOKE DETECTORS: Installed ____ Operating ____

Figure 8.4 Examples of PIO worksheets for varying incidents. *(Continued)*

HAZARDOUS MATERIALS INCIDENT WORKSHEET — Page 1

HAZARDOUS MATERIALS INCIDENT WORKSHEET

DATE: ______________________________ Fire _____ EPS _____ Hzmt _____

TIMES:
Dispatched: ____________ On Scene: ____________ Controlled: ____________
Tapped: ____________ Cleared Scene: ____________
Alarms: 2nd ____________ 3rd ____________ 4th ____________
Zone: 1 ____________ 2 ____________ 3 ____________ 4 ____________

DISPATCH:
Reason Dispatched: ______________________________
Address: ______________________________
Occupancy: ______________________________
Owner: ______________________________

OCCUPANCY:
Residential Structure: ______________________________
Multi-family: ______________________________
Commercial: ______________________________
Business/Function: ______________________________

APPARATUS:

Engines	_____	Aid	_____	Cmd Cars	_____
Ladders	_____	Hzmt	_____	Medic	_____
Amb	_____	PD	_____	WNG	_____
DOE	_____	EPA	_____	PUD	_____
RC	_____	Support	_____		

MUTUAL AID DEPARTMENTS/RESOURCES: ______________________________

____

____

____

SITUATION ENCOUNTERED/ACTION TAKEN: ______________________________

____

____

____

____

____

____

____

____

Quantity & State: ______________________________
Product Use: ______________________________

INJURIES: ______________________________

____

____

Figure 8.4 Examples of PIO worksheets for varying incidents. *(Continued)*

HAZARDOUS MATERIALS INCIDENT WORKSHEET — Page 2

COMPANY/INDIVIDUAL NAME (ADDRESS IF DIFFERENT THAN LOCATION)**:** _____
_____
_____
_____
_____

PRODUCT NAME: _____

SPECIAL CONSIDERATIONS - FIRE/WATER/AIR/HEALTH: _____
_____
_____
_____
_____

ROAD CLOSURES/EVACUATIONS: _____

CONTAINMENT/CONTROL: _____

ENVIRONMENTAL HAZARDS: _____

DECONTAMINATION: _____

CLEAN-UP: _____

FOLLOW-UP: _____

Figure 8.4 Examples of PIO worksheets for varying incidents. *(Continued)*

PIO WORKSHEET — Page 1

PIO WORKSHEET

Address ______________________ TOA __________ Arrival __________
Owner/Resident ______________________ Age __________
Alarm # __________ Type of Structure ______________________ Units __________

Unit _____ Arrival _____ Command ______________________
Unit _____ Arrival _____ Safety ______________________
Unit _____ Arrival _____ Liaison ______________________
Unit _____ Arrival _____ Operations ______________________
Unit _____ Arrival _____ City Safety ______________________
Unit _____ Arrival _____ City Public Relations ______________________
Unit _____ Arrival _____ Medical ______________________
Unit _____ Arrival _____ Investigations ______________________

MEDIA WORKSHEET

Unit __________ Name ______________________
Unit __________ Name ______________________
Unit __________ Name ______________________
Unit __________ Name ______________________
Unit __________ Name ______________________
Unit __________ Name ______________________
Unit __________ Name ______________________
Unit __________ Name ______________________

CAUSE OF FIRE

Location of Ignition ______________________
Cause of Ignition ______________________
Contributing Factors ______________________
Smoke Detector Present: Yes _____ No _____
Operate: Yes _____ No _____
Awoke Residents: Yes _____ No _____
Comments of Investigator ______________________

Figure 8.4 Examples of PIO worksheets for varying incidents. *(Continued)*

PIO WORKSHEET — Page 2

INJURIES

Name ______________________ Age ____ To/By ____________

Injury ____

Name ______________________ Age ____ To/By ____________

Injury ____

Other Information ______________________________________

____

____

____

AGENCY SUPPORT

Agency ______________________________ POC ____________

Type of Support ______________________________________

Agency ______________________________ POC ____________

Type of Support ______________________________________

Comments by PIO ______________________________________

Figure 8.4 Examples of PIO worksheets for varying incidents. *(Continued)*

Another critical thing to consider and become familiar with is the aspect of social media. Experienced PIOs can tell story after story of how bombarded they become when the media is looking for information or updates on current or recent stories. Providing regular "tweets" is one way to do this. There are many mediums for rapid information exchange. Blogs are another form, but these are generally not as fast. If you release instant bits of information quickly, you can stave off numerous calls as the media in your community subscribe to your sites and messages. The media now has access to real-time, up-to-date information they can act on instead of waiting to check with you to see if it is newsworthy or not. Sites such as "Facebook" can also be a great way of providing lots of information on your events, progress on items, or updates on general happenings. This type of site is generally for those who have time and gander your organization, while the other instant release mediums are for newsworthy or immediate events. Facebook or twitter and all the other mediums such as this require timely, accurate, and regular feeding of information. You cannot embark on using these mediums unless you are committed to provide immediate and timely information constantly. Inconsistency in transmissions will lead to disinterest and lack of trust.

We strongly encourage that you spend time familiarizing yourself with these various programs and sites and seek to determine who in your area uses which. This can be a double-edged sword. It can be very time consuming if you do not have enough personnel to maintain and manage it. However, it can save tons of time if you are able to keep up and provide regular data streams. In fact, many media use tweets as a way to craft their stories, saving them from sending out a reporter. Are you doing their work for them? Yes! Get over it! That's how it all works nowadays.

Written Communication

Proficiency in all forms of written communication is essential to public information functions. Press releases, letters, faxes, and e-mail have all become mainstays of communication. Be sure you can organize your thoughts and messages on paper in a fashion that will get your points across. Also become very adept at constructing the minuscule messages that are typical for tweets as previously discussed. Be able to adjust your style to accommodate a wide variety of audiences. In the fire service, we deal with a wide range of people from highly skilled engineers and political appointees to people from other countries and ordinary citizens who are unaware of our technical jargon. Be cognizant of your audience and deliver your messages appropriately. Be as grammatically correct as possible because good grammar shows professionalism. Know how your local media like to receive various types of information **(Figures 8.5 and 8.6, p. 188-189)**. You can learn this through early meetings while getting acquainted. These formats include:

- Reports of various kinds (annual, monthly, statistical, etc.)
- Public service announcements (PSAs)
- Fact sheets
- Media advisories
- Pamphlets, newsletters, brochures, and so forth

Use current publishing software to make attractive, quality documents that reflect your department's professionalism and dedication to good communication.

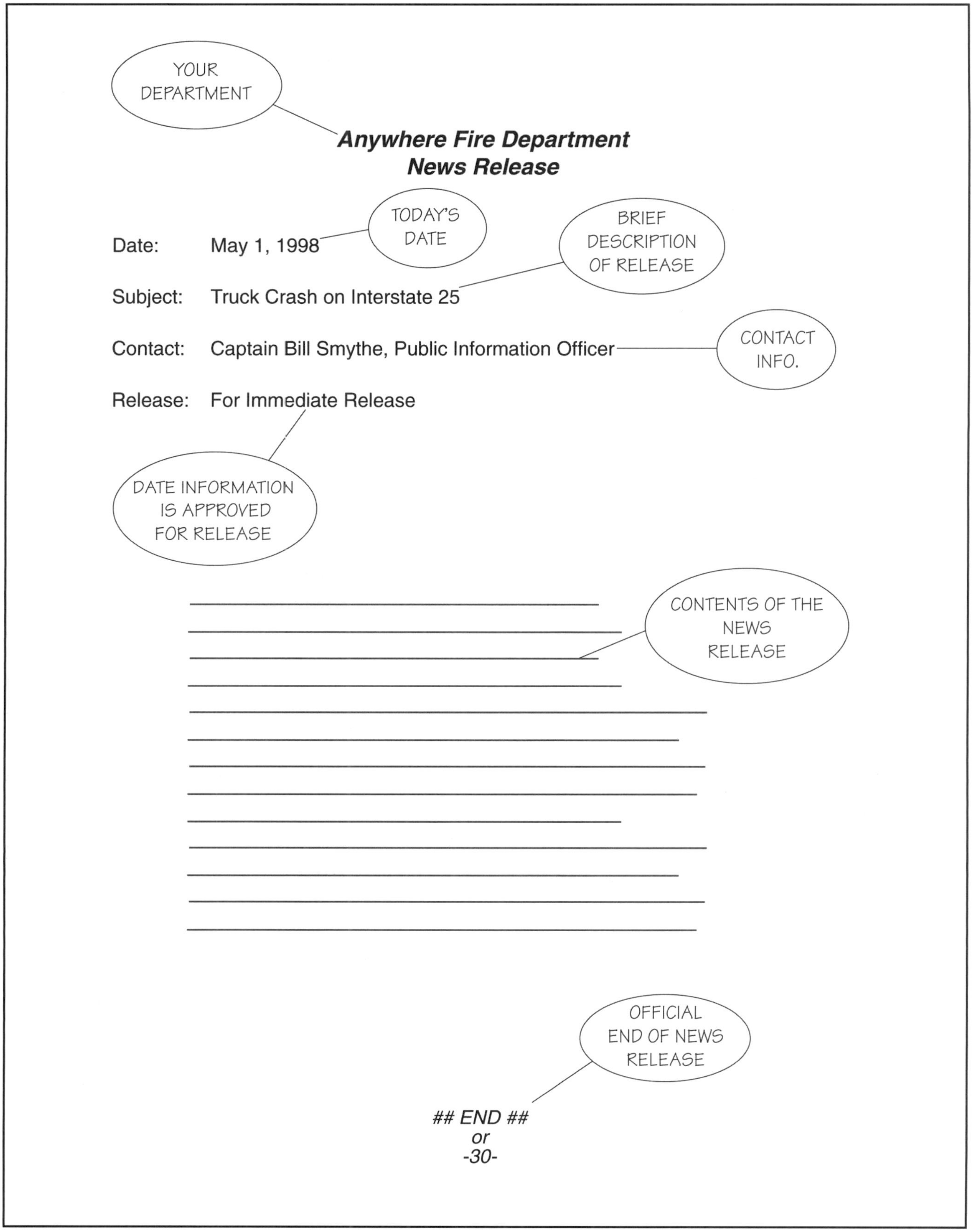
YOUR DEPARTMENT
Anywhere Fire Department
News Release
TODAY'S DATE
BRIEF DESCRIPTION OF RELEASE
Date: May 1, 1998
Subject: Truck Crash on Interstate 25
Contact: Captain Bill Smythe, Public Information Officer
CONTACT INFO.
Release: For Immediate Release
DATE INFORMATION IS APPROVED FOR RELEASE
CONTENTS OF THE NEWS RELEASE
OFFICIAL END OF NEWS RELEASE
END
or
-30-

Figure 8.5 News releases are among the public information officer's most valuable tools.

COLORADO SPRINGS FIRE DEPARTMENT
STEVE OSWALD
PIO CAPTAIN

CITY OF COLORADO SPRINGS

1398 Windmill Avenue Chimney Fire

Colorado Springs, CO – On December 12, 2014 at 7:16am, Colorado Springs firefighters were dispatched to a chimney fire at 1398 Windmill Ave. CSFD Truck 9 arrived within 5 minutes to the single family home and reported smoke showing from the roof. At 7:21am this incident was declared a working fire. After making access to the interior of the home, T9 reported all residents were out of the home and that the fire was possibly in the attic.

CSFD responded to this single family residence structure fire alarm with 2 fire engines, 2 fire trucks and a Battalion Chief on the first alarm. After declaring "working fire" and additional fire companies were added, there were 32 firefighters on scene working to setup water supply, vertical ventilation, fire attack and overhaul. Vertical ventilation (cutting holes in the roof) was an important tactic in stopping the progression of the fire in the attic.

The fire was extinguished at 7:53am and loss stopped declared at 9:04am. Upon arrival of the first fire companies, residents of the home had reported to firefighters that everyone was out of building and no injuries were sustained.

Fire investigators have ruled this fire as "Accidental." Investigators determined that due to the strategy and tactics used by CSFD Firefighters, we were able to save $197,174 of the homeowners' property.

Please visit out facebook page for any photos you wish to use.

#

Commission on
Fire Accreditation
International

Fire PIO
375 Printers Parkway
Colorado Springs, CO 80910-3191
TEL 719-385-7223 • FAX 719-385-7388

"Providing the highest quality problem solving and emergency service to our community since 1894."

Figure 8.6 Media or News Releases alert media outlets to important information coming from your department. *Courtesy of the Colorado Springs, CO Fire Department*

Public Speaking

People in our communities love to talk to and hear from fire department personnel, so PIOs should be groomed for this job. The types of public speaking engagements you may be called upon to participate in are numerous (**Figure 8.7**). These may include:

- On-scene emergencies
- Camera interviews
- Live radio broadcasts
- Panel discussions
- Seminars
- Presentations to business organizations, government agencies or policy-making groups, and neighborhood and building associations

Figure 8.7 Public information officers often use news or press conferences as a way to provide batch information to the media at emergency incidents.

The public generally asks many questions, some of which can be very challenging. If you have not done a lot of public speaking, offer yourself to as many civic organizations as possible for simple question-and-answer sessions. We also suggest joining organizations like Toastmasters. This organization is a marvelous way to enhance your oral communication skills, gain confidence, and provide connections with others in your community. In doing this, you do not always have to make a specific presentation, but can become comfortable standing in front of large groups passing on information. To speak well in public, you must be able to construct good written outlines that are easy to use for you or someone else in your organization. If you have accomplished this goal, and anyone in your field can use the same outline to communicate sufficiently, then you know you are preparing your presentation well. If you also have mastered the PIO's writing tasks, public speaking should be fairly easy.

Below is a list of "what not to do" when speaking in public:

- Pacing from side to side. This can be very distracting and makes everyone in the audience look like they are watching a tennis match.

- Reading to your audience. If you are going to read a speech to your audience, save everyone including yourself a lot of embarrassment and time and make copies of your speech and just hand them out.

- Holding or leaning onto the podium. Sometimes people uncomfortable with public speaking grab onto the podium — like they are in the middle of a tornado. This makes your audience tense and will make you tired.

- Using hypnotic fillers, words, or phrases like "uh," "okay," "ya know?" or "all right." We all have sat through presentations where people repeat these phrases or words so often that five minutes into the talk we are more interested in counting the repetitions than in listening to the speech. "Ya know" is our favorite because if the audience already knows, then why are you telling them?

- Looking sloppy. People look at the fire service as a professional organization. This does not mean in terms of paid versus volunteer but in terms of demeanor, appearance, knowledge, and application. Think of Shakespeare's phrase, "The clothes make the man"— or in this case, the person.

- Looking at incoming texts or emails on your phone.

- Answering your cell phone.

- Going beyond your allotted time. Practice, practice, and practice to make sure you hold to your time or less. Many a chief's favorite mantra is "Less is more!"

Try to take advantage of the media or someone else to videotape your presentations. If local television stations tape any of your presentations for their news coverage, ask if they will share dubs so you can critique yourself. This is becoming more and more difficult as their corporate attorneys avoid this type of exposure or liability; however, some smaller markets may be more cooperative.

Many reporters will help guide you on how to give the best sound bites. Do not ever be ashamed or afraid to ask them to do another take. They are likely more than willing to help you out if you do the same for them. When your presentation is over, review the tape for constructive criticism of yourself. Do not hesitate to ask them for input also. Learn by your mistakes and those of others. Likewise, when you see or hear a good presentation ask yourself, "What made this speaker so effective and how can I do the same?"

Audio /Visual Presentations

We recommend going to any number of the one- or two-day workshops on this topic that are frequently held in different parts of the country. Skills in photography, computer operations such as PowerPoint, desktop publishing software, and word processing applications can be used daily in this job position. Remember, your job is to communicate with any number of people. Using a variety of these media to deliver the same message to different audiences can have a profound impact on your success.

Understanding how newsletters (digital or in print) are constructed can be a huge asset to a number of organizations. For example, a local apartment or renter's organization may be happy to use preprinted or already laid out material. Various business organizations use newsletter formats. Nursing homes are another location where newsletters are popular. Web pages can be a critical source of information for everyone. Remember though, it must be current and maintained. Who in your organization is responsible for that?

When visiting local televised media, spend some time learning from them about video production. They love to show off, and the experience works wonders for your understanding of how to get the most from digital or taped formats.

Fire Department Operations and Functions

To do the best job possible of communicating fire protection or prevention messages or communicating on-scene media issues, familiarity with your department's tactical operations is crucial. Become intimately familiar with policies and procedures. If you are new to the fire service, read through relevant sections of the National Fire Protection Association's® *Fire Protection Handbook*® and become a diligent student of the fire service. It would be good to study IFSTA's validated training manual, *Fire and Emergency Services Orientation and Terminology,* so that you become familiar with the typical nomenclature. To communicate the emotional and factual sides of your messages, you have to become familiar with your business. The public looks to the fire service PIOs to provide comfort and confidence when things are going badly. It is not in our interest or theirs to provide critical information from an uninformed or ignorant perspective. Nothing could be worse than presenting critical information to a concerned public about an emergency incident and then falling on your face when you do not know the answer to a simple procedural question. The words, "I don't know," can be a terrible thing to say. However, tell the truth if you don't know; saying you don't is far better than lying. Make sure you give confidence that you will find out the answer and get back to them immediately.

The public and the media want authority and credibility. These are easy to establish if you are careful and diligent. True finesse comes through on-the-job training and years of experience or to very dedicated individuals who have spent countless hours learning their trade.

Legal Issues and Responsibilities

Be sure to develop thorough guidelines or policies that specifically itemize the information you can and cannot give out. Personnel matters are a good example. Typically, personnel or disciplinary issues are private and confidential and, therefore, cannot be disclosed in any detail. Make certain that you and your chief and legal counsel share a good understanding of your role. This may mean getting signoffs on press releases or anything else intended for publication before it goes to the media. Do not release the names of minors, particularly if they are involved in criminal activity, such as arson. Breaches of legal protocol, not to mention policy issues, can generate far more mistrust and discontent concerning the fire department than refusing to answer questions at all.

Summary

Be proactive. Establish relationships with your local media and be prepared for national attention. Remember, national news media can be anywhere within minutes. Media outlets are relying on individual citizens and their individual smart phones to photo and create real-time news casts. Know what your community expects of your department and be able to deliver it. Be familiar with the hazards and risks your department faces. Be able to articulate and explain to them, as well as how your department mitigates and responds to them. Plan ahead. If you live in a region of the country prone to wildfire, prepare burn restriction or ban release information and material, so you do not have to react spontaneously. Have information on evacuation preparedness and vegetation management ready to go. Holiday hazards, like Christmas trees and candles, are generally similar and typical. Do not wait until the last minute to prepare for these information releases. Another important aspect is to be timely in disseminating information. Be aware of your media's deadlines and help reporters to meet them. Remember, that everyone, including the police department, looks to the fire service for answers on many community emergency issues.

The job of a PIO is easy, except when things are most demanding, most important, and most challenging. People expect quick solutions to problems and factual information with which to react. They need professional reassurance and confidence to help them deal with tragedy and loss. Regular contact and information are critical to generate and secure public trust and calm. Recognize the difference between emergencies and nonemergencies. When commercial aircraft crashed into the World Trade Center on 9-11-2001, fear and anguish struck the hearts of Americans everywhere. In these types of incidents, the good PIO is immediately available and talking to the media.

Develop your delivery of product based on your public's needs. People make value judgments about your department during major emergencies or challenges, even those that do not occur locally. Be prepared to handle the stress, the questions, and the information professionally and meaningfully. You become the icon for your department. Dress the part, act the part, and garnish the confidence necessary to protect and inform the public you serve. We usually have a great story to tell about our services. Let your passion for the fire service be reflected in a professional and positive fashion.

The role of the public information officer is important in maintaining the image of the fire department. This position also is critical in presenting factual information to the public and others about public events, disasters, emergencies, or other vital information.

Years ago, transmitting or conveying information could take months. Today, CNN News can be at your doorstep in less than a few minutes broadcasting to the world. Technology has made instant communication possible; however, many fire departments have yet to realize the significance of this to departmental functions.

To provide the highest level of protection possible, reliance on our communities (our citizens) is vital, particularly when we are supposed to be providing more services for less. The best way to disseminate large amounts of good information is through our media. This includes print media such as newspapers, broadcast media such as television and radio, and social media such as Facebook and Twitter. To perform this task effectively, proper training, resources, and personnel should be provided. The role of the PIO is to disseminate information regularly and consistently. This involves community relations, media relations, and extensive written communication. Public speaking also is a critical function.

In the new age of video phones, computer presentations, and simulations, it is wise to have skills in creating quality audio/visual presentations. The youth are especially accustomed to high-quality graphics and fast-moving computer and video games. To properly communicate our messages, we must constantly seek ways to make our presentations more engaging. Using the Internet also provides good opportunities. However, be sure to consult with your legal advisor, as you do not want to violate any unknown copyrights or infringe on sharing or use agreements.

Finally, remember that many people will see you, and first impressions are very important. Dress the part, speak the part, and be professional at all times.

Chapter 8 Review Exercises

8.1   What is NFPA® 1035? _____

_____

_____

8.2   What is the role of the public information officer? _____

_____

_____

_____

8.3   What are six skills that a PIO should have?_____

_____

_____

_____

_____

_____

8.4 List four key aspects of community relations that a good PIO should know. _____

_____

_____

_____

_____

8.5 Create a job description for a news media person that lists qualifications, skills, and experience. _____

_____

_____

_____

_____

8.6 Create a job description for a PIO that lists qualifications, skills, and experience. _____

_____

_____

_____

_____

8.7 Compare your answers to questions 8.5 and 8.6.

_____

_____

_____

_____

_____

8.8 List ten common guidelines to use when dealing with the media.

_____

_____

_____

_____

_____

_____

_____

_____

_____

_____

_____

8.9   On a separate sheet of paper, create a sample PIO worksheet (cheat sheet) for collecting data on scenes.

8.10   On a separate sheet of paper, research the early reports of the Malaysian Jet crash that occurred in 2014 and create a press release that could better inform the families and relatives of victims as well as other government agencies.

8.11   List five "bad habits," or actions to avoid when doing work as a PIO.

_____

_____

_____

_____

8.12   On a separate sheet of paper, create a media blitz campaign (Twitter, Facebook, press release, fax notification to newspaper, media interview information, injury prevention materials) for the Fourth of July holiday.

8.13   Create a list of topics that may be legally sensitive. These should include one fire investigation aspect, one personnel issue, and one departmental political issue._____

_____

_____

_____

_____

_____

_____

_____

_____

_____

_____

_____

NOTES

1.   Mark E. Mathis, Feeding the Media Beast (West Lafayette, Indiana: Purdue University Press, 2002).

2.   Ibid.

3.   Ibid.

Fire Investigations

Table of Contents

Key Terms

Key Points

1. We should learn all that we can from every incident and use that information to help prevent similar events in the future.

2. Good investigative techniques secure the facts, thereby preventing or at least mitigating the degree of liability that a city or fire department may incur.

3. One of the greatest benefits of conducting fire investigations is the opportunity to gain insight into your community's fire problem.

4. Those involved in the fire investigation must be trained to the level of confidence needed for testifying to their findings in a court of law.

5. Investigators should be able to determine a fire's point of origin, its source of ignition, the material ignited, and the act or activity that brought the ignition source and materials together.

6. Document every scene as if you are preparing to testify in court the next day.

7. Allowing a fire's cause to remain undetermined is better than accusing people of something they did not do.

8. Determining that a fire was not accidental is not always the difficult part; prosecuting the individual who set the fire is!

9. Never be afraid to seek help from other authorities or agencies.

Learning Objectives

1. Define the national fire problem and role of fire prevention.*

2. Define the functions of a fire prevention bureau.*

3. Describe inspection practices and procedures.*

4. Identify and describe the standards for professional qualifications for Fire Marshal, Plan Examiner, FIre Inspector, Fire and Life Safety Educator, and Fire Investigator.*

5. List opportunities in professional development for fire prevention personnel.*

6. Identify the purpose of conducting fire investigations.

7. Understand the importance of fire investigations in a fire prevention program.

8. Identify the standards associated with fire investigations.

* FESHE Objectives (USFA)

Fire Investigations

Case Study

Shortly after noon on a warm day in July, a home oxygen-supply van (owned by LuvAir) entered a bulk-gas Breather Company facility to fill up from a bulk liquid oxygen (LOX) tank to make more home oxygen concentrator deliveries. Right after hooking up the transport LOX tank and initiating the refilling operation, a flash fire erupted out of and around the stainless-steel-tank shell and engulfed the Breather Company technician, spraying molten slag and piping components approximately 50 feet in about an 80-degree cone. This debris also started a secondary grass fire outside the plant that was quickly contained. Firefighters responded, treated the victim, and extinguished the remaining fire. The van was observed to have about 15 portable high-pressure cylinders and two oxygen concentrators. None of these appeared to be involved in the cause of the fire, but were instead determined to be an exposure to the fire emanating from the larger bulk LOX transport tank, which was manufactured by Cylandrasaurus. The fire had burned so severely that it cut through part of the van's side doorframe, the bottom of the van itself, and a side brace on the lower part of the van floor.

1. As the officer responsible for this investigation and your code enforcement responsibilities, why would a careful investigation be important with regard to the larger bulk storage tank?

2. If this is determined to be an accident, how many people or companies do you think could be involved in a subrogation action?

3. Overall, why would it be important for you to conduct a detailed investigation of this incident? Cite all reasons you can think of.

Why Investigate Fires?

We investigate fires for numerous reasons. We already have covered the importance of knowing the history of fires and how past fires have started. In this chapter we will explain in more detail the significance, purposes, and desired outcomes of good fire investigation.

Significance

Data gathered from fires are very useful fire prevention tools. Not only do we need to complete fire incident reports, but we must understand the importance of failure analysis and reconstructions of events. Most fire departments conduct critiques after incidents to identify the strengths and weaknesses of their operational and suppression efforts. This is an excellent method to determine areas where training is needed or to find ways to improve performance (**Figure 9.1**).

Figure 9.1 Catastrophic fires, such as in major power production facilities, can cripple cities and towns. Finding origin and cause of fires here that prevent a future occurrence can reduce power interruptions and save millions of dollars. *Courtesy of the Colorado Springs, CO, Fire Department*

The same type of origin and cause investigative approach should be taken after a fire to determine what steps could have been taken prior to the incident to prevent the event from occurring or to reduce the loss it caused. These steps could include fire or building code amendments or additions, behavior modifications, suppression modifications, or other relevant measures. Consider the outcome of the tragic events of 9-11-2001. Detailed fire modeling and analyses of how the World Trade Center buildings and the occupants reacted to the events have been conducted. Reluctance or failure to conduct a post fire evaluation or poor data collection after the fact would severely hinder good analysis and understanding of this event. We should learn all that we can from such incidents and incorporate our discoveries in reasonable modifications to fire code or other relevant solutions to prevent or mitigate similar events in the future. We should gain critical insight into future design and improved fire protection features. The examination of an incident should not focus just on failures or poor performance, but also on things done correctly. This type of study has resulted in improved designs and performance of fire protection systems and superior fireground operations.

Fire Investigation as Fire Prevention

Fire investigation is commonly assigned to the fire prevention division, predominantly because this group usually is designated to track the type of information collected during fire investigation, inspections, and the like. Basically, most fire prevention efforts collect data on all or most buildings within the community. Model codes can grant the fire code official the authority to investigate fires. For example, the 2015 *International Fire Code®*, Section 104.10 Fire investigations states: "The fire code official, the fire department or other responsible authority shall have the authority to investigate the cause, origin and circumstances of any fire, explosion or other hazardous condition. Information that could be related to trade secrets or processes shall not be made part of the public record, except as directed by a court of law."

Like almost all fire department tasks, fire investigations can be handled differently among various jurisdictions. Some locations have modified state or local laws or the adopted model codes allowing the state police or state fire marshal's office to handle some or all fire investigations. Others prefer to handle fire investigations at the county or city level. This is done either by the police or fire department. No one way is better than another, as long as everyone receives and shares the proper training, support, and data to accommodate the needs of the community. The goal should not be to protect turf or territorial issues, but to have a good investigative resource available to serve the public and accomplish the mission. In some situations, state and federal assistance can provide depth of resources that are not available at the local level. Typically, these resources will be in the form of equipment, staff, or laboratory services. Crime labs are a perfect example. Large metropolitan or state agencies can afford this type of facility, whereas a small town likely cannot.

Benefits of Fire Investigation

Fire prevention bureau personnel can use investigative information to try to prevent similar events in the future. One excellent method in using this information is through fire and life safety education. The information gained can and should be used to educate and inform the public of potential fire causes. For example, an unattended candle might start a fire in a home and cause significant damage to the home without injury to the occupants **(Figure 9.2)**. The investigation might then reveal that the occupants were awakened by their smoke alarm and followed their escape plan to safely exit the house. The cause of the fire could be used to inform residents about candle usage and safety. The positive results of having a working smoke alarm and an escape plan also could be used as educational messages.

Figure 9.2 Proper investigations should identify improper behavior, improper equipment installation, or manufacturing or code violations or deficiencies. *Courtesy of the Colorado Springs, CO, Fire Department*

If on a grander scale the fire received media attention, citizens might remember it during a fire safety education presentation. Tracking the causes of fires will benefit a fire prevention program that specifically addresses the "If." On a grander scale, if the fire received media attention, it may identify the community's fire problem allowing them to:

- Improve public awareness and education
- Implement more effective fire inspection practices
- Provide input into fire fighting tactics and operations
- Modify regulatory requirements to provide safer buildings or products
- Prevent or mitigate the impact of a similar occurrence

The value of good fire investigation is noteworthy. It provides benefits in many specific areas:

- Education
- Liability protection
- Code or legislative modification
- Research
- Consumer product improvement
- Identification and prosecution of the crime of arson

Incorporating the causes of fires into fire recruit training and required continuing education provides firefighters with knowledge to be better prepared during future events. The data on specific causes of fires by occupancy can prepare crews for the type of incidents they may encounter. Understanding how and why arsonists do what they do can protect firefighter lives by making them aware of potential hazards and enabling them to do a better job of scene recognition and preservation on their arrival. This in turn can help investigators do a better job of incident reconstruction and determination (**Figure 9.3**).

Figure 9.4 Trailers or indications of accelerant pours are important elements of evidence that are essential for criminal prosecution and the prevention of future hostile fires from the same individual. *Courtesy of the Colorado Springs, CO, Fire Department*

On fires where the fire department's performance may become a liability question, good investigative techniques secure the facts, thereby potentially preventing (or at least mitigating the degree of) liability that a city or fire department may be exposed to. For example, determining that an accelerant was placed in multiple locations of a building and ignited will explain why fire suppression crews were overwhelmed and could not prevent the building from being destroyed (**Figure 9.4**). Fact finding showing accidental causes may also protect business owners, thereby removing them from liability for a careless or negligent act.

The Cocoanut Grove fire discussed earlier in this book is a good example of how codes were modified and legislation was enacted to protect people who visit assembly occupancies.

Fire investigation research of the Apollo space program tragedy of 1967 revealed how an oxygen-enriched atmosphere in the command capsule led to a tragic fire, killing all three astronauts. This led to major changes in design, policy, and practice, preventing future mishaps of this type.

A — FDID ☆ State ☆ Incident Date ☆ MM DD YYYY Station Incident Number ☆ Exposure ☆ ☐ Delete ☐ Change ☐ No Activity — NFIRS-1 **Basic**

B Location Type ☆
☐ Check this box to indicate that the address for this incident is provided on the Wildland Fire Module in Section B, "Alternative Location Specification." Use only for wildland fires. Census Tract
- ☐ Street address
- ☐ Intersection
- ☐ In front of
- ☐ Rear of
- ☐ Adjacent to
- ☐ Directions

Number/Milepost Prefix Street or Highway Street Type Suffix
Apt./Suite/Room City State ZIP Code
Cross Street or Directions, as applicable

C Incident Type ☆
Incident Type

D Aid Given or Received ☆ ☐ None
1 ☐ Mutual aid received
2 ☐ Auto. aid received
3 ☐ Mutual aid given
4 ☐ Auto. aid given
5 ☐ Other aid given
Their FDID | Their State
Their Incident Number

E₁ Dates and Times Midnight is 0000
Check boxes if dates are the same as Alarm Date.
Month | Day | Year | Hour | Min
Alarm ☆ — ALARM always required
☐ Arrival ☆ — ARRIVAL required, unless canceled or did not arrive
☐ Controlled — CONTROLLED optional, except for wildland fires
☐ Last Unit Cleared — LAST UNIT CLEARED, required except for wildland fires

E₂ Shifts and Alarms
Local Option
Shift or Platoon | Alarms | District

E₃ Special Studies
Local Option
Special Study ID# | Special Study Value

F Actions Taken ☆
Primary Action Taken (1)
Additional Action Taken (2)
Additional Action Taken (3)

G₁ Resources ☆
☐ Check this box and skip this block if an Apparatus or Personnel Module is used.
	Apparatus	Personnel
Suppression		
EMS		
Other		
☐ Check box if resource counts include aid received resources.

G₂ Estimated Dollar Losses and Values
LOSSES: Required for all fires if known. Optional for non-fires. None
Property $ ☐
Contents $ ☐
PRE-INCIDENT VALUE: Optional
Property $ ☐
Contents $ ☐

Completed Modules
☐ Fire–2
☐ Structure Fire–3
☐ Civilian Fire Cas.–4
☐ Fire Service Cas.–5
☐ EMS–6
☐ HazMat–7
☐ Wildland Fire–8
☐ Apparatus–9
☐ Personnel–10
☐ Arson–11

H₁ ☆ Casualties ☐ None
	Deaths	Injuries
Fire Service		
Civilian		

H₂ Detector
Required for confined fires.
1 ☐ Detector alerted occupants
2 ☐ Detector did not alert them
U ☐ Unknown

H₃ Hazardous Materials Release ☐ None
1 ☐ Natural gas: slow leak, no evacuation or HazMat actions
2 ☐ Propane gas: <21-lb tank (as in home BBQ grill)
3 ☐ Gasoline: vehicle fuel tank or portable container
4 ☐ Kerosene: fuel burning equipment or portable storage
5 ☐ Diesel fuel/fuel oil: vehicle fuel tank or portable storage
6 ☐ Household solvents: home/office spill, cleanup only
7 ☐ Motor oil: from engine or portable container
8 ☐ Paint: from paint cans totaling <55 gallons
0 ☐ Other: special HazMat actions required or spill > 55 gal (Please complete the HazMat form.)

I Mixed Use Property ☐ Not mixed
10 ☐ Assembly use
20 ☐ Education use
33 ☐ Medical use
40 ☐ Residential use
51 ☐ Row of stores
53 ☐ Enclosed mall
58 ☐ Business & residential
59 ☐ Office use
60 ☐ Industrial use
63 ☐ Military use
65 ☐ Farm use
00 ☐ Other mixed use

J Property Use ☆ ☐ None
Structures
131 ☐ Church, place of worship
161 ☐ Restaurant or cafeteria
162 ☐ Bar/tavern or nightclub
213 ☐ Elementary school, kindergarten
215 ☐ High school, junior high
241 ☐ College, adult education
311 ☐ Nursing home
331 ☐ Hospital
341 ☐ Clinic, clinic-type infirmary
342 ☐ Doctor/dentist office
361 ☐ Prison or jail, not juvenile
419 ☐ 1- or 2-family dwelling
429 ☐ Multifamily dwelling
439 ☐ Rooming/boarding house
449 ☐ Commercial hotel or motel
459 ☐ Residential, board and care
464 ☐ Dormitory/barracks
519 ☐ Food and beverage sales
539 ☐ Household goods, sales, repairs
571 ☐ Gas or service station
579 ☐ Motor vehicle/boat sales/repairs
599 ☐ Business office
615 ☐ Electric-generating plant
629 ☐ Laboratory/science laboratory
700 ☐ Manufacturing plant
819 ☐ Livestock/poultry storage (barn)
882 ☐ Non-residential parking garage
891 ☐ Warehouse

Outside
124 ☐ Playground or park
655 ☐ Crops or orchard
669 ☐ Forest (timberland)
807 ☐ Outdoor storage area
919 ☐ Dump or sanitary landfill
931 ☐ Open land or field
936 ☐ Vacant lot
938 ☐ Graded/cared for plot of land
946 ☐ Lake, river, stream
951 ☐ Railroad right-of-way
960 ☐ Other street
961 ☐ Highway/divided highway
962 ☐ Residential street/driveway
981 ☐ Construction site
984 ☐ Industrial plant yard

Look up and enter a Property Use code and description only if you have NOT checked a Property Use box.
Property Use — Code
Property Use Description
NFIRS-1 Revision 01/01/04

The ☆ denotes a required field. 3–2 NFIRS 5.0 COMPLETE REFERENCE GUIDE

Figure 9.3 A National Fire Incident Report typically contains a wealth of important information.

K1   Person/Entity Involved

Local Option

Business Name (if applicable)   Area Code   Phone Number

Check this box if same address as incident Location (Section B). Then skip the three duplicate address lines.

Mr., Ms., Mrs.   First Name   MI   Last Name   Suffix

Number   Prefix   Street or Highway   Street Type   Suffix

Post Office Box   Apt./Suite/Room   City

State   ZIP Code

More people involved? Check this box and attach Supplemental Forms (NFIRS–1S) as necessary.

K2   Owner

Same as person involved? Then check this box and skip the rest of this block.

Local Option

Business Name (if applicable)   Area Code   Phone Number

Check this box if same address as incident Location (Section B). Then skip the three duplicate address lines.

Mr., Ms., Mrs.   First Name   MI   Last Name   Suffix

Number   Prefix   Street or Highway   Street Type   Suffix

Post Office Box   Apt./Suite/Room   City

State   ZIP Code

L   Remarks:

Local Option

Fire Module Required?

Check the box that applies and then complete the Fire Module based on Incident Type, as follows:

Buildings 111	Complete Fire & Structure Modules
Special structure 112	Complete Fire Module & Section I, Structure Module
Confined 113–118	Basic Module Only
Mobile property 120–123	Complete Fire & Structure Modules
Vehicle 130–138	Complete Fire Module
Vegetation 140–143	Complete Fire or Wildland Module
Outside rubbish fire 150–155	Basic Module Only
Special outside fire 160	Complete Fire or Wildland Module
Special outside fire 161–163	Complete Fire Module
Crop fire 170–173	Complete Fire or Wildland Module

ITEMS WITH A ☆ MUST **ALWAYS** BE COMPLETED!

More remarks? Check this box and attach Supplemental Forms (NFIRS–1S) as necessary.

M   Authorization

Officer in charge ID   Signature   Position or rank   Assignment   Month   Day   Year

Check box if same as Officer in charge.

Member making report ID   Signature   Position or rank   Assignment   Month   Day   Year

The ☆ denotes a required field.   3–3   NFIRS 5.0 COMPLETE REFERENCE GUIDE

Figure 9.3 A National Fire Incident Report typically contains a wealth of important information. (*Continued*)

Product safety recalls result from fire-related problems. Without proper fire origin and cause determination, faulty products would not be repaired or replaced, which could lead to millions of dollars in damage and countless life loss.

Arson crimes are unfortunately all too common. Proper investigations lead to the apprehension and prosecution of perpetrators, hopefully preventing these individuals from repeating their crimes. An example is the 1986 New Year's Eve Fire at the DuPont Plaza Hotel in San Juan, Puerto Rico, which killed between 96 to 98 people in just 12 short minutes. The individual who started the fire was apprehended and convicted, but only as a result of good fire investigation work. This fire investigation also led to fire safety improvements in hotel design and protection and to the Hotel and Motel Fire Safety Act of 1990. A good investigation program spawns many benefits that greatly improve our overall fire prevention and protection efforts.

Identifying Trends

The fire service is busier than it has ever been with regard to overall emergency response. While the number of overall fires is down, the need for accurate data and proper analysis of these data are becoming more and more important. As Tom McEwen points out in the *Fire Data Analysis Handbook,* "Balancing limited resources and justifying daily operations and finances in the face of tough economic times is a scenario that every department can relate to.[1] To identify our target and perform our work as efficiently as possible, not to mention to prevent future injury and harm, we must properly identify problems. Identifying trends is one way to do this. This will also be an important component for conducting risk analysis in the new NFPA® Standard 1730, *Standard on Organization and Deployment of Fire Prevention Inspection and Code Enforcement, Plan Review, Investigation, and Public Education Operations.*

The data to be collected on scenes are numerous. A typical National Fire Incident Report gives an idea of the amount of detail solicited, which just involves the victim.

- Date
- Day of week
- Time
- Victim's age or gender
- Type of casualty
- Severity
- Affiliation (civilian or firefighter)
- Familiarity of structure
- Location at ignition
- Physical condition before injury
- Cause of injury
- Activity at time of injury
- Body part injured
- Nature of injury
- What prevented escape

Another separate set of questions should include the following:

- Describe the type of situation.
- What is the property type?
- Where is the exact location?
- What was the method of alarm?
- What was the geographical area of the first-due responding apparatus?
- On which shift did the fire occur?
- How many personnel were involved?
- How many pieces of equipment were used and what type were they?
- Provide the number of injuries and deaths.
- What was the origin of the fire?
- What was the cause of the fire?
- What was the form of heat of ignition?
- How was the material ignited?
- What form of material was ignited?
- How was the fire extinguished?
- Where was the location of area of origin?
- What do you estimate the dollar loss to be?
- What is the extent of flame damage?
- What is the extent of smoke damage?
- How do you rate the detector/alarm performance?
- How do you rate the suppression system performance?
- Identify the smoke characteristics.

While company officers should be responsible for answering many of these questions, all are part of the fire investigation. The fire investigator is generally required to provide additional levels of detail for each question or item. This information then provides a good analytical picture of what occurred, how, and why. The resulting information can reveal trends that can be considered for future workload prioritization and problem correction. For example, a fire investigator might develop a graph that shows how much impact juvenile arson has in relation to other juvenile crime. This is good information for a fire chief if he or she plans on addressing the problem.

National and state findings are a good source of data, but the local fire data determine the specific elements relative to the community and of the local fire department's fire prevention efforts. One of the greatest benefits of conducting fire investigations is the opportunity to gain insight into your community's fire problem (**Figure 9.5**). The frequency of local incidents also will play a significant role in prioritizing fire-prevention activities.

Data trends can provide a wealth of information:

- Day of the week fires most often occur
- Time of day fires most often occur
- Most common type of fires
- Most common ignition sources causing fires
- Most common locations where fire occurs

Figure 9.5 This fire was determined to be arson and was connected to multiple other fires caused by a serial arsonist who was later arrested. *Courtesy of the Colorado Springs, CO, Fire Department*

Trend identification is also important for justifying requests for more resources. Remember not to lose the small data among the big data. Large headline events make it easier to procure resources and time for investigating, but many times it is the lower profile, inconspicuous fires that provide the real clues about what is going on. Statistically speaking, the big events are often thrown out as aberrations or outliers that do not factor into the real picture. Insurance claims, product liability, and subrogation claims (in which a third party assumes another's legal right to collect damages) are gathering more and more attention **(Figure 9.6)**. Although not necessarily significant for data collection, this does increase the workload.

Responsibility

As previously mentioned, the fire department is typically responsible for the investigation of fires. Fire departments conduct fire investigations at different levels. Some fire departments will simply investigate to determine where the fire started and what ignited it. This is referred to as origin and cause determination.

Figure 9.6 This fire became significant due to a failure of the installed fire hood and duct fire suppression systems. This will likely lead to a lawsuit where the building owner will seek damages from a failed extinguishing system installation.

The depth or detail of the investigation will also vary. Public investigators or investigation sections are not always responsible for going into as much detail as private investigators do for other interested parties, such as insurance companies, manufacturers or the like. Therefore, it is important that the public investigators' duties are clearly understood and codified in jurisdictional law, code, policies, or procedures.

Basic Investigation Training, Equipment, and Procedures

To do their jobs well, investigators must be properly trained in their tasks and duties. The nationally recognized standard of good practice for achieving this is NFPA® 1033, *Standard for Professional Qualifications for Fire Investigator*. The IFSTA *Fire Investigator* manual is a good resource for fire investigators, as it follows the training requirements established in the related NFPA® standards. Remember, education and training are the first order of importance. Determining the origin and cause of fires will most likely have some type of financial or legal outcome. Insurance companies as well as other attorneys involved in the litigation will most likely subpoena the investigation report. Those involved in the fire investigation must be trained to the level of competency with confidence sufficient for testifying to their findings in a court of law (**Figure 9.7**).

Figure 9.7 Fire investigators often are called to testify in court.

Another essential standard to become intimately familiar with is NFPA® 921, *Guide for Fire and Explosion Investigations*. Defense attorneys use this standard aggressively in their construction of cases. This reason alone is impetus to integrate the contents into your management of investigation programs. The requirements are fairly rigid, and much of its content is technical. Defense attorneys have been known to reference it in frequency and detail in an attempt to discredit investigators by proving things were not done according to the standard. As a result, we strongly recommend making this document a mainstay in your office, your organization, and at the kitchen table.

If a fire is determined to be deliberately set, it is called incendiary. However, an **incendiary fire** may not be arson. For example, a person who burns his or her shed simply to dispose of it will have created an incendiary fire. The fire is intentionally set, but there was no desire to defraud or cause harm. If the fire was deliberately and maliciously set with the intent to defraud or cause damage or injury, it is referred to as **arson**. The exact legal definition of arson may vary from jurisdiction to jurisdiction and is typically codified in State Penal codes. NFPA® 921 also provides specific definitions.

To begin a fire investigation, two questions need to be addressed:

- Is the focus of the fire investigation going to be strictly origin and cause?
- Is the focus of the fire investigation also going to include determining the crime of arson and subsequent arrest and prosecution?

In either case, if you are investigating fires to any degree, you need to meet minimum requirements.

Training

Before you start investigating fires, you must have a solid educational foundation. Any individual or group of individuals performing an origin and cause fire investigation should be prepared and adequately trained in:

Incendiary Fire — Any fire that is set intentionally.

Arson — The crime of maliciously and intentionally, or recklessly, starting a fire or causing an explosion.

- Basic fire science with emphasis on fire behavior
- An overview of insurance practices
- Report writing
  - Content of a good report
  - Sketches
- Interviewing
  - Verbal communication
  - Nonverbal communication
- Scene preservation and control
- Basic photography
- Origin determination
- Fire debris removal and evidence collection
- Cause determination
- Basic electricity
- Scene reconstruction
  - Burn patterns (low and high burns)
  - Depth of char
  - Spalling of concrete
  - Protected areas
- Working knowledge of resource materials (NFPA® standards, etc.)
- Basic building construction
- Basic understanding of incendiary fire indicators
  - Multiple fire sets
  - "Trailers" of flammable liquids
  - Absence of accidental causes
  - Crime scene
  - Murders
  - Theft
  - Fraud
  - Delayed ignition devices
  - Flammable liquid pours
  - Motives present
- Basic fireground tactics used in suppression
  - Ventilation
  - Interior attack
  - Exterior attack
  - Overhaul

Figure 9.8 Sometimes digging out a fire can involve lots of equipment, not just hand tools. Protection of fire investigators is equally important as shown by this heavy rescue shoring trailer. *Courtesy of the Colorado Springs, CO, Fire Department*

Equipment

Along with this basic training, you will also need the following tools and equipment to do a basic origin and cause investigation. Obviously, as you gain experience and determine your personal preferences, other tools and devices may help, particularly for more thorough investigations. The following may be required for arson investigation **(Figure 9.8)**:

- Personal protective equipment
  - Helmet
  - Coat
  - Gloves (heavy and light)
  - Boots (and a tub and soap with water for cleanup)
  - Respirator or self-contained breathing apparatus (SCBA)
  - Coveralls
- Flashlight and other portable lighting equipment and generators
- Portable heaters
- Cell phone or portable radio
- Pump can or air pressurized water extinguisher (for hotspot touchups and cleaning the floor after debris removal for better examination)
- Portable electrical lights
- Hydrocarbon sampling device
- "Get after it and dig for it" equipment
  - Trowel
  - Scraper
  - Shovel
  - Small saw
  - Chisel/hammer
  - Wire cutters
  - Multipurpose scissors
  - Needle-nose pliers
  - Vise grips
  - Gooseneck pliers
  - Pry bar
  - Garden tools
  - Battery powered reciprocating saw
- Measuring tape (laser works well)
- Ruler
- Camera and film (digitals work well)
- Barrier tape
- Paper
- Pen or pencil (pencils always work in a variety of climates!)
- Cleaning supplies and water to scrub tools, boots, etc.

While we cannot go into detailed instructions or standard operating procedures for using and maintaining this equipment, these items usually will provide you the basic tools to investigate the fire ground thoroughly.

Determining Origin and Cause

Investigators determine the origin and cause of most fires by first focusing on the area of origin, the heat of ignition, the combustible or flammable materials involved, and the actions of the occupants at the time of the fire and an evaluation of the properties of fire compared to the conditions present **(Figure 9.9)**. Remember, fire protection professionals must make every attempt to study all possible causes of fire to arrive at a complete understanding of what took place to help determine fire prevention solutions that address the community's overall fire problem. Investigators should be capable of evaluating a scene to determine the point of origin, source of ignition, the material ignited, and the act or activity that brought the ignition source and materials together.[2] They should also be able to secure the scene properly and protect potential evidence until it is no longer

Figure 9.9 Sometimes the investigation is slow moving. Here, crews are shoring up the structure around the area of origin to protect investigators as they carefully sift through the debris. *Courtesy of the Colorado Springs, CO, Fire Department*

necessary. They should provide a good scene survey, both inside and outside the building or area involved. Investigators should be able to recognize burn patterns and understand how the fire affected structural components. Analyzing burn patterns and fire behavior with regard to ventilation and fluid dynamics becomes instrumental in evaluating pieces of this puzzle. Proper examination of fireground debris requires the use of various tools and equipment. Using the tools and equipment provided, the investigator should be able to remove insignificant debris and preserve those elements of the scene that correlate to the origin and cause. Investigators must make certain that no evidence destruction takes place in the event the determination is suspicious.

Reconstructing the area of origin is important. This can be done any number of ways, such as witness information, verified debris, and statements from fire crews. The inspector should be able to verify building mechanical equipment and any other special equipment to rule in or out accidental causes. Verification by experts in various fields, such as mechanical or electrical engineers is also helpful. Determining explosive effects may also become important. The investigator should know how explosions can be caused and what impact they will have on a structure, vessel, or other area of origin.

Documenting the scene is critical. Photographs, diagrams, and measurements all are important parts of the investigation. Notes and documentation will be fair game for evaluation and submission as evidence, so doing this thoroughly and accurately is very important.

Interviews with witnesses, residents, or workers provide invaluable information. Carefully document all relevant information and make sure that you know where to find these people again after you leave. Information such as social security numbers, driver's license numbers, and dates of birth is important. Various courses are available to provide good advice and explain techniques for interviewing witnesses or suspects. We strongly recommend anyone doing investigations take these courses if at all possible. Again, seeking courses on NFPA® 921 can be very worthwhile.

Report writing, computer data entry, filing, and document preparation are all important follow-up duties. The better you are organized and prepared, the better you will be able to handle future inquiries on the events, particularly three years down the road when the case finally goes to court. That late date is not the time to try to remember what happened, what you saw, or what you did. Document as if you are preparing to testify in court the next day. Voice-recognition software can be a powerful tool in dictating detailed reports quickly.

Arson Investigation

The FBI's Uniform Crime Reporting (UCR) Program defines arson as any willful or malicious burning or attempting to burn, with or without intent to defraud, a dwelling house, public building, motor vehicle or aircraft, personal property of another, etc. You should obtain specific definitions from various state or local statutes where you work. Additionally, information can be acquired from various sources such as NFPA® 921 and various other texts. In essence, arson is harming someone or their property by hostile fire.

Training and Equipment

The basic techniques of arson investigations are no different than those of simple origin and cause investigations with regard to scene processing. Investigators must follow the same procedures and practices for properly determining origin and cause, as that is still the key to prosecuting a crime. To properly capture and preserve evidence for use in criminal trials, however, they must perform other tasks that require a higher level of training and additional equipment.

Arson investigators typically need training in the following law enforcement concepts and procedures:

- Court decisions
- Burden of proof
- Warrants
- Arrest procedures and rights
- State statutes related to arson
- Federal statutes related to fire investigation
- Evidence collection and processing

Keep in mind that different jurisdictions may have different requirements depending upon their policies and procedures.

In addition to the basic equipment for origin and cause determination, arson investigations generally will require the following items to gather evidence and document the fire scene (**Figure 9.10**):

- Unlined paint cans with secure fitting, airtight lids
- Paper and plastic (sealable) bags
- Tape
  — Duct
  — Masking
  — Clear

- Glass jars with caps
- Numbered marking tents or triangles
- Permanent marker (fine and chisel tip)
- Engineering paper (gridded)
- Awl or diamond-tipped scribing tool
- Small voice recorder (nice to have)

Once an investigator has the appropriate training and equipment, the additional work can begin. Keep in mind that collaboration with district attorneys, police departments, and all who process and store evidence is essential. Changes occur regularly as a result of court cases and improvements or changes in technology. These changes can dictate equipment, handling, storage, and the processing of evidence.

Figure 9.10 The typical equipment used for collecting evidence.

Procedures

We do not intend the following sections to be all inclusive in an arson investigation, but simply to give you a perspective and overview on what is involved. This should facilitate informed decision-making and planning in respect to starting or evaluating this function in your fire prevention bureau. If you want to conduct fireground investigations, you should look into other courses, training, and books on the subject.

Size-Up

Determining the origin of a fire requires a systematic approach. Your starting point should be your initial arrival on the fire scene (assuming that you arrive on the scene during the incident or shortly after mop-up operations). It will be very beneficial in your investigation if you can reconstruct in your mind what the scene looked like prior to the fire department's arrival and where the first signs of fire were located. Barricading and securing the scene as early as possible, even if mop-up operations are taking place, will help manage evidence, control, and disruption of the scene. If the building has been completely consumed by fire, you will need to reconstruct what it looked like and the location of its contents when the fire started. It is also important to know what the building looked like during the course of the fire and how it behaved. Previously, we mentioned the importance of evaluating burn patterns and the evaluation based on ventilation aspects. Lots of research is underway regarding small-scale-fire replication. Monitoring the progress of this research may be invaluable, if it progresses to the point where it becomes practical and available for reconstruction of fire behavior events.

Visit with fire crew members, and accurately document what each member saw upon his or her arrival. Keep a record of what suppression tasks were performed. Also, make sure that each crew member documents in his or her report what was observed upon arrival, particularly, ventilation components such as open or broken doors and windows, or did crews force entry. Ventilation plays a key part in fire behavior and knowing these details early, can be very helpful to the investigation. Keep in mind that what they initially saw might have been consumed by fire and will no longer be visible to you. They will be able to save you time during your reconstruction process. It may even be helpful to have them draw a sketch. Fire suppression crews may have left the scene or have been relieved by another crew. It is always a good practice for the fire investigator to attend the fire suppression crew's critique of the incident.

Talk to whomever reported the fire and document the time of the alarm and how it was transmitted. If it was a large fire, there may be many people to interview. You should ask questions or solicit information about a number of concerns, such as:

- Are there any injuries or fatalities? If yes, does anyone dislike the victims?

- If a fatality occurred, could the fire have been started to cover up evidence? Most likely, law enforcement agencies will be involved if a death has occurred in a fire.

- What were the weather conditions? Do all the things you see and hear from crews and witnesses correlate with the weather? (Closed or open windows, doors, etc.)

- Was lightning a factor?

- What was the wind direction and speed?

- Were there any unusual delays or obstructions to the fire suppression crews?

- Where did fire suppression crews make initial entry?

- Do you notice anything unusual about witnesses or bystanders at the scene?
- Did broken glass seem to be from a break-in or as a result of the fire?
- What color of smoke did fire crews observe? This question can be important as certain types of materials produce certain colors of smoke. However, given the types of fuels available today, distinguishing them by the color of their smoke is getting more and more difficult.

Examining the Exterior

After you have sized up the scene, the next critical step is to determine whether the fire started on the exterior or the interior of the building. Safety is always a primary concern **(Figure 9.11)**. During the scene examination and during the interview with the suppression crews, it is important to determine approximately how much water was used to extinguish the fire and what parts of the structure are not safe to enter. The firefighters will most likely be at the incident prior to the fire investigator's arrival. They will be able to explain what they encountered and how the fire scene initially looked prior to suppression activities and further consumption of the combustibles by fire. If the firefighter has some insight into the cause of fires and what has taken place in the past, he or she may be able to provide a critical piece of information that will help the investigator determine the cause of the fire. Essentially, this method utilizes the fire suppression crews as preliminary fire investigators. In fact, some fire departments do not have a designated fire investigator. Instead they use on-duty fire personnel to conduct the fire investigation.

As you examine the four sides of the structure, determine if the fire started or communicated to the structure from the exterior **(Figure 9.12)**. Note if the fire ventilated through the roof on its own or if the suppression crews opened the roof as part of the ventilation process. Are there any unusual prints, tracks, or debris on the outside? Were the windows broken in or is glass on the exterior? Are doors and windows secured? Are any utilities cut off or disrupted other than for fire suppression? Look for anything that seems out of place, either with what you see or what witnesses state.

Examining the Interior

Determine where the most significant damage occurred based on the fire's highest temperature and longest duration. Trace the burn patterns back to the fire's origin and rule out those patterns that can be attributed to other fuel sources or affects due to ventilation. Look for uneven burning or localized deep char. Where is the most ceiling and roof damage? Find the lowest point of the burned area within the area of origin. Was

Figure 9.11 Examining the exterior of this building and talking to first-arriving crews, it looks readily apparent this fire started on the outside of this apartment building. *Courtesy of the Colorado Springs, CO, Fire Department*

Figure 9.12 This photo shows a home starting from a combustible roof ignition as a result of embers deposited by a wildfire. *Courtesy of the Colorado Springs, CO, Fire Department*

the floor burned through, and if so, from the top or from the bottom? Once you have localized this fairly well, evaluate potential air movement or drafts that could have influenced the fire. Also try to determine the types of furnishings or objects in the area before the fire to see if they had any effect that would be inconsistent with the apparent burn **(Figure 9.13)**. Look for heat flow paths. Are there multiple ignition locations? Do you see a traditional V-pattern pointing to the low point like the one in the Figure 9.13?

Figure 9.13 The V-pattern can be seen behind this grill and limited damage to the front of the equipment shows how the fire originated behind the grill. *Courtesy of the Colorado Springs, CO, Fire Department*

Always examine the lowest point of burning. Accidental fires typically burn vertically upward leaving the floor and lower areas much less damaged than the upper levels of the room (fires do not typically burn downward, as heat travels upward). After you have dug to the lowest burn level, it may be a good time to use the pump can to clean the floor and lower levels away to see what is actually burned. Do not be afraid to dig **(Figures 9.14 and 9.15)**. We have seen valuable clues missed and incorrect decisions made simply because the investigators were too lazy to work their way through the entire scene and surrounding area. To get to the bottom of the fire, you need to get to the bottom of the fire!

Indicators of fast, propagating fires can be severe overhead damage in comparison to other parts of the room and distinct, sharp V-patterns. The IFSTA publication, *Fire Investigator* (Second Edition), explains how glass crazing has more to do with rapid cooling, such as by the application of water, as opposed to heat or fire behavior. Small, alligator-scale charring-patterns in wood is not a significant indicator of fire behavior. It has more to do with moisture content of wood and should not be used as a definitive indicator of fire behavior based on accelerants. Spalling of concrete is not indicative of flammable liquid pours or accelerant deposits. Spalling occurs when moisture in concrete cannot escape fast enough as it heats up, causing steam pockets that break the concrete. This is typically an indication of an area that sustained a significant amount of heat that can result from a multitude of sources.

Figure 9.14 Fire investigators need to sift through debris looking for evidence.

Figure 9.15 It is important to completely remove and examine the debris to check for signs of the origin of the fire.

Once you have isolated the point of origin, the next step is to determine the cause. To do this, you must eliminate any and all natural or accidental causes before concluding that a fire is suspicious or incendiary. Many times the determining factor will be the process of elimination, which must be based on evidence. For example, you could eliminate any energized electrical sources if your investigation revealed the power to the building was disconnected at the time of the fire.

Look for any potential sources of accidental ignition, such as electrical heating devices, smoking materials, pinched electrical cords, overloaded electrical supplies, tripped breakers, or the like **(Figure 9.16)**. Were any candles being used? Was painting, cleaning, or other activity taking place prior to the fire? How does the information from your initial scene interviews correlate to what you see with your own eyes?

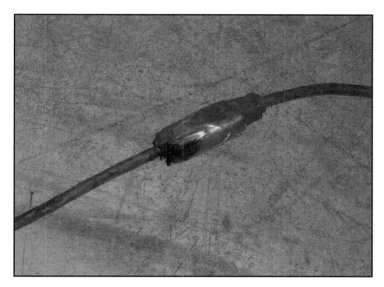

Figure 9.16 This extension cord is beginning to burn and if left unchecked, it would have started a fire. *Courtesy of the Colorado Springs, CO, Fire Department*

The investigator must consider and eliminate all of these possibilities before making any other determination. Sometimes the fire's cause remains undetermined. Although not desirable, it is certainly better than accusing people of something they did not do or determining the fire to be something that is later disproved. For example, you might "suspect" an electrical cause and document the finding as an "electrical fire" so you have a cause. Then, years later, someone comes forth and confesses to setting the fire. Attorneys on both sides will have fun with your written report and testimony in court.

Write the report to show the exact cause is undetermined, while ruling out any potential causes as in our electrical example above. Occasionally, it may be undetermined due to two or more possible sources. These can be cited, just be sure to clarify it is not finite thereby remaining undetermined. A benefit to leaving the final cause of the fire as undetermined and the case as "open" may allow for reevaluation later if additional evidence becomes available. If the cause was intentional, the person who set the fire may set another and not necessarily in your jurisdiction. When the arsonist is finally caught, he or she may confess to setting other fires, including the one you investigated. If you have already listed its cause as electrical, modifying your report will be difficult, even with the arsonist's confession. During the suspect's prosecution, the defense attorney will stop at nothing to discredit you and your initial findings, if only to confuse the issue. Remember, criminal issues must be proven beyond a reasonable doubt. To quote the defense attorney's famous admonition regarding a glove produced in evidence at the O. J. Simpson murder trial, "If it doesn't fit, you must acquit!"

If you suspect accelerants were used, you must mark their locations, photograph the area, take control samples and good, secured evidence samples and send them to a qualified lab. Document all of your findings. Many times, the collection of accelerant samples and other evidence may be better left to law enforcement agencies trained in this task. Fire departments do not always have the expertise to handle this type of situation. It is important that all investigators know their limitations and seek help from others. It is rare to find someone who can be an expert in all aspects of fire investigation. Investigators may need the expert advice of electrical engineers, appliance professionals, or bomb experts, to name just a few.

Getting the results from evidence samples can take weeks to months in some cases. Some jurisdictions may be able to secure the use of accelerant dogs. This is another example of using experts as collaborative partners. These dogs are phenomenal to watch and very good at what they do. If you have the opportunity to work with one, make sure that you ask the dog's handler for special instructions before you do anything. The handlers are highly skilled at working their animals (partners) and the team operates very methodically. Most state agencies have access to these experts.

Remember, determining that a fire was not accidental is not always the difficult part; prosecuting the individual who set the fire is! Some of the biggest mistakes we have seen investigators make on fire scenes include:

- Failing to take enough photographs (Film is cheap and digital is even cheaper)
- Not digging and scraping to the bottom of the fire scene
- Not digging and scraping with care (Don't just shovel stuff out the window! Some of that "stuff" may be evidence.)
- Not documenting where the evidence samples were taken
- Failing to get and document accurate statements from witnesses

- Assuming something happened without making certain
- Failing to write a comprehensive report
- Failing to seek the assistance of outside agencies
- Failing to seek the assistance of experts

After Origin and Cause, What's Next?

If you have determined where and how the fire started and that it is suspicious or incendiary, you must proceed through another entire process. You need to determine if crime specialists are needed. It is a good idea to already have developed a solid working relationship with your district attorney. Prosecutors may want to be called to see the scene firsthand. They can often give you hints on what to do and what not to do, avoiding pitfalls later if the case does go to court. If a fatality is involved, what did the coroner or medical examiner determine as cause of death? Involve local, county, state, and federal officials if you need more help or if leads keep popping up that are beyond your capacity to handle. Do not ever be afraid to get help. Often, this networking among law enforcement agencies yields far more than going it alone.

Smaller fire departments usually do not have the experience and training of more experienced law enforcement agencies. Fire department personnel are generally pretty good at determining what caused the fire, while law enforcement personnel are experienced at collecting evidence, preserving the crime scene, and working to prosecute the offender. Establishing a working relationship between law enforcement and fire personnel enables a much more thorough and efficient investigation team.

Looking Toward a Long-Term Solution

The crime of arson is interesting and complex. It is a killer. Set fires burn quickly, trapping occupants, and endangering firefighters. Arson is a local problem. It is a crime with high potential for financial and emotional gain and very low risk of detection. Two major categories of arson are those motivated by financial gain, which is constantly weighed against the risk of loss vs. gain, and those motivated by psychological gain. The latter are far more difficult to prevent and control. What we are finding is an increasing trend of juvenile fire setters whose motives fall into the second category. We must continue our efforts toward juvenile firesetter intervention prevention programs. To make policymakers at all levels aware of these trends, we should provide them with better data so they can see the problem clearly. As a fire service professional, you should make it your mission to provide these data as clearly and effectively as possible, allowing decision makers to properly measure the problem's magnitude and our progress in controlling it.

The best way for you as a fire service professional to address a good fire investigation program is to strive to achieve the following goals:

- Develop good policies and procedures for everyone to follow.
- Integrate fire investigation functions into the mission of the fire department.
- Train everyone on the job about what it means to do good fire investigation.

- Get management's commitment to this process.
- Provide proper training and equipment.
- Develop arson task forces where appropriate.
- Gather and track good data.
- Develop an early arson warning system.
- Develop a progressive public education program citing all aspects of fire.
- Partner with community organizations.
- Develop relationships with law enforcement agencies and prosecutors.
- Support advances in fire investigation technology.

By following these goals and attending to the details and complexities of fire investigation, you will have the basic tools to form a comprehensive program, providing a solid foundation for future fire prevention efforts. Keep in mind the need to ensure that staff and equipment are available for your use. You must do this well before an event takes place. We cannot and should not be solely responsible for fire investigations. We need to use all of the resources available and to share information in our mutual endeavor to prevent fires or control their spread.

Summary

The purpose of a fire investigation is to determine how the fire started and why it behaved as it did. The data collected from the investigation of fires can be a key element in addressing a community's fire problem. The data gathered in a fire investigation can be used for fire prevention.

Fire prevention bureau personnel can use fire investigation information to try to prevent similar events. One excellent method of using this information is through fire and life safety education. The information gained from a fire investigation can and should be used to educate and inform the public of potential fire causes. The best opportunity for presenting this message is when the media attention is greatest—just after an incident takes place.

The information can also be used to identify code modifications that will reduce the potential for similar fires. This is very similar to the case studies we reviewed in Chapter 2, "History of Fire Prevention." The only difference is that the lessons learned can be community specific. If many communities experience the same type of incident, then it gains more attention, and more people benefit from finding the cause of the fire. At times this may even lead to a product recall on a national level.

An important level of fire investigation is determining if arson was the cause of the fire. Arson directly impacts the community where it takes place. There is the potential for the community to lose revenue as well as for the death of a firefighter or innocent citizen. To combat arson effectively, it is essential for fire investigators to pool their resources and work with other agencies, such as law enforcement.

Chapter 9 Review Exercises

9.1   List four reasons for investigating fires? _____
_____
_____
_____
_____

9.2   How does fire investigation prevent fires?
_____
_____
_____

9.3   What is a cause and origin investigation? _____
_____
_____
_____

9.4   What are nine of the common mistakes made by fire departments
        during a fire investigation? _____
_____
_____
_____
_____
_____

9.5   What is arson?_____
_____
_____
_____

9.6   How does arson affect the citizens in a community? _____
_____
_____
_____

9.7   What NFPA® standards apply to fire investigations?_____
_____
_____
_____

9.8 How are the methods of conducting a fire investigation similar to those of conducting a fire inspection? _____

_____
_____
_____

9.9 How can fire investigation be used for code development? Cite an example. _____

_____
_____

9.10 Provide a list of information that could be collected during an investigation. _____

_____
_____

9.11 Provide a list of tools and equipment that could be used during an investigation. _____

_____
_____
_____
_____
_____
_____
_____
_____
_____
_____
_____
_____
_____
_____
_____
_____
_____
_____
_____
_____
_____
_____

9.12   What could be important data trend points to look at if you wanted
       to maximize your resources?_____

       _____

       _____

       _____

9.13   What are three safety concerns for fire investigators?

       _____

       _____

       _____

       _____

9.14   Outline twelve goals that a fire investigations unit should strive to
       achieve.

       _____

       _____

       _____

       _____

       _____

       _____

       _____

       _____

       _____

       _____

       _____

       _____

       _____

       _____

       _____

       _____

       _____

       _____

       _____

       _____

       _____

       _____

Notes

1. Tom McEwen, *Fire Data Analysis Handbook* (Washington, D.C.: U.S. Fire Administration, n.d.)

2. NFPA® 1033, *Professional Qualification for Fire Investigator* (Quincy, Mass.: National Fire Protection Association, 1998).

Construction
Document Review

Table of Contents

Key Terms

Key Points

1. The construction document review is an opportunity to begin building a fire prevention coalition with the owners of the building.

2. The individuals involved in the construction document review process must be fair, accurate, and consistent.

3. The owner of a facility will be around to deal with the fire inspectors for years to come, long after the architect and contractor have completed their work.

4. The building official's role is to ensure protection of the structure's current and future occupants and owners and of the surrounding property.

5. Becoming actively involved in the construction document review process is the best opportunity for fire departments to eliminate hazards or protect those hazards that cannot be eliminated.

6. Fire department personnel who are contacted by a building owner who is contemplating a renovation should always seek the advice of an architect.

7. Because the construction document review can create conflict, it must be performed precisely and systematically.

8. The results of the construction document review must be communicated in writing.

9. Communication breakdown will always result in poor product and time delays and will likely upset customers and staff.

10. On-site fire inspections ensure the construction site is safe from fire and verify the work being performed conforms to the approved drawings.

11. The construction document review process should create sufficient documentation to revisit the project in 10 or 15 years and clearly understand the code requirements when the building was constructed based upon its original design.

Learning Objectives

1. Define the national fire problem and role of fire prevention.*

2. Define the functions of a fire prevention bureau.*

3. Describe inspection practices and procedures.*

4. Identify and describe the standards for professional qualifications for Fire Marshal, Plans Examiner, Fire Inspector, Fire and Life Safety Educator, and Fire Investigator.*

5. Describe building construction as it relates to firefighter safety, buildings codes, fire prevention, code inspection, firefighting strategy, and tactics.*

6. Classify occupancy designations of the building code.*

7. Identify the types of construction documents reviewed by fire departments.

8. Understand the development process and the fire department's role in it.

9. Understand the importance of the fire department's participation in the construction document review process.

* FESHE Objectives (USFA)

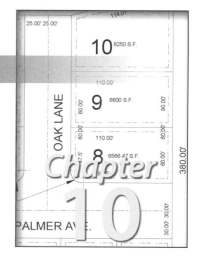

Construction Document Review

Case Study

The local youth ministry organization has purchased a foreclosed industrial building. The large nonsprinklered building, constructed of wood timber, is vacant and has limited parking. The youth organization purchased the building without consulting a design professional or the local fire department. The organization intends to make the facility a youth center. Construction is limited to the installation of cooking equipment in one of the existing break areas. The center will have basketball courts, game rooms, a kitchen, and large meeting rooms. The facility will be used on weekends for concert venues.

During a routine fire inspection of the facility, the new owner meets the fire inspector and explains his future plans for the facility, which will take place next week. What fire department concerns would need to be portrayed to the new owner? Who would need to be involved in the project?

What is a good course of action?

Introduction

Whether a municipality or fire district is seeing rapid construction growth and development or none at all, the fire department responsible for providing its fire and emergency services also should provide construction document review services (**Figure 10.1**). There is only one opportunity to ensure a building's construction features are adequate to prevent the spread of fire and include all of the required built-in automatic suppression systems. That opportunity occurs when the building is constructed or remodeled. The fire department should not see the construction document review process as just another unfunded mandate. The fire department should view construction document review as an opportunity to begin building a fire prevention relationship with the owners of the building and as an opportunity to identify potential hazards and risks that will impact the safety of the occupants and the firefighters who may be called there in the middle of the night to mitigate an emergency incident. This is one of the greatest examples

Figure 10.1 Construction document review is an important aspect of fire prevention.

where fire prevention impacts fire department operations. Even small volunteer fire departments need to take an active role in construction document review at some level. The fire department should consider construction document review one of the essential services it provides and a critical element of the fire prevention program. Making the right construction review decisions pertaining to fire protection features and design will impact the safety of firefighters and building occupants for many years to follow.

Unfortunately, during tough economic climates, this is frequently a service that is reduced or even eliminated. Many fire departments have chosen to relieve the economic burden of the personnel cost associated with construction document review by utilizing a third-party plan review agency. The cost of the service is then placed on the burden of the developer. Even when fire departments utilize a third-party construction document review firm, they can still need to have a role in the construction document review process and ensure their operational needs are addressed, while building a relationship with the new building owner.

The construction document review process is one of the few fire department activities that incorporates all three of these principles of fire prevention—education, engineering, and enforcement. If the process is worked properly, this is a great way to build trust, the lack of which is a common complaint among regulatory agencies. Fire department personnel involved in the construction document review process educate building owners, developers, and architects on the importance of fire protection. Many times, fire department personnel can explain the code's intent and the reasons for its development. They can accomplish this through comparing the occupancy under construction with similar occupancies that have had fires. Some fire department personnel use engineering skills during their review of the construction documents. For example, fire department personnel may review site drawings to ensure fire department access or review hydraulic sprinkler calculations. Various departments have a range of staff, such as engineers, fire plans examiners, and inspectors, to do these tasks.

The enforcement of the codes and standards is the foundation of the construction review process. One reason fire department personnel participate in construction review is to ensure that buildings are designed and constructed in accordance with the codes and standards the fire department has adopted.

In many areas of the country, the fire department may not have the authority or staff to oversee the enforcement of nationally recognized codes and standards (see Chapter 3). Architects and builders can then construct buildings as they see fit without the benefit of oversight. Architects do, however, have some legal, if not ethical, obligation to ensure a building's design is safe. Constructing a building to meet the adopted building code indicates it meets the minimum safety requirements.

Difficulties arise when a developer, who is unaccustomed to having a fire department provide input during construction, constructs a building in a location where the fire department actively reviews construction documents. The fire department providing construction document review faces confrontation from a frustrated developer or builder who indicates his construction method was approved in another location. Why it is not approved in this location? As with most municipal ordinances and laws, most municipalities do not have the same building and fire codes. They may have adopted identical nationally recognized codes, but most municipalities make amendments to the model codes. This also places the builders in a difficult situation. It is for this reason we recommend municipalities require and issue permits for construction only after the project has been through a construction document review process.

The construction document review process in most municipalities can become very political, which in turn may create pressures on the individual responsible for conducting the construction document review. Even the third-party construction document review firm can face pressure during the review process. The individuals involved in the construction document review process must be fair, accurate, and consistent in their review practices. Following these philosophies will not relieve all of the political pressures applied to the construction document reviewer but will provide a means to alleviate potential conflicts and accusations of selective enforcement. The construction document review should be treated in a similar fashion as a judge passing on sentencing. The results of review are based upon adopted codes and ordinances not based upon political ties or other extenuating circumstances or personal bias.

The demands on architects and builders to meet construction deadlines and to be under budget are enormous coupled with the municipality or fire district's desire to have the new facility open to generate revenue. In almost every construction situation, the owner, contractor, and architect encounter significant economic penalties if the building's scheduled completion date is missed or if additional costs are incurred that are outside the initial scope of the project.

Methods are available for contractors to ensure that they complete a project within budget and on time. Unfortunately, not all contractors or architects follow these methods. Some individuals will cut construction costs by using substandard material or by cutting corners and not using approved construction methods. Some builders, untrained in the benefits of requirements for certain materials to be UL-listed, can improve their profit margin by substituting the type of material used. This practice can save the builder substantial costs if products appear to be similar. The owners of the building may not be getting what they actually intended to purchase and the product will not perform as needed. For example: A builder who substitutes a normal caulk (found in most hardware stores) can save the builder substantial costs. The owners of the building may not be getting what they actually intended to purchase (caulking that is UL-listed for fire-stopping). This is where the fire department's service to the owner is very beneficial.

In most situations, the owner of the building wants a safe structure, architects want to design a building that complies with codes, and the contractor wants a quality product. Fire departments can honestly state, "They are working on behalf of the owner of the building." The fire department is basically a consumer advocate in this instance. However, more often than not, the owner or user of the building is frequently not present during the construction process. If your process of plan review involves the owner, you must be a collaborative participant in decision making. You will improve your ability to gain a trusted ally and substantiate your reputation as "owner advocated."

Providing the owner the opportunity to learn this early in the project design process helps to establish a rapport with the owner. Keep in mind, the owner of the facility will be around to work with the fire inspectors for years to come, long after the architect and contractor have completed their work. This may be the only other time the building owner has interaction with the fire department.

Architects do not often see first-hand the devastation of fire and may not always comprehend the significance of constructing a building to comply with nationally recognized standards. Architects usually do not have the opportunity to see the impact of a simple change of occupancy or see the implications that performance-based design can have on the fire inspection process until well after the building is complete and occupied. The building's occupancy may change from its original design, and the parameters of a performance-based design can impact the fire department throughout the life of the building.

The Construction Document Review Team

The person responsible for the fire department's construction document review needs to establish a relationship with all of individuals involved in the construction document review process. These people all have vested interests in the process, and understanding them will help fire personnel to work with them as a cooperative team. As already mentioned, third-party firms may provide this service. It is imperative that the third-party firm understand and portray the fire department's vested interest in the process in a professional and diplomatic fashion. Third-party firms need to be involved in the conceptual phase of the project and communicate the needs of the fire department to the entire construction document review team. The selected third-party firm needs to be a constant collaborative partnership with the fire department.

Building Officials

Building officials are involved in the construction document review process and have a number of concerns beyond compliance with the building code. Of all the people involved in construction review, the building official's participation and close working relationship with the person responsible for fire department construction document review is critical. The building official is usually responsible for enforcing a building code that contains many requirements to assist and protect firefighters during a fire. The scope of the building code is to ensure that the safety of a structure's occupants is provided for during fires and similar emergencies. Many of the model codes' requirements were developed as the result of significant fires. A major intent of the building code, which is enforced by the building official, is to prevent the spread of fire and provide for safe egress of occupants. The building official's role at times parallels the role of the fire official, and frequently their duties may overlap. The building official's role is to ensure protection of the structure's current and future occupants and owners and of the surrounding property and to oversee changes in the building design.

Insurance Rating Bureau

In some instances, the insurance rating bureau and or insurance companies can be involved in construction document review. They have an interest in the classification of risk and evaluation of the construction to determine the insurance premiums. Many of the insurance carrier's requirements during the construction document review process will be more restrictive or specific than those of the local fire or building department. The insurance requirements are based on the risk or loss potential of the building or its operation. Insurance providers develop their requirements based on national code as well as on the industry's loss experience with similar occupancies and buildings. Many times, the loss potential is so great that the insurance company will require protection above and beyond the national codes.

State Fire Marshals

State fire marshals enforce codes adopted by their state. Their role in the construction document review process will vary from state to state, depending on the type of occupancy and on jurisdictional constraints upon their own offices as well as upon the local fire department. Typically, state fire marshals are responsible for enforcement of state codes in occupancies such as state owned buildings, schools, nursing homes, and hospitals. Some state laws may exclude fire departments from having jurisdiction in these types of occupancies. In these instances, local fire departments may have the opportunity to conduct a cursory review, but the final approving authority will be the office of the state fire marshal.

Fire Department

Traditionally, municipal building departments are actively involved in construction document review, and the fire department's fire prevention role continues after the building is constructed. The fire prevention codes enforced by fire departments are considered *maintenance codes*. Once a building is constructed, the fire department is left with enforcing the maintenance of the built-in fire protection systems, egress components, and processes with which that occupancy may operate. The fire department also is usually responsible for ensuring that the building is used as designed.

Becoming actively involved in the construction document review process is the best opportunity to eliminate hazards or protect those hazards that cannot be removed. During this process, fire department personnel can make the biggest impact on the building's overall fire protection features. Long-lasting decisions made during construction document review most likely will be in place for the life of the building or the duration of the occupancy of the facility. It is critical, therefore, that fire departments employ individuals who are capable of conducting thorough and accurate construction document reviews. As stated previously, we have only one opportunity to ensure the building is constructed in accordance with adopted codes and standards. This opportunity is during construction document review. The selection of a third-party firm for construction document review must take into account the firm's ability to work with the fire department, to understand its needs, and to identify its fire protection philosophy for construction document review.

Progressive fire departments will ensure fire department personnel or their designated third-party firm is involved in the construction document review process for a number of reasons including:

- Enforcement of codes for construction practices (elimination of construction deficiencies)
- Firefighter safety
- Occupant safety
- Property conservation
- Environmental conservation
- Preincident planning
- Opportunity to build a fire prevention coalition with the owner of the facility
- Identify performance-based-design criteria which requires fire department monitoring during the life of the building

Other entities that may be involved in the process as members of the construction document review team are:

- *Engineering*. The municipal civil engineering department typically has construction document review responsibility for items pertaining to drainage and easement issues. This department may also be responsible for flood control.
- *Public works*. The municipal public works department typically has review responsibility for issues such as water, sewer, roads, and improvements on the public right-of-way.
- *Zoning*. The municipal zoning department focuses on ensuring buildings and their associated uses are constructed in a permitted area zoned for such use.
- *Health*. The health department in a municipality is typically responsible for issues that directly impact its citizens' health. This may include those health items associated with restaurants or sanitation issues.
- *City manager or Mayor's office* (city official's office). This office is responsible for protecting the interest of the entire municipality and will take an active role in a variety of issues of which some may likely become political.

The Construction Document Review and Permit Coordination Process

Although the procedural details of the construction document review process may vary from municipality to municipality, some elements of the process are consistent. Municipalities may have procedural guidelines on how to submit the documents for review or who is responsible for reviewing them. For instance, some municipalities have a department responsible for economic or community development that may oversee the entire construction document review process. In most municipalities, construction documents are submitted to one department and then routed for review comments to the other departments that comprise the construction document review team. The easiest way to explain

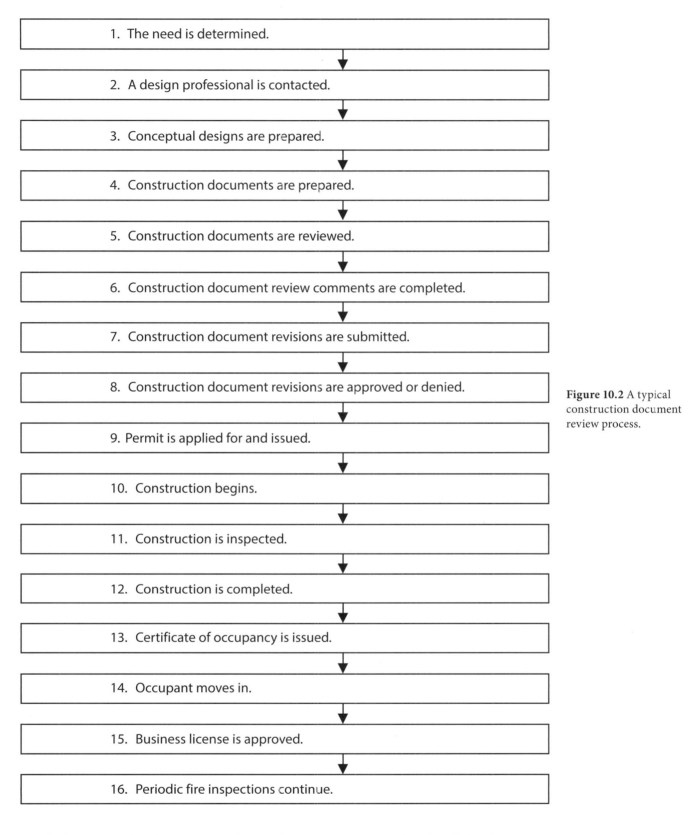

1. The need is determined.

2. A design professional is contacted.

3. Conceptual designs are prepared.

4. Construction documents are prepared.

5. Construction documents are reviewed.

6. Construction document review comments are completed.

7. Construction document revisions are submitted.

8. Construction document revisions are approved or denied.

9. Permit is applied for and issued.

10. Construction begins.

11. Construction is inspected.

12. Construction is completed.

13. Certificate of occupancy is issued.

14. Occupant moves in.

15. Business license is approved.

16. Periodic fire inspections continue.

Figure 10.2 A typical construction document review process.

a typical construction document review and permitting process is to break it down into steps as shown in **Figure 10.2**. The sixteen steps briefly summarize how a building is constructed and how the fire department and other construction document review team members are involved.

Step 1: Determine Need

The entire construction document review process actually begins when the owner, potential owner, or occupant of a building determines the need for constructing a new building or renovating an existing building. This sometimes can stem from a potential change in the use of an existing building or desires to expand. Many times, what may be viewed as a simple building expansion or occupancy use change can have the greatest impact on the model building codes.

Step 2: Contact Design Professional

After determining a need, the owner or occupant contacts a design professional, such as an architect, to develop conceptual drawings based on that need. However, this may not always be the path that owners or occupants choose. Owners or occupants may simply hire contractors to begin the work and not realize that the facility occupancy expansion or change in facility affects many codes and standards. The project can be delayed by not using a design professional. The design professional is responsible to ensure that the design of the project meets all required codes and is constructed safely with quality workmanship. The architect may also monitor the construction process to ensure it is being constructed in accordance with the design. Architects also are available to solve any issues that may arise as part of the construction process. This is extremely important during renovations of existing buildings. Typically, these issues become labor intensive among the AHJ, the contractor, and the designers. Architects are trained professionals and should always be consulted by the building owner in the design process. Fire department personnel who are contacted by a building owner who is contemplating a renovation should always seek the advice of an architect and seek input from other construction document design review team members.

Step 3: Preparation of Conceptual Design

The first document presented to the municipality for comment may be only a conceptual design. **Conceptual designs** are preliminary documents and are not intended to be used for construction or permit approval. These drawings may include a sketch of the site and the proposed building, along with sketches of exterior elevations and rough outlines of the interior space. Conceptual drawings typically serve as a means to facilitate a meeting of the construction document review team with the owners and their design professional. These conceptual design review meetings, or "**preconstruction meetings**," provide the architect, other design professionals, and owner the opportunity to ask the construction document review team specific questions regarding the project. The fire department can respond to the civil engineer's questions regarding fire department access needs. If a civil engineer is part of the construction document review team, you could discuss other specific design questions pertaining to the actual road construction. The preconstruction meeting is an excellent opportunity for everyone involved in the project to meet all of the individuals from the design team and construction document review team. The architects and designers benefit from such a meeting because they will reduce or eliminate

Conceptual Design — A preliminary document that is not intended to be used for construction or permit approval.

Preconstruction Meeting — A meeting during which the people involved in a building project review the conceptual designs.

design mistakes. The construction document review team benefits because ensuring their concerns are addressed in the initial design will most likely shorten the review time.

Many times, the experienced design professional understands the importance of seeking the input from fire department personnel and other review team members early in the design phase. Depending on the nature of the project, the fire department may comment on the conceptual site plan or the interior layout. These comments would be very preliminary and even somewhat vague. For instance, the fire department may indicate that the drawings do not reflect a fire lane along one side of the building. The interior details may not show a separate room for the fire command station. Seeking the construction document review team's input avoids project delays and begins the establishment of a working relationship. When the construction document review team can offer a service to meet with the owner and design professional before final construction documents are completed, the entire review process is most often greatly improved.

Step 4: Construction Document Submittal for Review

Fire department personnel need to review the entire construction document package, which includes:

- Architectural drawings
- Structural drawings
- Mechanical drawings
- Electrical drawings
- Site, landscaping, civil, and utility drawings
- Plats
- Fire protection drawings
  — Sprinklers
  — Standpipes
  — Fire detection and alarm
  — Other fixed special fire suppression systems

The construction documents provide details of two types of fire protection systems: active and passive. Passive fire protection systems include the roof, floors, walls, ceilings, doors, egress, and vegetation management. Active fire protection includes fixed fire suppression systems (such as automatic sprinklers), standpipes, special fire suppression systems, fire detection systems, and smoke control systems.

Some elements of the construction documents require more fire department involvement than others, and the construction features reviewed can vary from fire department to fire department. This variation occurs because some fire departments may only review fire-suppression and fire-detection related issues and leave the remaining items to another member of the construction document review team, such as the building department. Many of the items overlap in the construction document review process. This only reemphasizes the need to establish a cooperative working relationship with other construction document review team members. Such a relationship allows the fire department to express its concerns informally to the other team members who then can communicate the issue formally to the design professional.

Architectural Drawings

Architectural drawings provide a significant amount of detail that may require fire department review. In some situations, the architectural drawings may be an inclusive submittal that contains all of the construction documents. However, the actual architectural drawings provide details of nonstructural elements such as fire-resistive-construction features, exit components, occupancy classification, occupant loads, occupant load calculations, travel distance to exits, door schedules (door details), door hardware, and the building's fire suppression features. The details of the fire suppression and detection systems are not provided in the architectural drawings. These systems are usually listed among the architectural features to be provided and submitted later.

Floor Plan — A detailed drawing that indicates the dimensions and use of each room.

The **floor plans** are also included in the architectural drawings **(Figure 10.3)**. Floor plans indicate the room uses or occupancy of each room. Important elements such as generators, fire pump rooms, hazardous storage rooms, and transformer vaults are just some of the typical details shown on the architectural drawings. The drawings are very detailed and even show the location of furniture and equipment as well as reflected ceiling plans highlighting the locations of lighting fixtures, air ducts, and other items found on ceilings.

Fire Department's Concern during Architectural Document Reviews — Comprehensive architectural drawings offer the greatest amount of construction detail. The fire department representative responsible for construction document review will spend a great deal of time reviewing the architectural drawings.

Those fire departments responsible for reviewing egress components also will spend considerable time reviewing the architectural drawings. Fire departments are concerned with items such as the building's occupancy classification (This classification is noted on the architectural drawings and identifies the building's intended use). Fire departments ensure compliance with items such as flame-spread ratings for interior finishes, fire stopping details, construction of hazardous storage rooms, and details of any proposed hazardous operations.

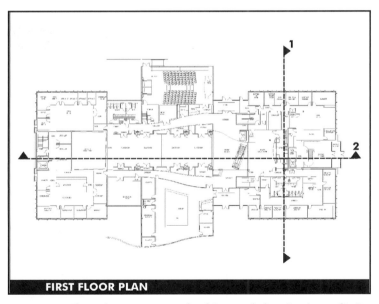

FIRST FLOOR PLAN

Figure 10.3 Floor plans are a type of architectural plan. See Appendix B, page 365 for a larger view of this drawing.

Structural Drawings

Structural Drawing — A drawing that provides details on how a building is put together.

Structural drawings provide details on how the building is put together **(Figure 10.4)**. Essentially, the structural elements are the critical components of the building that hold it up or keep it from falling down. The structural drawings provide details of the load bearing walls, roof, floors, and ceilings, structural steel, and other components.

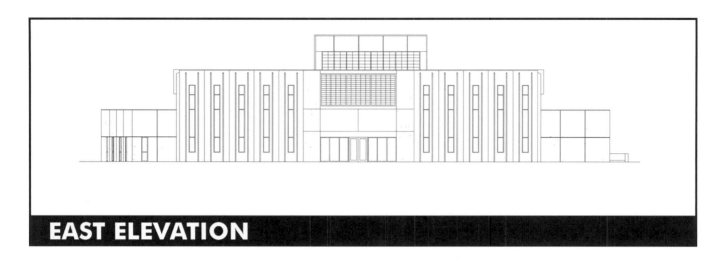

EAST ELEVATION

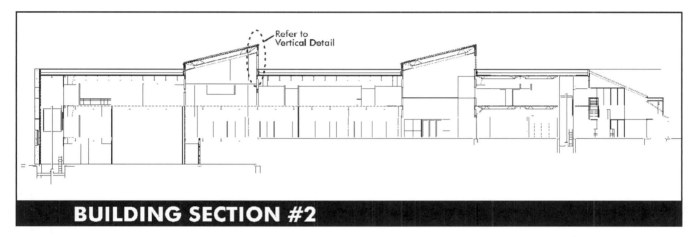

Refer to
Vertical Detail

BUILDING SECTION #2

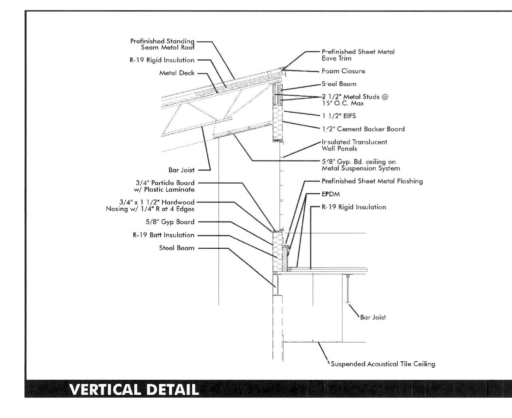

Prefinished Standing
Seam Metal Roof
R-19 Rigid Insulation
Metal Deck

Prefinished Sheet Metal
Eave Trim
Foam Closure
Steel Beam
2 1/2" Metal Studs @
15" O.C. Max
1 1/2" EIFS
1/2" Cement Backer Board
Insulated Translucent
Wall Panels
5/8" Gyp. Bd. ceiling on
Metal Suspension System

Bar Joist
3/4" Particle Board
w/ Plastic Laminate
3/4" x 1 1/2" Hardwood
Nosing w/ 1/4" R at 4 Edges
5/8" Gyp Board
R-19 Batt Insulation
Steel Beam

Prefinished Sheet Metal Flashing
EPDM
R-19 Rigid Insulation

Bar Joist

Suspended Acoustical Tile Ceiling

VERTICAL DETAIL

Figure 10.4 Three types of construction drawings are (a) elevation views, (b) section views, and (c) detail views. See Appendix B, page 366 for a larger view of these drawings. *All three courtesy of C.H. Guernsey & Company*

Fire Department's Concern during Structural Document Reviews — Fire departments review construction drawings for code compliance as well as for preincident planning and firefighter safety. Elements of the structural documents pertain to both firefighter safety and preincident planning. Fire department personnel will review items shown on the structural drawings, such as firewalls, parapets, openings in walls, roof construction, and type of construction and roof access. These passive fire protection elements are critical during a fire. When a fire does occur, this type of information can help the fire officer make tactical decisions. These decisions could include determining how long to continue an interior fire attack or whether to place firefighters on the roof for ventilation duties.

Mechanical Drawings

The **mechanical drawings** reflect many of the elements of the building's mechanical systems, such as plumbing, heating, air conditioning, and sometimes specialty features such as refrigeration, compressed air, oxygen systems, inert gas systems, swimming pools, kitchen hood exhaust, or others (**Figure 10.5**). In some instances, the mechanical drawings show the building's fire suppression system. For the purpose of this section and for clarity, we will consider the fire protection drawings to be a separate document.

Mechanical Drawing — A drawing that indicates the elements of the building's mechanical systems, such as plumbing, heating, air conditioning, etc.

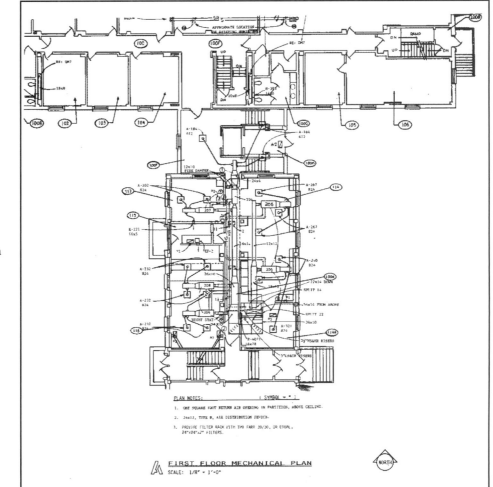

Figure 10.5 A typical mechanical plan. See Appendix B, page 368 for a larger view of this drawing.

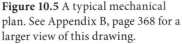

Fire Department's Concern during Mechanical Document Reviews — Some items on mechanical drawings significantly affect fire fighting activities, and fire departments must consider them even if mechanical reviews are not among their responsibilities. For example, the building's ductwork may contain provisions for smoke removal or smoke control. The ability to control a building's ventilation system can be critical to mitigating structural fires, particularly in large buildings, such as malls or high-rises.

Another important element on most mechanical drawings is the building's plumbing details. Fire department personnel will review the plumbing details if the building has automatic fire sprinkler protection. The water service that supplies the automatic sprinkler protection is shown on the building's plumbing drawings. Other automatic sprinkler details, such as the backflow prevention device, may also be shown on the plumbing details.

Examination of the various mechanical and plumbing chases that run throughout the building also provides invaluable information on how fire or products of combustion may travel or affect the rest of the building.

Electrical Drawings

The **electrical drawings** reflect the details of the building's electrical system (**Figure 10.6**). The power supply for the building and the associated wiring diagrams are typical details in these drawings. Some electrical drawings include the building's fire alarm drawings, and this is typically done for subcontractor bids rather than for formal fire detection and alarm system reviews. Since this is not normal practice, we will discuss fire alarm drawings in further detail later in this chapter when we explain the fire protection drawing submittal.

Electrical Drawing — A drawing that reflects the details of the building's electrical system.

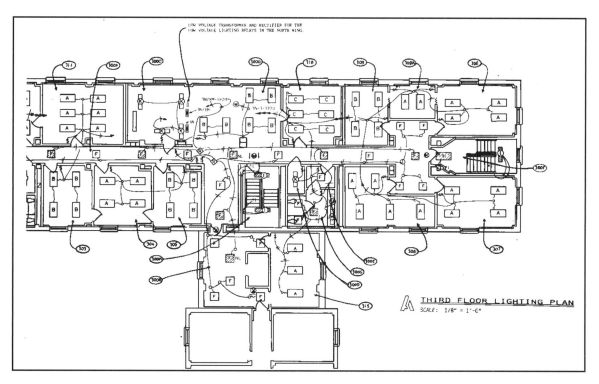

Figure 10.6 A common electrical system drawing. See Appendix B, page 369 for a larger view of this drawing.

Fire Department's Concern during Electrical Document Reviews — The fire department may review the electrical drawing details pertaining to the installation of the electrical fire pump, emergency generator, emergency lighting, exit lighting, egress lighting, and location of transformer vaults. Fire department personnel assigned to construction document reviews typically do not have the expertise of an electrical engineer. However, they need an understanding of how to locate these items on electrical drawings and understand basic electrical drawing symbols and details.

For larger, complex facilities or those with special electrical systems, fire department personnel assigned to the construction document review should seek the assistance of the team member who is responsible for the comprehensive electrical review. The *National Electrical Code®*, which is adopted by many municipalities, is a published by the National Fire Protection Association®.

Site, Landscaping, Civil, and Utility Drawings

The **site drawings** for the building are composed of a variety of details pertaining to a number of areas **(Figure 10.7)**. Most details on the site plan concern civil engineering. For example, the site plan will include details pertaining to the grade of the land for storm water retention and detention needs. Road construction details are included as part of the building's site plan as well. Details pertaining to the building's utilities, such as water, sewer, natural gas, and electrical supply from the public utility, are shown on the building's site plans.

Site Drawing — A drawing that indicates a variety of details concerning topography, landscaping, and civil engineering details.

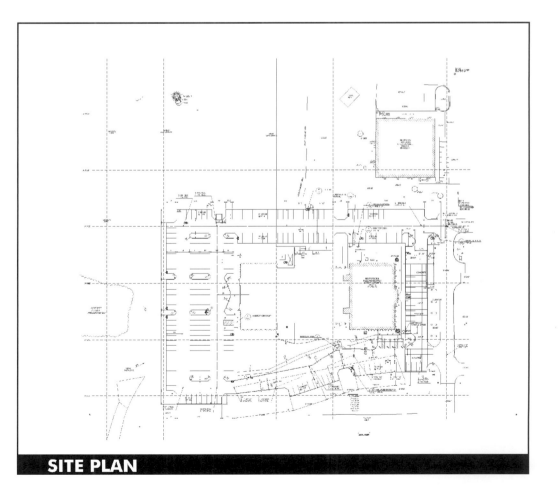

Figure 10.7 A site plan. See Appendix B, page 370 for a larger view of this drawing. *Courtesy of C.H. Guernsey & Company*

SITE PLAN

The site plan also shows comprehensive details of the topography of the land and site of the facility, including the landscaping features. The only building details shown are the structure's shape and location on the site. Although the site plan does not reflect any building construction details, it does contain two of the most critical elements of the building's fire protection features: access and water supply. Fire departments must be provided with the necessary tools to do their job in an emergency. Those tools include access to the building and an adequate water supply. It is important that newly constructed buildings contain these features. Fire department personnel should not be hindered by poor access to the building or an inadequate water supply.

Fire Department's Concern during Site Document Reviews — The fire department will focus on fire department access, hydrant locations, and vegetation management. During the review of the site drawings, details such as street widths, dead-end streets, and turning radius all pertain significantly to the fire department's access. In addition to building access, considerations may include access to fire hydrants, ponds or other bodies of water, and fire department sprinkler or standpipe connections. During this portion of the review process, the fire department may determine that the building needs fire lanes to ensure fire department access during an emergency operation. The person responsible for the construction document review should understand the fire department's apparatus to ensure that access will be adequate to accommodate them; fire lanes are of no benefit unless fire department personnel and their apparatus can use them.

Another important aspect of the site drawings is the utility details pertaining to the water supply and hydrants. The fire department will review the documents to ensure the provision of an adequate number of hydrants designed in accordance with the adopted codes. The fire department may consider discussing the water supply details with the Insurance Services Office (ISO). The Insurance Services Office evaluates the water supply as part of its grading of the municipality. The required fire flow for the structure determines the size of the water mains. It is important to consider not only the proposed structure undergoing construction document reviews but future structures anticipated as well. Fire department personnel need to work closely with the construction document review team's civil engineering and public works members. They review the site plans in great detail and can offer expertise in areas pertaining to road construction and utility installation.

Other important features involve the landscape plan, which can be part of or separate from the site plan. This is particularly important if the building is within the wildland/urban interface. Vegetation, landscaping materials, steep slopes, grade, and so forth all play parts. For example, if you are looking at a large apartment or townhome plan, rescue may be severely hampered if slope and vegetation prevent the placement of ground ladders. Also consider the ability to advance handlines for fire suppression, particularly if a wall is blocking access.

From a wildland fire perspective, vegetation types and proximity to the structure are very important, which may involve examining the slope adjacent to the structure. In some cases, when the structure is located high on the property, 200 feet of clear space or well-managed vegetation may be necessary if possible. Other issues may be the types of trees, proximity of bushes and shrubs, and so forth. It may also be beneficial to suggest the types of plants best suited for planting, such as those that are high in water content and naturally resistant to fire. The proximity of decks to combustible vegetation is also important.

Plats

Plat — A legal document illustrating the legal description of a property as well as any legally binding easements.

Plats are legal documents prepared by a land surveyor. The plat contains the legal description of the property as well as any legally binding easements (**Figure 10.8**). The dimensions of the lot lines are shown along with the associated property lines.

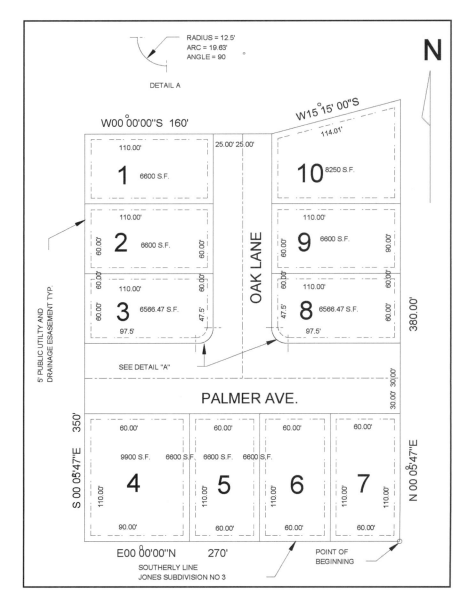

Figure 10.8 A typical plat. See Appendix B, page 371 for a larger view of this drawing.

Fire Department's Concern during Plat Document Reviews — The fire department should review the plat to ensure that it reflects any easements pertaining to fire department access. When property is being subdivided to construct additional buildings, the fire department may review the plat to assess the new structures' impact on the department's ability to access the site and provide fire protection. In some situations, the plat may reflect existing farmland subdivided into a major development that will have many implications on the fire department's ability to provide service.

> **Fire Protection Drawing** — A drawing that indicates the systems and elements pertaining to a building's fire protection systems.

Fire Protection Drawings

The **fire protection drawing** submittal will contain a variety of different systems and elements pertaining to the building's fire protection systems (**Figure 10.9**). The types of fire protection drawings submitted may include the following:

- *Sprinklers.* Detailed drawings and calculations pertaining to the building's automatic sprinkler system and fire pump where applicable. Sprinkler piping details, sprinkler locations, and sprinkler riser details are all reflected on the sprinkler drawings. Depending on the design of the sprinkler system, fire pump details may also be reflected on the sprinkler drawings.

- *Standpipes.* Detailed drawings and calculations pertaining to the building's standpipe system. They show the system's piping arrangement and hose valve locations. Depending on the design of the standpipe system, they may also show fire pump details.

- *Detection and alarm.* Detailed drawings and calculations of fire detection initiating devices and indicating devices including power load and performance and battery calculations. These drawings will show location and placement of devices, types, numbers, and the like.

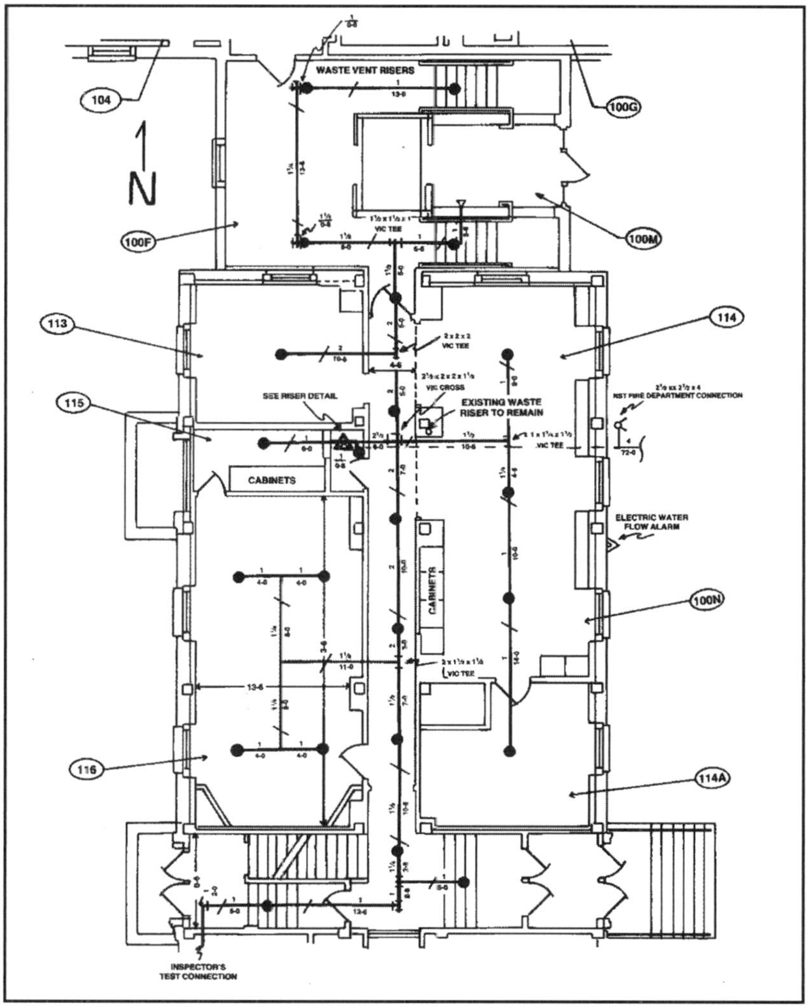

Figure 10.9 A sprinkler system drawing. See Appendix B, page 372 for a larger view of this drawing.

Fire Department's Concern during Fire Protection Document Reviews — The fire department may have the greatest involvement in the review of fire protection drawings. The review will focus on the drawings' compliance with the adopted codes, standards, and ordinances. The National Fire Protection Association® publishes fire codes pertaining to fire protection systems. In addition to reviewing the fire protection construction documents for compliance, the fire department will need to review the location of the fire protection system components, such as sprinkler risers, fire alarm control panel, and voice evacuation panel. The location of these system components must accommodate fire department operations. For instance, most fire department personnel will enter the front of the building for fire alarm activation. Ensuring that the fire alarm panel or a remote annunciator is in the front entrance of the building will help responding personnel locate the fire alarm panel.

Step 5: Construction Document Review

Taking part in the construction document review provides the fire department the opportunity to ensure that newly constructed or renovated buildings meet the adopted codes. Because the construction document review is one aspect of code enforcement that can create conflict, it must be performed in a precise and systematic fashion. Fire department personnel responsible for reviewing construction documents must be able to read blueprints, be knowledgeable of the adopted codes, and be able to communicate fire department concerns professionally and diplomatically. Many fire departments employ fire protection engineers to review their construction drawings. In some municipalities, an outside agency or third party performs the construction document review.

Step 6: Generating Construction Document Review Comments

After the construction documents have been reviewed, the next step is to communicate the results to the owner, tenant, design professional, general contractor, and key construction document review team members. Again, individual design team members often have overlapping responsibilities. For example, a review of hydrant placement may have some impact on the public works review team member. Depending on the nature and scope of the project, communicating the review comments to the entire construction document review team may be prudent. Some organizations send all of the review comments from every team member as a single communication. This may work for some municipalities but not for all.

Most important is to communicate in writing the results of the construction document review. The review comments must be clear and concise to enable the design professional to respond if requested. An effective review comment will identify the following three elements:

1. The deficiency. What is the problem with document under review?

2. The code and code section of the deficiency. Where can the design professional go to find more information?

3. What is needed to correct the deficiency? What must the design professional do?

The review comments will also indicate if the construction drawings have been approved for permit or if they have been denied and must be resubmitted. If there are no review comments, communicate this. Many fire departments will include the code requirements for ensuring the fire department is present for witnessing the acceptance testing of any fire protection systems.

Just how important an effective preconstruction meeting can be becomes apparent during the development of the construction document review comments. If the design professional and other individuals associated with the project have had the opportunity to discuss the project and meet the review team, many of the review comments will have been addressed in the submittal. If either the design professional or review team members have concerns, they then can contact an individual from the preconstruction meetings to help answer their questions. Many times, the ability to communicate verbally can reduce the bureaucratic snares associated with the construction document review process. If the issue is resolved verbally, note this in the construction document review process. This will avoid potential conflict in the future. Documentation is very important.

Rest assured, communication breakdown will always result in poor product and time delays and will likely upset customers and staff. Make every effort to communicate as effectively and completely as possible.

Step 7: Construction Document Revision Submittal

If the construction documents are not approved for permit, then the design professional may have to submit revised drawings and begin the review process again at Step 4. This typically means that the designer will get the plans back with appropriate comments and notes as well as a rejection notice or stamp of some type. The designer must then address the discrepancies, make the necessary changes or provide the missing information, and resubmit the plans. Keep in mind, generally speaking, the fire department's job is not to design but to review and ultimately to approve the design.

Step 8: Construction Document Revision Approval or Denial

When the construction documents have been approved, review comments are generated that indicate approval for the permit. In some cases, the construction documents may have been conditionally approved, or what we term "red lined." This may include approval based on trust and assurance that appropriate changes will be made, or it may be based on an item's being addressed at a future date prior to completion of the project.

Step 9: Permit Application and Issuance

When all of the construction document review team members have approved the construction of the project, a permit will be issued allowing work to begin. The format of permits varies from municipality to municipality. Some jurisdictions issue a separate permit for individual construction stages or systems. For example, **Figure 10.10, p. 250**, reflects a permit to begin construction of the building. **Figure 10.11, p. 251**, shows a permit to begin the installation of the automatic sprinkler system. Most municipalities assess fees with building permits. They may range from development fees to construction document review fees. Once the fees are paid, a permit is issued for construction.

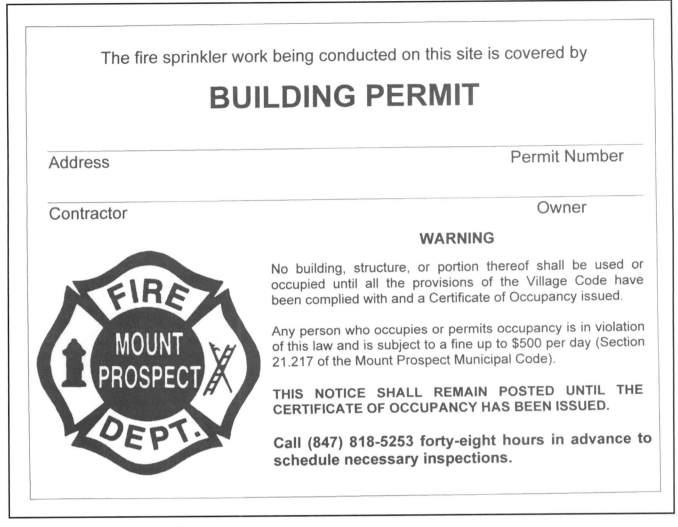

The fire sprinkler work being conducted on this site is covered by

BUILDING PERMIT

Address   Permit Number

Contractor   Owner

WARNING

No building, structure, or portion thereof shall be used or occupied until all the provisions of the Village Code have been complied with and a Certificate of Occupancy issued.

Any person who occupies or permits occupancy is in violation of this law and is subject to a fine up to $500 per day (Section 21.217 of the Mount Prospect Municipal Code).

THIS NOTICE SHALL REMAIN POSTED UNTIL THE CERTIFICATE OF OCCUPANCY HAS BEEN ISSUED.

Call (847) 818-5253 forty-eight hours in advance to schedule necessary inspections.

Figure 10.10 A permit to allow building construction.

Step 10: Construction Begins

The fire department's involvement does not stop when the construction documents have been approved and a permit has been issued. Fire departments will continue to be involved with the project while it is under construction. If the fire department is using a third party review firm, the firm should either be involved in the project under construction or have a fire department member involved. The involvement does not end when the construction documents have been approved. Assurance that water supplies or temporary standpipes are provided as soon as combustible materials are present or construction is started is one example of why continued involvement is important. The fire department will want to ensure it has access to the construction site should the need arise for emergency personnel to respond to the site. If the review process has successfully identified items to be corrected, very few changes for noncompliance should be needed during construction. Prewire tests, prehydrostatic tests, or any number of other inspections may also be required before the real "guts" of the construction inspection take place. Another aspect to consider is familiarity with the facility by stations and engine companies.

STILLWATER
FSD
1895

Heritage • Service • Pride

PERMIT CERTIFICATE

Date / Time: 09/08/04   08:32

Occupancy Name: Headquarters Fire Station

Address: 1506 S Main ST
Stillwater, OK 74074

Phone:

The following permit has been issued:

Permit No. 001280

Issued To: John Doe, USA Sprinkler Company

Type: SP INST. Automatic Fire Sprinkler Installation (New)

Issued: 09/08/04

Effective: 10/08/04

Expiration: 10/08/04

It is the contractor's responsibility to ensure that conditions are in accordance with applicable State and Local Fire Codes and Standards. Please contact Stillwater Fire Department for more information.

_____   _____
Installer / Contractor: **Date**

_____   _____
Inspector: Trent Hawkins   **Date**

* This permit must be available on-site while inspection, testing, or maintenance is being performed.

Office of the Fire Marshal · 1510 South Main Street · Stillwater, OK 74074
Phone (405) 742-8308   Fax (405) 747-8050

Figure 10.11 A permit to allow sprinkler installation.

Step 11: Construction Inspection

While the building is under construction, fire department personnel and other construction document review team members will conduct a variety of inspections. On-site fire inspections ensure the construction site is safe from fire and verify the work being performed conforms to the approved drawings. Most municipalities require a job copy or approved copy of the construction documents to remain on the construction site. If a discrepancy should arise between what is being constructed and what was approved, the fire inspector can view the approved set of construction documents as well as the construction document review letter. Conflicts can arise during the inspection of the facility while it is under construction. The field conditions may have necessitated changes that were not noted on the drawings, such as as-built deviations from the initial design. If the contractor or design professional did not consult the construction document review team prior to making the changes, there may be issues to address later that can cause project delays and escalate construction costs. This is just one of the many reasons that frequently visiting the construction site is important and conducive to the project's progress. The construction site visits also allow the parties involved to discuss issues that may have arisen. The fire department as well as other members of the construction document review team may need to inspect parts of systems before they are concealed behind walls or ceilings. This is especially true for some components of the fire protection systems.

The last inspection performed is the final acceptance testing of the fire protection systems. The fire department does not conduct the acceptance tests. They are performed by the contractors with the fire department as a witness. The fire protection system installer is responsible for conducting the tests. The fire protection system installer should have conducted troubleshooting tests before contacting the fire department to come and witness the final test. The fire protection system installation should be 100 percent complete when the contractor asks the fire department to be present at an acceptance test.

Step 12: Construction Complete

When construction is completed, the fire department and other members of the inspection or construction document review team may be called to conduct a final inspection. The final inspection will verify the project was completed according to the approved drawings. If the fire department had not visited the site as indicated in Step 11, there could be a number of very costly issues to address after the work is completed.

Step 13: Certificate of Occupancy Issued

A certificate of occupancy is issued to the occupant when the building has been determined to be safe to occupy and has been inspected to meet the code requirements. The certificate of occupancy ensures the building's fire protection features are operable. The building department typically issues this.

Step 14: Occupant Moves In

The occupants can move into the building once they have received the certificate of occupancy. In some situations occupants move into the facility prior to final approval by the municipality. These occupants may face legal action by the municipality. Moving into a building without final approval from the municipality is an unsafe practice. On some occasions, temporary certificates of occupancy may be issued to allow this, but typically only after all fire/life safety systems are installed and tested. This is also an opportunity to document existing conditions such as storage limitations, etc. for future reference.

Step 15: Business License Approval

Some municipalities require occupants of commercial occupancies to obtain a license to conduct business. The requirements for when and if a business license is required will vary from one municipality to another.

Step 16: Periodic Fire Inspections

When the building has been completed and occupied, periodic fire inspections are conducted to ensure modifications have not been made without the benefit of the construction document review process and to identify any changes in hazards associated with the operation of the facility. The frequency of the periodic inspections will vary depending on the hazard and its potential to cause large loss of property or life. The documentation generated in the construction document review process becomes a very useful reference source for fire inspectors when they conduct their periodic inspections.

Significance of a Good Construction Document Review Process

One of the most important aspects of the construction document review process is to ensure the building is constructed in accordance with the adopted codes, standards, and ordinances. Periodic fire inspections are needed to ensure the building and its operation continue to meet the fire codes. It is expected that the owner of the facility will need to make changes to both the building and the operation of the building. In fact, during the life of the structure a variety of uses may take place in the building. With this in mind, providing thorough documentation is paramount to enable any member of the fire prevention bureau to determine what the building was originally designed to be used for and what fire protection systems were provided. We know that sprinklers have a success rate of over 96 percent. One reason for the limited failures that do occur is that the system is not designed specifically for the hazard it protects. This usually does not result from an error in the construction document review process but from a change in occupancy. The use of digital photography makes an excellent record of approved conditions at the time of occupancy. This is very common in the documentation of kitchen equipment locations and storage practices.

For anyone in the construction document review process, the important task is to create sufficient documentation to revisit the project in 10 or 15 years and clearly understand the code requirements when the building was constructed based upon its original design. When fire department personnel become active in the construction document review process, they have the opportunity to build a relationship with the building owner or tenant. This should lay the foundation for building a positive partnership in fire prevention. The contractor will leave the site after the project is complete, but the owner or tenant will be involved with the fire department during fire inspections for years to come.

Summary

The fire department's role in construction document review is another essential component of its overall fire prevention efforts. The construction document review provides an opportunity for the fire department to begin building a fire prevention coalition with the owners of the building. It also offers an opportunity to identify potential hazards and risks that will impact the safety of the occupants and the firefighters who may be called there in the middle of the night to mitigate an emergency incident.

The fire department is part of a team of individuals with a vested interest in the construction document review process. The individuals may include representatives from the city manager's office, public works, building department, zoning department, economic development department, engineering department, and many others depending on the size of the project and the organizational structure of the municipality. The fire department provides input to an array of construction documents that may include site or civil drawings, zoning drawings, plats, architectural drawings, mechanical drawings, fire protection drawings, and electrical and mechanical drawings.

For the construction document review process to prove successful, all of the parties involved must work together during the developmental process that begins with the conceptual drawing and ends with the occupant obtaining a business license. An important element for fire departments to remember during the construction document review process is that the building's occupants as well as firefighters will rely heavily on the effectiveness of its life safety features long after the first certificate of occupancy is issued. As part of the construction document review process, the fire department will help to ensure that the life safety features are in place during construction. The final step of the construction document review process is to ensure the fire and life safety features are maintained by conducting fire inspections at the facility for the life of the building.

Case Study Results

The owner needs to be informed in a diplomatic fashion that the new use of the facility is a change in occupancy classification from an industrial occupancy to assembly occupancy and certain codes for egress, fire protection, etc. will be impacted. Then, the owner should contact a design professional and meet with the entire construction review team. The fire department will invite its third-party-review firm or plan reviewer to the meeting. Others with concern are planning and zoning (parking and zoning), engineering (traffic issues), health department (kitchen and food sales), and village manager's office (potential political impacts).

The best course of action is to meet with the new owner and identify all the issues that will need to be addressed prior to a certificate of occupancy.

Chapter 10 Review Exercises

10.1 Explain the fire department's role in the construction document review process and explain its importance. _____

_____

_____

_____

10.2 Identify three individuals who may have a vested interest in the construction document review process and identify why each would be involved in construction document review.

_____

_____

_____

10.3 Identify five reasons why fire departments are involved in construction document review._____

_____

_____

_____

_____

_____

10.4 List each type of building construction drawing typically found in the construction document review package and describe the fire department's possible concerns with each document._____

_____

_____

_____

_____

_____

_____

_____

_____

_____

10.5 What are preconstruction meetings? _____

_____

_____

_____

_____

10.6   Why are preconstruction meetings beneficial?_____

_____

_____

_____

10.7   What is the intent of conceptual drawings? _____

_____

_____

_____

10.8   What is a certificate of occupancy?_____

_____

_____

_____

10.9   What three elements must good plan review or construction review comments have? _____

_____

_____

_____

10.10 What are fire protection drawings?_____

_____

_____

_____

10.11 Identify the steps in a thorough development review process?

_____

_____

_____

_____

_____

_____

_____

_____

_____

_____

_____

_____

Fire Inspection Procedures

Table of Contents

Key Terms

Key Points

1. Two primary reasons for fire inspections are (1) to identify and correct hazards or conditions that could cause a fire or contribute to its spread and (2) to determine conditions that will endanger occupants and firefighters during a fire.

2. One thing is certain: the fire inspector must be professional and credible in the performance of his or her duties.

3. The fire inspection program must be established for the protection of the community served by the fire department.

4. The ability to explain why correcting a problem is important will help significantly in having the occupant remedy the deficiency and in establishing goodwill.

5. One of the best ways to build fire safety partnerships is through the education that takes place during a fire inspection.

6. Just as operational personnel should spend a great deal of time preplanning for fire incidents, inspection personnel should take ample time to prepare for an inspection.

7. It is a good practice to follow-up on fires in regulated occupancies with a top-down inspection "while the embers are still hot."

8. The inspector's ability to deal with the general public is just as important as his or her knowledge of the applicable codes.

9. No matter who is responsible for conducting an inspection, it must be organized, systematic, and methodical.

10. The most important thing to remember is to inspect every interior and exterior area of the building as well as the grounds of the facility.

11. Voluntary compliance is always better than pursuing legal action.

12. The inspector's most valuable tool for achieving voluntary compliance is the ability to develop a rapport and level of trust with the building contact.

13. The legal ramifications of following improper inspection procedures can be costly and painful.

Learning Objectives

1. Identify the goal of a fire inspection program.
2. Define the elements of a fire inspection.
3. Identify methods to conduct a fire inspection.
4. Determine what constitutes a "good fire inspection."
5. Identify methods for compliance of inspection findings.
6. Define the national fire problem and role of fire prevention.*
7. Define laws, rules, regulations, and codes and identify those relevant to fire prevention of the authority having jurisdiction.*
8. Define the functions of a fire prevention bureau.*
9. Describe inspection practices and procedures.*
10. Identify and describe the standards for professional qualifications for Fire Marshal, Plans Examiner, Fire Inspector, Fire and Life Safety Educator, and Fire Investigator.*
11. List opportunities in professional development for fire prevention personnel.*

* FESHE Objectives (USFA)

Fire Inspection Procedures

Case Study

A relatively new fire inspector receives an assignment to visit a local day care facility to do an annual state inspection. The inspector arrives and approaches the facility with his clipboard and inspection forms in hand. He makes contact with a male worker (Ben) and finds out that the owner/operator is out for the day. Ben states that he is just filling in and has no responsibility for decisions made at the facility. Actually, he is not sure if he should let the inspector inside the facility. After some conversation, the inspector tells Ben that he "will be allowed in or he will issue him a summons for obstructing a fire official." The inspector enters, walks around, and finds only one significant violation: a dead-bolt with a thumb-turn on the rear door, which is a marked and dedicated exit. Ben told the inspector that previously a fire company with three firefighters did the inspection and told them they could have it. Ben stated that the owner got some kind of "okay or variant or something." Ben said he knows this because the owner of the day care asked him to do the work and install the thumb-turn lock. The inspector says it is not allowed; he was writing it up; and it needed to be removed immediately.

- Was the inspector prepared for this inspection? Why?
- Was his approach done in a legal or customer service manner? How?
- If you were to choreograph this inspection, how would you rewrite the whole event?

Why Conduct Fire Inspections?

One of the fundamental tasks of a fire prevention bureau is to conduct fire inspections. When the term *fire prevention bureau* is mentioned, the average citizen usually thinks of fire inspections. The purpose of conducting fire inspections is **not** to write violations or issue citations for court appearances. Two of the most significant reasons for fire inspections are to (1) identify and correct hazards or conditions that could cause a fire to start or contribute to the fire's spread and (2) to determine conditions that will endanger occupants and firefighters during a fire. This last one has the most potential to directly impact firefighter safety on the fireground. It is also a task that many firefighters feel as insignificant to the overall mission of the fire department. Once hazardous conditions have been identified, they are then communicated to the property representative in a manner conducive to educating the individual, not reprimanding him or her. Fire inspections are a means to educate the public on the importance of fire prevention and to eliminate hazards or conditions that will contribute to the development or spread of fire. The inspector represents the fire department and needs to ensure he or she always performs in a professional and diplomatic fashion. Fire departments provide fire inspections as a service to the community. Inspections should be conducted so that excellent service is provided to the customer. Inspectors need to be aware that being heavy-handed or badge-heavy is detrimental not only to the fostering of relationships but to the reputation of public safety.

The ability to explain the history or the reasoning behind an unsafe act or condition helps the occupant understand the danger and its significance. In most cases, once fair-minded people understand the reasons, they have an easier time addressing the issue. Nobody likes to do things just because the book says so! Not that fire prevention is an easy sell; it can definitely be tough at times. Depending upon local "flavor," simple things, such as using the term *deficiency* instead of *violation,* may help achieve compliance. Explaining how a deficiency impacts occupant safety and firefighter safety is another strong selling point. Whenever you can, always try to keep word choices positive rather than negative. Use action statements! Keep in mind, what you do and say will impact the safety of not only the occupants but of visitors and of the firefighters who will be called there in the middle of the night to extinguish a fire or deal with any other emergency.

For the most part in the United States, people presume when they enter a public or commercial building that it is safe to do so. The citizens rely on safe buildings, just as they rely on a fire engine or ambulance to be available to take care of their emergency needs. Governmental agencies should address this presumption held by citizens from a liability or perception standpoint.

Implementing Inspections

How does a fire department begin to conduct fire inspections? The establishment of a fire inspection program may result when the fire department's governing body (city government, for example) annexes property or when a fire department's strategic planning identifies the need to take a more active role in fire prevention by conducting fire inspections. Because of economic conditions, fire departments may experience a period of time where they have to curtail inspections. The dilemma a fire department or fire prevention administrator then faces is that once time has elapsed without conducting inspections, it can be a difficult task to resume. As an inspector, you may be faced with numerous deficiencies as well as building owners who are not accustomed to an inspection. Building owners may have to spend considerable amounts of money to rectify larger, more complex violations that have evolved over time.

Whether the fire department is new to fire inspections or has been conducting them for some time, it is essential that the fire inspector be professional and credible in the performance of his or her duties.

Do not equate the term *professional*, as some do, in the sense of whether a firefighter is paid or volunteer. In the context of this discussion, professional is the quality of expected performance. Many "professional" fire inspectors in volunteer fire departments conduct inspections without pay. The fire inspector continues to evolve as one of the most technically competent and professional members of the fire service. It does not matter whether you are a volunteer or paid firefighter; full-time or part-time inspector, sworn or civilian. One of the fire department's most important functions is conducting fire inspections and must be regarded accordingly. The success of a fire inspection is sometimes difficult to measure. However, you must conduct a fire inspection thoroughly and accurately — just as a paramedic would provide emergency medical treatment to a patient. Many still argue that you cannot measure the value of inspections and fire prevention activities. Do not get discouraged — technology is improving every day. You can justify the worth of an inspection if you dedicate time and resources to good data collection and statistical evaluation.

Step 1: Analyze

Why are we starting a fire prevention inspection program? Chapter 4 explains the need for the fire department to involve the fire prevention bureau in its strategic planning process and to ensure that fire prevention activities are part of the overall mission of the organization. One of the fundamental planning steps is to analyze the level of service to be provided. Obviously, this is initially started based on the needs and initiatives established through the strategic planning process which identifies what, when, and how to provide the proper level of service. The planning process is generally finalized through the policymakers who oversee the governmental entity. These officials may be either elected or appointed which, believe it or not, can affect the way the process is accomplished. Elected officials can frequently be swayed by constituency opinion, which may fly in the face of what may be good for the community. Appointed folks may be more solidly grounded in principle but they too have a boss or bosses who may give direction that is contrary to our professional position.

The establishment of the fire inspection program must be derived from the need to provide the service. The need to provide the fire prevention service is often scrutinized during the budget process. It may be controversial because businesses do not want to be saddled with additional expenses and barriers to their enterprising efforts. Again, this is why you must be able to sell your program and the worth harder than ever.

During the recent economic downturn, the ability to have tracked and documented fire prevention successes was an important element for demonstrating to the elected officials the need for fire prevention. The need for fire prevention is sometimes recognized as a result of the strategic planning process or a catastrophic loss. The inspection program, however, must be structured appropriately for the protection of the community and firefighters.

Step 2: Determine the Enforceable Regulations

The second step in establishing a fire prevention inspection program is to identify the community laws, policies, and procedures that assign the fire department the duty to conduct fire inspections. The fire department's authority may originate from local or state government agencies (see Chapter 3). Once the fire department has determined its authority to conduct inspections, the next logical step is to learn what codes and standards can and should be enforced (**Figure 11.1**). The codes are the inspector's rulebook. A look at the codes adopted by the governing entity may reveal that they are not the most recent editions published, or it may reveal that less stringent state or federal codes can be enforced. These issues may require the fire department to adopt a different edition of the code, another code, or a combination of both with amendments (This is something that should be brought forward at the next fire department strategic planning meeting!). For the purpose of this discussion, the code itself is not necessarily as important as *what* code is or is not enforceable.

Figure 11.1 A number of codes and standards are needed for proper guidance of code enforcement activities.

Step 3: Determine Inspector's Level of Training

The next step is to determine the level of training required for conducting inspections. The National Fire Protection Association® offers courses online that individuals can complete at their own pace. IFSTA's *Fire Inspection and Code Enforcement* manual, published by Fire Protection Publications (Oklahoma State University), is also an excellent resource for learning about fire inspections and code enforcement. The principles of this text create a good foundation for the inspector or fire prevention bureau manager to build upon. Many junior colleges and professional organizations offer courses and seminars in code enforcement. Company officers are commonly being assigned to some or all of the duties of fire inspection. In many cases, this practice has become a direct result of budget restraints and the loss of fire prevention staff. Also, it is simply an effective practice. The company officer has the potential to have the greatest impact on the Operations Division fire prevention efforts. The text, *Fire Prevention Applications for the Company Officer*, provides an overview of how this can be accomplished. The text is written for existing or future company officers.

Conducting a fire inspection for the first time in any occupancy takes time and extra effort. This is true even if other inspectors have inspected the occupancy previously. An inspector may conduct a fire inspection and identify conditions that are in violation of the adopted codes. The condition may have existed for years without being identified by previous inspectors. Situations like this make educating the occupant about the severity of the deficiency difficult. If it was not serious enough for other inspectors to identify, how has it become such a serious issue now? This should also be a good indication to the fire prevention manager that previous inspector training was inadequate. Depending on the severity of the deficiency, the occupant may not be able to correct the problem immediately, and the inspector should be prepared to allow a reasonable amount of time for the occupant to work on it. Having the ability to explain why correcting a problem is important will help significantly in having the occupant remedy the deficiency and in establishing goodwill.

For example, the *Life Safety Code*® requires exits to be enclosed with rated walls and doors. If the occupant of a building removed the doors on a basement exit stair entry to move material easily to a storage area in the basement, an inspector might assume that this was approved as part of the construction process and was acceptable. More inspections might have occurred without anyone ever identifying or correcting the deficiency. When a new inspector visited the occupancy and identified this condition as a deficiency, the owner/manager might be upset that it was not previously identified. Instead of confronting the occupant, the inspector could offer to research the question and get back to the owner. The inspector would then determine that doors were originally required by code and installed.

Upon returning to the occupancy, the inspector would provide the following reasons why the doors were needed. Doors would:

- Prevent fire spread from the basement to the second floor.
- Provide a safe passage for the occupants to exit.
- Ensure firefighters had an enclosed and protected area in which to extinguish a fire.

The inspector has now explained the dangers of the deficiency and the benefits of correcting it, instead of just reciting verbiage from the code book or giving the impression of enforcing some made-up version of the code. It may take time to have the violation corrected, but the end result will be compliance.

Remember, too, that inspections provide an opportunity to educate (remember, be careful with that word) or make others aware of what you are trying to do as a fire service professional. "Sell" your issues. It is easy to be a badge-heavy thug and make people do things just because you can. Don't do that! Find issues that need to be corrected, explain why correcting them is important, and be flexible when you can. Remember, any time you put pen to paper or open your mouth, you are costing this business money. It is one thing to tell warehouse owners they have to take down the top row of crates because they are stacked too close to the ceiling. It is quite another for the owners to lose over a third of their inventory capacity. We are not saying that you should not enforce the code but that you should think about what you are doing and the impact your decision has on the overall operation of the business. Sell the reasons for the requirements. Sometimes the decision needs to be shared by more than just you and the business owner. Work with your clients, don't browbeat them. You have authority and responsibility, but you must not abuse your position.

The goals of the fire inspection are to ensure that the building is safer when the inspector leaves than when the inspector entered it. The building should be in compliance with the adopted codes and standards, and the owners/managers should have been made aware of fire safety practices. Achieving these goals may not happen overnight. Educating the occupant about practicing fire safety can help to ensure the property owner/manager will eliminate potential hazards without requiring the fire inspector's presence. If the occupants gain an understanding of the hazards associated with their property, they may be more willing and able to take corrective action on their own. Fire safety is not just the fire department's responsibility. It is best accomplished through partnerships with the community. One of the best ways to develop these relationships is through the fire safety education that takes place during a fire inspection.

To achieve compliance, the fire inspector needs to be able to help the owner/operator prioritize any list of deficiencies to be corrected. The challenge for the inspector is determining if the deficiency identified is serious enough to warrant immediate attention or if the corrective action may be delayed a reasonable time. As an inspector, this takes time and experience to master.

Step 4: Establish Organizational Fire Inspection Policies and Procedures

Now that the need and the legal right to conduct fire inspections have been determined, it is prudent to develop organizational policies and procedures for conducting fire inspections. The level of detail in the policies and procedures will vary dependent on the size of the fire prevention bureau, the level of technical training, and the size of the community served. In addition, the policies may have to be tailored to accommodate inspections conducted by the operations division to coincide with their duties.

Most fire departments have developed standard operating guidelines (SOGs) for suppression and emergency medical services. Similar types of guidelines are needed for fire inspection activities. An important consideration is that the guidelines are not developed as a means to "police" or "control" the inspection process within the fire department, but as a means to provide guidance for the inspection staff. It also provides and promotes continued focus on mission and purpose. Some of the guidelines may result from ordinances or laws. For instance, a local ordinance or state law may require all day-care occupancies to be inspected twice a year. A corresponding guideline for the frequency of conducting inspections would list day-care centers as being inspected twice a year.

Procedural guidelines also assist in the enforcement of codes in situations where it becomes necessary to pursue legal action. The guidelines may state that a regular inspection form is used on first inspections. If the violation persists, then an order notice is issued. If the violation again continues, the guidelines would indicate the forms to use when issuing a citation or summons and when it could be issued; then they would detail the court appearance procedures.

Fire Inspection Priorities

To determine *when* to inspect *what*, you need a good understanding of your local fire problem. Good fire protection practice would dictate inspecting all commercial and public code-regulated occupancies or properties a minimum of once a year. Smaller departments can generally accomplish this if they have the staff because the inventory is manageable, but relatively few large departments can do so even with an adequate level of staffing. The answer to success for all fire departments is inspection prioritization.

To prioritize inspection targets, first identify your fire problem **(Figure 11.2)**. As an example, in Colorado Springs, Colorado, in 1997 William H. Wallace analyzed community risk.[1] Wallace concluded that the number-one cause of fires was unattended cooking in private dwellings followed closely by the same problem in apartments. Since formal fire inspections do not typically address single-family homes, the department placed the inspection focus on apartments. Apartments are structures the department has the authority to inspect. The issue surrounding detached single family dwellings was shifted to an educational message versus an inspection posture.

As a target is identified as previously shown, a department may decide to mount an aggressive fire and life safety education campaign and offer free home inspections to single-family homeowners while targeting multifamily residences or apartments for mandatory inspections.

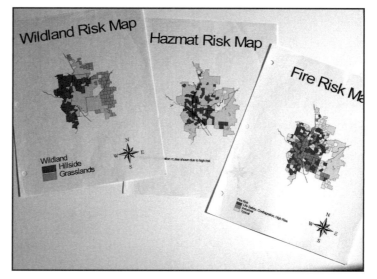

Figure 11.2 Good fire prevention programs identify the location of community risks.

The goal of fire prevention activities is to mitigate the threat of fire. In the interest of achieving that goal, determine the type of occupancies in your jurisdiction where fire incidents are most prevalent. If they are regulated by your department, prioritize them from highest risk to lowest. Finally, identify the number of occupancies in each group and devise a plan or schedule to attack the problem in the most efficient and cost-effective manner with the available resources.

Some departments place occupancies on inspection cycles. The first year they may inspect apartments and schools, the second year hospitals and care facilities, and the third year high-pile storage warehouses and malls. The following or fourth year they may start the cycle over again or plug in another occupancy if trends show a problem in a specific area.

For initial targeting, follow the same order as for operational considerations. What is most important? Life safety. What is second? Property conservation. With that in mind, identify locations with high life loss potential. Next, identify the occupancies with potential for high value property loss or the potential for significant economic impact to the community. Next, identify locations documented as high hazard such as hazardous materials storage or process locations. The remaining inspectable occupancies will rank low, so fire and life safety education may be appropriate and sufficient to address the fire problems.

Figure 11.3 Training is essential to prepare for fire inspections.

Preparing for Fire Inspections

Most times, the fire inspector prepares for a methodical and organized preparation before leaving the fire station or the office. To prepare for a fire inspection, fire inspection personnel should take ample time — just as fire department operational personnel spend a great deal of time preplanning for fire incidents. Fire inspectors should spend time researching background information about potential hazards or deficiencies that they may encounter in a particular type of occupancy (**Figure 11.3**). Likewise, firefighters should spend time learning about the hazards that they may encounter in a particular building. Taking the time to perform this exercise will ensure the inspector is prepared to perform his or her duties and enable the inspection process to be performed smoothly. Most all of the code requirements are based on the occupancy classification of the building.

Occupancy Hazards

One of the first items to consider is the potential hazard associated with a particular occupancy. Where can you find such information? The following are a number of references available:

- Factory Mutual (FM) Property Loss Prevention Data
- *Fire Protection Handbook*® — National Fire Protection Association®
- NFPA® — *Fire Code Handbook*®, National Fire Protection Association®
- IFSTA *Fire Inspection and Code Enforcement*, 8th edition, manual — International Fire Service Training Association
- *Industrial Fire Hazards* — National Fire Protection Association®

These resources provide a guide to some of the typical hazards that inspectors may encounter during an inspection. They offer insight into the conditions that may warrant corrective action by the owner. The inspector's research should not only identify typical hazards associated with the occupancy, but also determine why these conditions create fire hazards. Much of the needed information will provide a short synopsis of fires that started or rapidly spread as the result of a condition identified as a hazard. Keep in mind, the inspector will have to sell the need to correct the condition. Convincing the occupant that a condition can contribute to the spread or ignition of a fire will be easier if the inspector can clearly cite examples of past fires that the condition caused in similar occupancies. It is also important to consider how these hazards can impact fire department operations and the potential to cause injury to firefighters. The reason something needs to be corrected is more convincing if the inspector can cite details of how it can contribute to a fire's starting or spreading. The inspector can gain this information either through experience, research, or a combination of both. Do not be too upset if the occupants still are not happy with your explanation of why something needs to be corrected. You are costing the building owner money to fix the problem. You are not paying for the changes, they are. Do not shirk your duties, but do not take it personally if they listen with empathy and still think your requirement is stupid and an unwarranted expense. Besides, the occupant will likely tell you they "never had to comply with it before" or the "last inspector did not tell me to do it," and they very well could be right. This situation tends to become more prevalent during tough economic climates when budget restraints forces inspection frequency to be decreased. Hazardous conditions within a building may escalate without notice, resulting in a perception that the condition has always existed.

Specific code requirements may address the hazards of an occupancy. Remember that most codes are derived from an incident that has occurred and is brought forward for consideration in the code proposal process to prevent the incident from occurring again.

Determining process operation hazards before the inspection can help prepare the inspector for his or her evaluation during the inspection. Process hazards may also require the inspector to review codes or standards applicable to the process hazard. For instance, if an occupancy contains a dipping and coating operation, the inspector should become familiar with NFPA® 34, *Standard for Dipping, Coating, and Printing Processes Using Flammable or Combustible Liquids.*

History of Occupancy

Previously, we identified the importance of selling the need to correct a deficiency and overcoming the response "it has always been that way." An invaluable tool for doing this is the ability to review what has occurred at the occupancy in the past. Before conducting the inspection, the inspector should know as much history as possible about the occupancy. It may have been a considerable length of time since the last inspection. Additionally, after fires occur in regulated occupancies, it is a good practice to follow up with a top-down inspection. The purpose of the inspection is not meant to cause further stress to a business owner. However the inspection is a means of working with the investigators to determine origin and cause, evaluate prefire and postfire code effectiveness, and to identify and evaluate other deficiencies to assist the owner in the future. Some aspects of the occupancy's history to consider include:

- History of fires at the occupancy
- Date of last inspection and contact person (Knowing who the department interfaced with can be useful in defusing the argument "It has always been that way.")
- Number of fire alarms or requests for service
- Past fire inspection and fire prevention deficiencies noted

- Outstanding fire prevention deficiencies not corrected
- Construction permits for building alterations and what it entailed (Additional alterations may have since taken place without the benefit of a permit, construction document review, or inspection.)
- Use of the facility in the past (Is the use of the occupancy different than at the time of the last inspection?)
- Results of most recent annual fire alarm testing
- Results of most recent annual fixed fire suppression system testing
- Results of most recent annual fire pump testing
- Written correspondence, such as complaints from citizens, suppression personnel, and letters of inquiry
- Fire prevention permits

As inspectors become more experienced, they realize the importance and benefits of documentation. Develop enough documentation as if you were the inspector assigned to the inspection for the first time. What level of history inspection details would you like to have when you conduct an inspection at a facility for the first time?

Fire inspectors must constantly provide documentation that will enable any inspector to pick up where another left off. This documentation may not seem important at the time, but it may play a critical role later down the road when a deficiency needs to be corrected, particularly by someone else.

Fire Inspection Public Relations

Most citizens encounter fire department personnel through fire inspections or emergency medical service responses and not through fire responses to their home or business. In fact, a busy fire prevention division will likely have more face-to-face contacts than the emergency operation's division will. In some circumstances, fire prevention inspection personnel will be the only fire department personnel who have any opportunity to make a lasting impression on citizens. The fire inspection provides the fire inspector with an excellent opportunity not only to educate the general public about the importance of fire prevention, but also to sell the fire prevention division's excellent services **(Figure 11.4)**. When the fire inspector performs in a diplomatic and professional manner, the entire fire department is portrayed as professional and diplomatic.

Figure 11.4 It is important for fire inspectors to sell the importance of compliance.

The image of the inspector and his or her fire department will be dictated by how the inspector talks and dresses and how he or she produces the inspection report.

The inspector's ability to deal with the general public is just as important as his or her knowledge of the applicable codes. The inspector must communicate his or her knowledge of the code to the occupant in a fashion that is neither condescending nor overly authoritative and therefore obtain voluntary compliance. The fire inspector should constantly seek voluntary correction of any deficiencies found during the inspection. This holds true even with trying to achieve voluntary compliance for minor deficiencies such as a missing or malfunctioning exit light. If the violation can easily be corrected at the time of inspection, then it is usually wise to continue with the inspection without even writing the violation. If the violation is a common occurrence at the premise, it is better to go ahead and note the violation and then document it was corrected at the time of inspection.

Government entities in general can have difficulty enforcing compliance issues simply because they are part of a governing body. Many individuals resent being forced to do anything mandated by government. As the inspector begins to build a rapport with the occupant, the ability to see from the occupant's viewpoint helps to achieve this goal of voluntary compliance. For example, conducting an inspection when the occupant's fiscal year is ending may contribute to the occupant's resentment and resistance to voluntary compliance because his or her money is running low. If the inspector is able to discuss the occupant's concerns of fiscal restraints, the occupant may comply voluntarily at the beginning of the next fiscal year. Of course, delaying compliance on a fire code issue requires careful evaluation and approval from the fire department's highest-ranking officer. However, even the best fire inspector will still encounter conflicts with the occupants. Resolving these issues requires conflict management skills that will improve over time as the inspector encounters them.

Managing conflict can be difficult. There are three common ways of dealing with conflict:

- Domination
- Compromise
- Integration

One expert on resolving conflicts, Mary Parker Follett, has suggested that **compromise** is the popular way to resolve many issues.[2] Each side gives up something, and they reach an agreement. While sometimes necessary in other situations, this likely would dilute fire code requirements, which is not always a preferred choice among code enforcement professionals. **Domination**, Follett believes, is a victory over an opponent. One side is the victor, and one side is the loser. In this instance, she proposes nothing is ever resolved, it is merely postponed. This likely would be a poor choice for fire code enforcement. Follett goes on to say that the most sophisticated approach to resolving conflict is **integration**. This is where a mutual solution is found and both sides achieve their goal to some degree with neither being wrong or bad. Integration is probably the best choice for resolving conflict; this allows the inspector to enforce the code and the owner to accomplish his or her objective.

Compromise — A method of resolving conflicts in which each side gives up something in order to reach an agreement.

Domination — A method of resolving conflicts in which one side is the victor and one side is the loser.

Integration — A method of resolving conflicts in which a mutual solution is found and both sides achieve their goal to some degree with neither being wrong or bad.

Some fire departments have implemented means for occupancies to conduct a self-inspection program. These departments set the criteria for determining if a business is suitable to participate in the self-inspection program. The criteria may be based on the size of the facility, occupancy of the building, or results of previous fire inspections. Some departments set a time limit for how long the occupancy can participate in the self-inspection program and provide mandatory training for participants. The occupant of the business conducts the inspection and provides a copy of the inspection report to the fire department. The fire department may then stop by to verify the occupant's findings or may accept the inspection report as submitted without visiting the location.

Conducting the Inspection

The general principles of conducting an inspection are the same regardless who is responsible for the task. An inspection must be organized, systematic, and methodical. Beyond that, a variety of methods or techniques are used to conduct an inspection. The inspector needs to choose a systematic method that fits his or her needs and needs to follow the same format for every inspection. Repetition helps avoid mistakes and will assist in keeping the inspector on track during the inspection. Depending on the complexity of the inspection, it may become easy to be overwhelmed or distracted.

Whenever possible, the fire inspector should call the contact person for the facility and schedule the inspection. This will help to ensure that the contact person can spend time with the inspector. The inspector can also use the phone call making the appointment to explain exactly what he or she will be looking for. If deficiencies from the last visit are still outstanding, the inspector should let the contact person know that he or she will want to see if the deficiency has been corrected or that progress has been made to address it. The inspector also should indicate any documentation, such as fixed suppression testing, fire alarm testing, or material safety data sheets, that will be useful or required for the inspection,

The inspector should consider if any other individuals should also be present at the inspection. Many times, the actual owner of the property will not be the contact person for the inspection. In a number of instances, the building is leased to a tenant, who may not be responsible for all areas of the building's maintenance and upkeep. This is particularly likely in situations where multiple tenants occupy a single building. In circumstances such as these, the inspector should schedule the inspection so that both the owner and the occupant are present.

Some building owners may become frustrated if they are inspected by multiple entities over a short period of time. In some communities, the health department will inspect restaurants; the housing or property maintenance division will inspect multifamily occupancie; and so on. Fire inspectors may wish to coordinate their inspections with those of any other entities that may be required to inspect the occupancy. This demonstrates the desire to limit interruptions to the business and not have multiple inspections at the same level of government (i.e., city, town, county, etc.). The potential drawback of conducting simultaneous inspections is that the occupant may become overwhelmed and frustrated dealing with multiple inspectors, each of whom has his or her own specialty. The occupant may even have a sense of being "ganged up on."

Once the inspector has made an appointment for the inspection, being punctual is an important element in establishing credibility. Very possibly, inspectors may be delayed in their appointments or may be called to assist at an emergency incident. During these situations, it is imperative that the inspector at least make a phone call to cancel or reschedule the appointment for a later date. In many cases, a department secretary can make the phone call to the occupant and explain the circumstances. When scheduling, inspectors need to ensure adequate travel time between inspections and allow enough time to conduct a thorough inspection. Hurrying to finish one inspection in order to proceed to the next is usually not conducive to providing a quality inspection service.

Some circumstances permit just dropping in for the inspection. For example, engine companies generally do not schedule those assigned fire inspections. If an engine company must still be in service to respond to emergency calls, it cannot know exactly when it can conduct the inspection. Often, when fire personnel assigned to an engine company visit a location unannounced for an inspection, the owner may ask them to come back at another time or day. Likewise, it is not uncommon for fire inspectors to visit a location unannounced. Frequently, fire inspectors will drop in on small businesses to conduct inspections, especially when a number of businesses are in close proximity, and the inspector can go from one occupancy to the next.

Fire inspectors today are equipped differently than their predecessors. The age of electronics assists them tremendously but also adds to their equipment list. For example, in addition to the traditional fire department uniform and identification, an inspector's equipment typically includes:

- Tablet or portable computer
- Electronic ruler or measuring tape
- Rechargeable flashlight
- Cellular phone
- Pen
- Clipboard
- Inspection forms
- Digital camera

And of course, there are coveralls for extremely dirty areas.

The inspector can choose between several systematic inspection approaches. Remember to inspect every interior and exterior area of the building as well as the grounds of the facility.

Upon arrival at the site, spend a few moments to drive around the facility and become familiar with the building's layout. This is also a good way to determine whether the facility is accessible for emergency vehicles. Please note that this is not the exterior inspection but only a tool to become familiar with the building and surrounding areas. It allows personnel to make some mental notes regarding the height of the building, exterior housekeeping practices, and notable construction features.

Unless otherwise requested by the owner of the property, the inspector should enter the premises at the front door or the main entrance to the facility. The inspector should identify him or herself and proceed to meet with the contact person. In most all situations, inspectors will want a building representative to accompany them during the inspection. Often more than one person will accompany an inspector. Greeting these people with a positive attitude is important. The contact person accompanying the inspector may have had a bad experience with an inspector in the past as well as naturally being a little apprehensive about what is going to take place. During the initial contact or opening remarks, the inspector explains what he or she is going to do. This is also a good time to emphasize the philosophy of working together to resolve any deficiencies. Good inspectors will also indicate that they understand it can take time to correct any deficiencies and that together they will establish a "game plan" to address concerns. Voluntary compliance is always better than pursuing legal action.

Some inspectors start on the exterior of the building; others start on the interior. Where the inspector starts does not matter, as long as he or she is thorough and systematic. During the inspection, the inspector must take the time to explain any deficiency to the occupant and to explain any corrective action needed. In some cases, the inspector may recommend the occupant contact an architect or engineer to help in correcting the deficiency. The inspector should offer to assist by meeting with the architect or engineer to discuss the deficiency further and in greater technical detail. The inspector can indicate to the owner how the fire department will provide this service and that the fire department is looking out for the best interest of the occupant. In most cases, the occupant may not have the technical understanding needed to correct the deficiency.

Interior Inspections

Three common systematic approaches are used to conduct an interior inspection. They include:

- Start from the roof and work down to the lowest level.
- Start from the lowest level and work up to the roof.
- Follow the manufacturing process from the point where the raw goods enter the facility to the point where the finished product is placed for shipping or storage.

Some fire inspectors prefer to inspect each area in a clockwise direction, a counter-clockwise direction, from front to rear, or vice versa. No one method is necessarily better than another. The important point is that the inspection is systematic. Occupants might not show the inspector an area of the facility unless the inspector specifically asks to see it. As the inspector visits an area, he or she should look for doors, stairs, or fixed industrial access ladders and ask what is located in the areas to which they lead. The inspector should request to enter these and all other areas. In documented circumstances, building occupants have used restrooms to store hazardous materials, not considering that the inspector would ever look in a restroom!

If the inspector has elected to follow the manufacturing process from start to finish, a systematic approach is still warranted. This method is conducive for the building's contact person to rush the inspector through the process. The inspector needs to stop along the way and ask questions about the process and the facility. For instance, questions pertaining to the types of materials used in the process may lead to where the materials are stored and handled.

The inspector's most valuable tool for achieving voluntary compliance is the ability to develop a rapport and level of trust with the building contact.

Exterior Inspections

The exterior inspection of the facility should be done with the same systematic approach as the interior. The building's exterior is an important fire protection feature that is the first to be utilized by responding emergency personnel. For example, fire department vehicle access, fire hydrants, sprinkler connections, and building addresses are all exterior features that, if not maintained and in good working condition, may cause difficulties when fire personnel first arrive for an incident. This is just one of the many examples of how fire prevention activities support fire department operations.

Inspectors may need to examine structures or conditions that are not usually thought of as fire department concerns. These could be inspections of single-family developments, making sure roadway width is appropriate as designed and pavement or overlay meets the required specifications or inspections of vegetation management in the wildland/urban interface. These inspections must be carried out according to local laws or regulations but should be handled in the same fashion as described above.

Fire prevention personnel provide a service not only to citizens, but also to fire suppression and emergency services personnel. A significant but sometimes overlooked function of the fire inspection is to ensure fire department personnel can handle any emergency as safely as possible. Of course, fire fighting will continue to be a dangerous job; however, inspectors must evaluate the building's conditions and hazards to minimize risk to firefighters. Items, such as open floors or doors or stairs fixed in a permanently closed position, may pose a threat to firefighters in a smoke-filled environment. Most model fire codes specifically address signage requirements for open hatches, etc. that may endanger firefighter activities. There are now code requirements for identifying dangerous buildings. If your department chooses to enforce those provisions, you must be prepared to execute the proper procedure for doing that task.

Fire Inspection Evaluation

The types of inspection forms used will vary from a multicopy paper form to electronic documents (see Chapter 14). Some fire departments will use a checklist. Others will use some other type of simple form. Some feel that check sheets may keep the inspector too focused on checking boxes rather than looking for and discussing fire prevention deficiencies. The use of either usually has a direct correlation to the experience level of the inspector.

Fire Inspection Steps

The inspection process is relatively simple, but it should not be underestimated. The legal ramifications of improper procedures can be costly and painful, if not for you, then for your boss or your boss's boss. We recommend the following guidelines to complete a good fire inspection:

- Contact the business owner if possible to schedule an appointment for the inspection. Sometimes surprise inspections are beneficial and necessary, but avoid them when you can. No business owner likes to be surprised.

- Arrive on time. Time is money to a business owner or operator. Treat this time as a very valuable commodity to them. Show respect.

- Look professional, act professional, and be *nice*. You are representing not only yourself, but also your organization and the fire service as a whole. Generally, people will react the same way they are treated.

- Introduce yourself and explain the purpose and objectives of your visit. If the business owner hands you off to someone else, never leave until you make an effort to close your inspection with the owner or operator of the business. They are responsible and accountable; let them share in your efforts.

- Ask questions if you have them. Important information can come in bits and pieces. In fact, you will likely find problems you did not know about, as long as you are methodical.

- Be systematic. Whether you start inside and work your way out or start at the top and go down or vice versa does not matter. Just be consistent, every time. Conducting inspections is like playing an instrument; you will develop a rhythm. Do it the same way everytime, and you will find being more consistent and organized from one visit to the next.

- Never stop at verbally requesting a correction of a code violation. Make every request in writing. We cannot begin to tell you the number of times we have had to start over because someone forgot to record exactly what was wrong or what needed to be fixed. If it is a dangerous violation, it is worth writing down. If it is a violation worth noticing, it is worth writing down. This not only helps you, but also helps the business owner, the inspector who has to follow you, and the attorney who has to defend you three (3) years from now. Write everything. You can never have too much documentation. Is this hard? Yes, but protecting everyone involved is worth the effort.

- *Sell, sell,* and *sell.* Sell every fix you have to a problem. Let the occupant know why and how it affects everyone. Generally, the problem exists for you because no one in the occupancy understood that it was a problem to begin with. You are the messenger of fire-safe behavior and issues. Be thoughtful.

- Always take the high road. Why get into a spitting contest? If you are right and the occupant is resistant, write it up, and explain how to appeal through your local appeals process. Smile and say thank you. Do not be confrontational.

- Make sure you praise the positive conditions. Everyone likes to be told good things or to be complimented. Take the time to share with occupants the things they are doing right.

As we have discussed in previous chapters, one of the fundamental concepts for a successful fire prevention program is the continued development of coalitions and building a community partnership for the implementation of a balanced fire and life safety program. One of the successful tools being used by many fire departments is self-inspections. In an effort to address the needs of our customers, we have taken their input and are launching our self-inspection reports.

A number of circumstances dictate the need for a self-inspection process and a variety of self-inspection reports are being used today. It is obvious that no matter which circumstances apply to your community or which type of report you are using, the end result is the same....... stronger fire and life safety coalitions, enhanced relationships as well as reduced workloads for fire prevention staff. However, the most notable result is improved efficiency in achieving compliance through the establishment of a positive relationship.

What Are Self-Inspections?

Many fire prevention bureaus have been faced with staff reductions without a reduction in the levels service needs such as inspections. Fire departments have been successful implementing a self-inspection program for certain occupancies determined by the fire department. The fire department's selection of the occupancies eligible for a self-inspection varies from occupancy classification, the results of the community risk evaluation, occupancy classification fire loss history, past violations, and complexity of facility size and operations. The methods for inspection implementation noted previously can apply to self-inspections as well.

What Are the Types of Self-Inspections?

Self-inspections are performed by the completion of a customized self-inspection form designed for the specific inspection tasks. For example:

Basic facility fire and life safety survey or specific inspection tasks:

- Fire damper testing (visual and or functional)
- Exit lights
- Emergency lights
- Fire extinguishers (visual)
- Fire suppression systems (visual inspections)
- Egress (door operational, clear paths, etc.)
- Fire hydrants (visual for obstructions removal of snow)
- Fire department connections (visual)
- Update of SARAH Title III reporting requirements to the AHJ
- Update of the Hazardous Material Inventory Statement (HMIS) required by the International Fire Code®

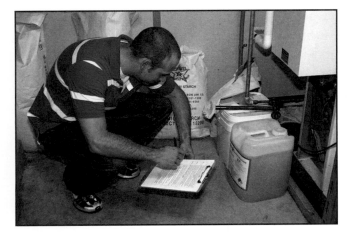

Figure 11.5 It is becoming more common for property owners to conduct self-inspections.

How Are Self-Inspections Used?

The self-inspection is usually completed by the building occupant or building owner (**Figure 11.5**). Once the form is completed, it is provided to the authority having jurisdiction to document the completion as well as note any areas to be corrected. It serves as a guide to walk the person through the inspection process.

How Are People Trained to Perform Self-Inspections?

Some fire departments have elected to provide self-inspection training seminars as a means to begin the coalition building process. Others have chosen to use web-based self-inspection training programs (See Chapter 14). Completion of the web-based training ensures the person is adequately prepared to perform the task. Most importantly, the person learns when there is a need to contact the local AHJ for assistance and to meet with them for a consultation. The most important emphasis of the training is the establishment of a coalition working for a common goal of enhancing the level of safety for building occupants and responding emergency personnel.

Does this eliminate the need for inspectors? The purpose of self-inspection is to provide a tool for the fire departments or building owners to enhance communication for an effective fire and life safety inspection process. The purpose is not to eliminate positions. This tool does not eliminate the need for inspections but allow a diversion of resource allocation to target hazards or emergent priorities. Some fire departments use the self-inspection tool for occupancies typically inspected by fire company operations. The fire company officer and crew use the completed self-inspection from the occupant as a discussion and learning mechanism with the occupant when they follow up with an onsite meeting. This allows the occupant to meet the first responders and learn from what was identified in the self-inspection process. The benefits include compliance through coalition building and the opportunity to divert limited resources to emergent tasks. The process serves as an excellent tool for compliance dialogue and education of the importance of ensuring a fire safe facility for occupants and emergency responders.

How Is a Self-Inspection Process Implemented?

As a mandatory requirement, some fire departments have elected to implement the self-inspection process. In some locations, a qualified individual who completes a self-inspection eliminates the requirement for the local AHJ to conduct a fire inspection. Some fire departments place limitations on the number of self-inspections that can be considered as a fire department inspection, and others perform random follow-ups.

Some jurisdictions make the self-inspection process voluntary with the goal of improving the coalition building process. Fire departments encourage the self-inspection as a means for the occupant to ensure the deficiencies are addressed prior to the fire department's inspection. As an incentive for self-inspection, some fire departments will not have a fire department representative visit the location for an inspection for a determined amount of time. In some locations, the option of a self-inspection by the occupant is based upon the fire inspector's discretion and violation history. For example, if the facility is relatively small with no complex hazards and there is no history of violations, the inspector may tell the occupant, "thank you for your continued effort and partnership for a fire-safe building." "Next year you will be eligible to participate in the fire department's self-inspection program.

Self-inspections are not intended to replace Fire Department Inspections; their intent is to compliment the process. Many jurisdictions simply don't have the resources to inspect every property every year; however, they have the authority to mandate a self-inspection annually. The inspection process noted above can also be applied for self-inspections.

In today's fire department culture of increasing costs and decreasing resources, it is our responsibility to the citizens we serve to find alternative solutions to ensure the public and firefighters are protected from the devastations of fire. New technologies offer increased efficiency and effectiveness. Including the property owner as part of the solution rather than part of the problem only makes sense. Combining a web-based self-inspection reporting system with a community partner is an excellent solution to enhance a community's fire and life safety program while building a strong community risk-reduction coalition. (See Chapter 14 for further web-based reporting discussion.)

Record Keeping

Documentation is very important. The inspector should document all of his or her activities in a format that allows easy recall. Inspectors can use paper for record keeping, although most departments use a computer database to maintain their documents. Most inspections require some type of signed verification of notice received, and many jurisdictions require a copy of this signature. Your department's legal authority can tell you how to handle these records. Records are an essential component of the fire prevention bureau's work. Organize these records in a systematic fashion that enables easy retrieval. The volume of records continues to grow throughout the years of the occupancy. It may be beneficial to archive the documents or to separate the documents into categories.

Chapter 14 further examines the use of information technology for managing the fire prevention bureau's records. Most fire departments have implemented an electronic means to collect data and perform inspections. Utilization of a web-based system for record management is becoming more and more prevalent.

Summary

Inspections are one of the most important fire prevention bureau functions. Inspections are important because they provide an opportunity to identify and correct hazardous conditions that could start a fire or contribute to the severity of a fire, endangering not only the occupants but firefighters as well.

During tough economic conditions, these services are often scrutinized and subjected to cuts in lieu of maintaining emergency response capabilities. Inspection functions are important to maintain, whether by dedicated fire prevention personnel or operations folks.

For a fire department to serve the public properly, it must establish and implement the key elements prior to beginning an inspection program. If a program is already established, review it to determine if any of those elements are lacking or need adjustment. The elements include:

- Analyzing the level of service the community needs and wants
- Determining and codifying enforceable regulations
- Determining the level of training inspectors will need
- Establishing organizational fire inspection policies and procedures

Next, the inspection priorities must be established. This means identifying those risks and hazards that are the biggest threat to your community and fire department and dealing with them at the appropriate level of detail and frequency to address the risk. Then, personnel must prepare for conducting the inspections. This may include very detailed research of various occupancies, processes, loss history, past inspection history, and so forth.

Once the appropriate research has been conducted, the inspection can begin. One important thing to remember is that any time you open your mouth or put pen to paper during an inspection, you cost the business or property owner money. That is your job and it is your responsibility. You are there to identify problems and have them corrected. However, never forget that what takes two minutes to document may take up to a hundred thousand dollars to fix.

Working with business owners often requires a series of stiff negotiations and reasonable compromise. We must always honor the intent if not the letter of the code; however, we must be empathetic and understanding enough to work with individuals in achieving compliance. When initiating the inspection, make sure your appearance is neat, your demeanor is professional, and your attitude positive. As far as the technical aspects of the inspection, you may follow any number of systematic approaches to going through a property or building. The key is to remain consistent so that you do not overlook or forget things. Also, make sure you allow a reasonable amount of time for the owner to abate the hazard or correct the violation. The decision on when something must be done is really the inspector's. However, it should be a negotiated agreement with the owner/manager unless there are extenuating circumstances, such as a resistant occupant or owner.

Document, document, and document. Never conduct an inspection without following up on the proper documentation. This will be vital if an issue ever comes up in court, and it will be equally important if the occupancy suffers a fire. All records will be pulled and examined. Any department that does not have good record keeping not only will be viewed as an embarrassment but could incur legal or civil responsibility due to litigation.

Chapter 11 Review Exercises

11.1 List the elements of a fire inspection. _____

_____

_____

_____

11.2 Is credibility important for a fire inspector? Why or why not?

_____

_____

_____

11.3 Why should a community have a fire inspection program?

_____

_____

_____

11.4 Identify the steps (discussed in this chapter) to conduct a fire inspection.

_____

_____

_____

11.5 What constitutes a good fire inspection? _____

_____

_____

_____

11.6 What is a self-inspection program?_____

_____

_____

11.7 What are the advantages and disadvantages of a self-inspection program?

Advantages: _____

_____

_____

Disadvantages: _____

_____

_____

11.8 Why is an efficient method to document inspections an important element of a fire prevention program? _____

_____

_____

11.9 What skills should a fire inspector possess? _____

_____

_____

11.10 How should a fire inspector prepare for an inspection?

_____

_____

_____

11.11 What equipment does a fire inspector need to conduct an inspection?

_____

_____

_____

11.12 Discuss the fire inspection guidelines identified in this chapter.

_____

_____

_____

11.13 Define a reasonable time frame for gaining compliance.

_____

_____

_____

11.14 Why is it important to have a good set of operational policies or guidelines for an inspection program?_____

_____

_____

11.15 What is the goal of an inspection program? _____

_____

_____

11.16 Explain what it means to, "Sell fire prevention?"

_____

_____

_____

11.17 How would you prioritize occupancies or facilities you choose to inspect? _____

_____

_____

11.18 List ten facility historical items that are important to identify prior to conducting the inspect of the history of a facility. _____

_____

_____

11.19 Is technical knowledge of the codes more important than customer relations? Why? _____

_____

_____

11.20 Identify three ways people tend to deal with conflict?

_____

_____

_____

11.21 Is it better to conduct your inspection from the outside in or from the inside out? Explain your answer._____

_____

_____

_____

_____

_____

_____

Notes

1. William H. Wallace, *Community Risk Issue: Structure Fires* (Colorado Springs, Colorado: Colorado Springs Fire Department, 1997).

2. This discussion of Mary Parker Follett's ideas is based on Donald F. Favreau, *Fire Service Management* (New York: Donnelley, 1969)

Identification and Protection of Hazards

Table of Contents

Key Terms

Key Points

1. Because the perception of dealing with events in a particular type of occupancy differs among departments, the perception of the hazard will differ, too.

2. Identifying hazards begins with a good origin and cause determination.

3. No matter how small the department, developing a target protection plan requires a high-quality team.

4. Without following through by reevaluating your target protection plan you will lose credibility and likely waste energy doing unproductive things.

5. Choosing the wrong fire protection systems can have disastrous results: large property loss and potential loss of life.

6. Suppression and control is the best all-encompassing method of achieving a fire protection and life safety goal.

7. Automatic fire protection systems only work as well as they are engineered: "Bad design, bad performance."

8. When not asleep or incapacitated, people do a good job of discovering fires; they just don't know what to do about them.

9. Controlling hostile fires before you have to rely on response truly gives the best bang for the buck.

10. From a code enforcement perspective, fire prevention efforts should take into account all important aspects of fire fighting.

11. As a fire safety professional, you are the one responsible for determining whether or not things are safe.

12. Much as a chain is only as strong as its weakest link, fire barriers are only as effective as their performance integrity.

13. Think of your options, think of the expectations and the possible outcomes and use good practical judgment with the tools you have available.

14. Considerable training and behavior modification must take place to give people the necessary confidence that they can and will survive a fire even though it may seem to be bearing down on them.

15. When applying the performance-based design approach, it is paramount to strive for a solution whose results will be equivalent to or more effective than the prescribed codes.

Identification and Protection of Hazards

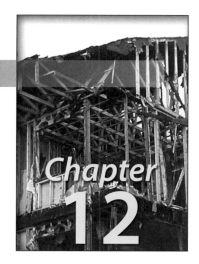

Chapter 12

Case Study

A large fire department experienced a serious fire in its community. The fire was in a structure in which one part of the building was used as a senior independent living facility and the other part was used as a hotel. The 12-story, unsprinklered structure was built in the 1950s with a basic fire alarm and a class II standpipe system installed. When the fire department personnel arrived on the scene, they were notified that there were two fatalities with over 80 rescues. Many people (156) were hospitalized — some serious, some treated, and some released. Several occupants were not expected to live the night. The standpipes did not work and fire hose had to be hand-jacked up to the fire floor on the 8th floor. Six apartment units were damaged. Everyone (197 people) from the 6th floor up would have to be relocated to shelters or other locations as the upper floors of the building were untenable for the next few months. While talking to the crews, all of them stated this was going to happen. Senior officers knew that it was a worn-out building and that it had no reliable fire protection equipment. They stated that it had been this way for the last 20 years.

1. Was this a facility that should have been worked on more diligently by fire prevention? How?

2. Given the very basic understanding you have of this facility and its occupants, what type of fire prevention or life safety plan might have helped?

3. How could you have worked with the owners to upgrade the fire protection systems?

Types of Hazards

The term **hazard** has a number of definitions. None of the definitions are especially specific, particularly in the context of fire prevention. To frame this chapter's discussion, we will define a hazard as a condition or element that provides a source of ignition for a hostile fire or that contributes to the spread and severity of a hostile fire. **Hostile fire** is a term for any unwanted or destructive fire.

Historically, the distinction between common hazards and specific hazards has been a subject of much discussion. **Common hazards** are typically referred to as those hazards that are common among many occupancies or locations.

Hazard — A condition or element that provides a source of ignition for a hostile fire or that contributes to the spread and severity of a hostile fire.

Hostile Fire — Any unwanted or destructive fire.

Common Hazard — Hazard that is found among many occupancies or locations.

An example would be the storage of flammable or combustible liquids, such as paints, lacquers, or cleaners, which can be found nearly anywhere (**Figure 12.1**). **Specific hazards** are those hazards that are isolated to specific operations or locations, such as cryogenic liquid storage tanks at an industrial facility or quench tanks as found in heat treating facilities. These installations are not common to all buildings, but instead are specific to occupancies or locations with particular needs or processes (**Figure 12.2**).

Figure 12.1 This photo shows a common fire hazard created by numerous extension cords that are being used for permanent wiring, and too many being fed from one circuit and various wires being supported on fire sprinkler piping. *Courtesy of the Colorado Springs, CO Fire Department*

Figure 12.2 This photo shows a special fire hazard that consists of a liquid oxygen tank located immediately outside a hospital, along a fire department access route. *Courtesy of the Colorado Springs, CO Fire Department*

Another method of identifying hazards is by occupancy type. While this is more typically used to classify hazards, identification can be accomplished based on the general subcategories as well.

However, this method can be inefficient, as different localities and jurisdictions may perceive types of hazards differently. Hazards can be items that are small and specific within a building, or they can be an entire building, plant, process, or geographical area.

An example could be found in a small town with a single high-rise protected by only a couple of fire stations. This high-rise could be the community's most significant hazard or target hazard. New York City, however, may not consider high-rises with much significance, as they have a large number of these buildings that have incorporated various types of fire protection systems as well as a large resource-supplied department that can handle such fires. Because the perception of dealing with events in these locations is different, the perception of the hazard can be different, too.

Because this book is created to be more of an administrator's philosophical approach to fire prevention applications, this chapter will not delve into the specifics of hazards within buildings or processes. Numerous other texts already address these issues.

Examples include the *Fire Protection Handbook®*, from the National Fire Protection Association® (an excellent resource); the *Fire Inspection and Code Enforcement* manual, from the International Fire Service Training Association; and *Fire Prevention Applications for Company Officers*, from Fire Protection Publications. These all provide expert-level explanation and analysis down to an apprentice-level look at these types of hazards and issues.

Instead, we will use global or macro-level examinations to look at hazards. This holistic approach is an overall community hazard identification and protection and is not merely a review and analysis of the single exit sign that is nonfunctional in a motel hallway. The study is to be more global in the perspective of how the entire motel is viewed and protected as a hazard and risk within the community.

Identifying the Target Hazards

Assume for a moment that you are the fire marshal of a mid-sized American city and that you recently have been placed in this position. This new position is the opportunity of a lifetime. You have an excellent staff, you work for a progressive department, and the community has liked your organization's performance for a number of years. What do you do now?

One of the first things you must do is to define exactly what fire hazards you are trying to protect people from. That's a big task. If you are stuck, list the top five causes of structure fires in your community. For purposes of illustration, list five causes that are common to many communities;

1. Unattended cooking

2. Arson

3. Discarded smoking materials

4. Electrical shorts

5. Children playing or experimenting with matches or lighters

Now that we know the cause or rather how your structure fires are starting, let's determine the types of structures or occupancies where these fires are occurring. Once more, for purposes of this example, let's list the following occupancies that are common to many communities:

1. Single-family dwellings

2. Apartments (multifamily)

3. General business offices

Now let's readdress the question. What are you trying to protect others from? The answer to this question can define your community's target hazards. The data in the example may indicate that your community does not know how to cook safely, that you have bad people running around town using fire as a weapon, and that your public is at risk both at home during the night and at work during the day. This type of approach makes identifying the hazards fairly straightforward.

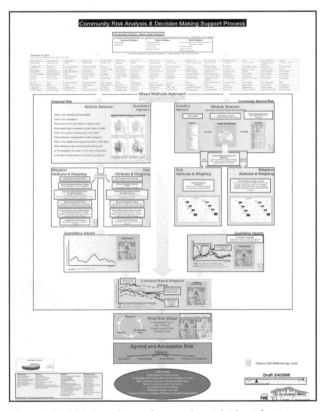

Figure 12.3 This is a photo of a complex risk identification process flowchart. While detail cannot be seen, you can observe the complex number of steps involved in identifying risk. *Courtesy of the Colorado Springs, CO Fire Department*

Colorado Springs conducted a detailed survey of various fire service industry professionals and others who were considered Subject Matter Experts (SMEs). These SMEs included battalion chiefs, company officers, firefighters, and others. They were asked various questions like what types of occupancies are "worse" than others relative to risk, what is the realistic threat, how many of a given type of occupancy are there, and what was the difficulty in extinguishing a hostile fire in those locations. This involved a process similar to a Delphi weighting process. The process involved asking comparative, weighted questions of one item relative to another to all of the SMEs cited above as well as citizens, business leaders, which provided weightings of opinions based on knowledge from everyone interviewed, driving answers that focused on fewer and fewer elements (**Figure 12.3**). This eventually developed decisions or choices on topics or points, basically occupancies. These choices then provided specific information relative to real or perceived risk, depending on who provided the answers. This established the areas for concern relative to others. The SMEs then examined the locally adopted fire code and evaluated how well it addressed or regulated these selected choices. This collective data was then analyzed, categorized, and placed in a risk matrix like the example shown in **Chapter 6**, **Figure 6.5, page 135**.

The higher weighted risks are placed in the red box, while the lowest risks are in the green box. The gradients of color from green to red represent various increases in risk moving from green to red. One of the interesting outcomes of this study was apparent in the representation of lowest risk being educational occupancies or schools. Intuitively this may not ring true; however, this was derived from the fact that educational occupancies are typically occupied only during the day, have fire alarm systems installed, and have a population that routinely practices fire drills and knows how to respond to these emergency signals. It is also evaluated and inspected many times through the year, frequently by different agencies which assist in keeping hazards to a minimum. All of the newer schools in this community are also sprinklered. Based on all these findings and the relative comparisons of other occupancies, it was apparent that educational facilities were the best protected facilities and trained group of occupants regulated by the fire code. Therefore, it was placed in the lowest risk category.

Many references list specific conditions related to hazards in particular operations or occupancies. While these are vital to your overall fire prevention mission, you must first deal with your immediate problem: The community signs and symptoms you have identified.

Without accurate data collection and analysis, you will frequently be shooting in the dark. How can you look at a problem if you cannot see it?

Good origin and cause investigation is a crucial part to good data collection relative to effective or ineffective fire prevention efforts. So, identifying hazards should include thorough origin and cause determination which must be factored into your overall risk analysis **(Figure 12.4)**. Strengthen your data collection if you want to hit the target hard and truly make a difference. In Chapter 14, we will discuss using technology to your advantage when collecting and using data.

Think Strategically in the Target Identification Process

Think back to Chapter 4, "Development and Implementation of Fire Prevention Bureaus." It discusses the importance of strategic planning in establishing the fire prevention bureau. What does it mean to think strategically in the target identification process? Some people have compared it to viewing the world through a telescope, looking way down the road to see where you are going, and understanding why you need to be there. So, thinking strategically in identifying your target means knowing very clearly your community's unaltered direction, the objectives needed for course correction, and the methods and resources you will use to attain your goals and objectives. Planning for the resources you need to have available to accomplish your goals is essential for your success and future program sustainability.

Figure 12.4 This photo emphasizes the need for good origin and cause investigation. Many aspects of fire code compliance, code mitigation aspects, and identification of deficient regulation are important in modifying risk. *Courtesy of the Colorado Springs, CO Fire Department*

Done regularly, strategic planning allows you to deal dynamically with changes in identified targets. What was it that Winston Churchill said? "Plans are of little importance, but planning is essential." People, technology, and hazards change all the time. As such, we also need to be flexible and adjust to those changes by thinking long term. Thinking strategically will keep our fire prevention efforts appropriate to our jurisdiction's ever-changing circumstances.

Develop a Target Protection Plan

How do you get started? Let's return to our previous scenario, in which we determined that single-family dwellings, multifamily dwellings, and businesses were our target. With this in mind we start our plan.

The planning process consists of the following seven main steps:

1. Create or gather the team who will do the work. The team should include stakeholders from the identified hazards as well as SMEs.

2. Create, modify, or reaffirm with the team your mission statement based on the task at hand.

3. Establish the strategic goals you want to accomplish for the specific hazards identified. These goals should be realistic and appropriate.

4. Work with the team to establish the specific strategies you will use to accomplish the identified goals.

5. Print the document and have everyone sign it to affirm their agreement with and commitment to its content.

6. Implement and evaluate over time, just as you would any program management cycle.

7. Make adjustments as necessary based on the evaluation, and evaluate again.

Little is possible without an engaged and high-quality team. Even if your fire department is small and you are the chief, fire marshal, inspector, or training chief, you will require help reducing the risks with a team, small though it may be. Involving representatives of the potentially affected stakeholders in the process from its inception is the best way to develop good synergy and cooperation. If you have naysayers, give them the opportunity to "own" part of the solution. That way, the community and the stakeholders all have some ownership in the success of the plan.

Figure 12.5 This photo shows strategic planning in process that involves community members stakeholders and fire department staff. *Courtesy of the Colorado Springs, CO Fire Department*

Identify Your Resources

Once you have worked out all of your strategies, you need to identify how to work on them. Think outside the "box." Do not lock yourself into looking only within your department **(Figure 12.5)**. What other stakeholders in the community will benefit from this program? What stakeholders can be more effective at doing certain tasks?

We identified apartments as a target hazard. Do you think the local apartment management association has something to say about their members' property getting burned up every week? Sure they do! Involve them. Consider all potential interested parties and those who benefit or can assist you.

Look for funding sources, such as grants and philanthropic organizations, to help you. Include local civic centers, YMCAs, and any place where owners or occupants of your target hazard may live, work, or play. People will help if they know what to do and have direction. By establishing your strategic plan, you will show them you are serious and assure them they can participate knowing they are part of an organized effort. Leverage the other people on the team. Generally, people love to help and if you build relationships and create coalitions, you will discover just how powerful enlisting other people can be.

Involve Your Team

Brag about your team. Use as much publicity as possible. Get your fire and life safety educators, your public information officers, and your chief involved. Good press gives the effort a great boost **(Figure 12.6)**. Let the team see their progress regularly. Show them off, and they will do wonders for you and the effort. This is especially true for organizations outside the fire department. They love to be recognized standing by a fire truck or next to the fire department chief or you, the fire marshal. Have fun and celebrate the successes, no matter how slight. Once the ball gets rolling, the path is generally downhill. It will pick up speed.

Figure 12.6 This photo shows a media highlighted event where children were recognized for performing lifesaving skills that they learned from fire department fire and life safety educators. Publicizing these events provide lots of community recognition and support. *Courtesy of the Colorado Springs, CO Fire Department*

Reevaluate the Plan

The saddest way for planning to end is getting everyone motivated to do the work and take off, only to see them fall flat on their faces because no one did any follow-up. Planning and execution are just like swinging a bat at a baseball: follow-through is very important. Without it, you will lose credibility and likely waste energy doing unproductive things. No plan is perfect. Be ready to make mistakes. Always look for better ways of doing things. Don't be afraid to blunder. Those who do not blunder at some point usually are not doing anything.

Solutions for Control

When you have finished your plan and have provided for reevaluating and revising it as needed, the hard part is done. Now you use your knowledge of fire protection issues as a tool to solve the problems you have identified, within the targets of opportunity you chose. When we examine methods to protect a hazard or a series of hazards, we are assuming a fire will occur. We need to think of where the fire will most likely occur and how the fire will progress. We also need to consider the fire's impact on any occupants who are present, including our crews.

Engineering is the best, although not always the cheapest, method of dealing with identified hazards. Eliminating as much of the "human factor" as possible will keep incidents to a minimum. Remember the three *Es*: Engineering, Education, and Enforcement. As a species, human beings have engineered solutions to natural or man-made problems ever since we came into existence. Down through the ages, engineering has been in the forefront of history. Its task is to put knowledge to practical use.[1] The objective of fire protection engineering is to find ways of preventing or mitigating hostile fires. In this chapter, we concentrate primarily on mitigation as we previously addressed prevention issues. Because of our limited resources for prevention and the recognition that prevention will never be 100 percent successful, we have to plan and design so as to mitigate damage and injury when fire occurs.[2]

You can use many different approaches and systems, and some will work in more than one scenario. However, a select few generally will work in any given situation. It becomes paramount, then, that as a fire protection professional and as a manager of a fire prevention bureau, you must become the resident expert in understanding the best approach for a given application. In recent economic climates, business supportive solutions and economic impact are key to whatever system you choose. Being keenly aware of your local political environment as well as evaluating your professional expectations can sometimes be a struggle. Stick to the plan, make sure you have reasonable buy-in, and you will succeed. Let's analyze the ways we can mitigate fire situations. For our purposes, we will use Webster's definition of the term **mitigate**: to make less harsh or hostile or to make less serious or painful. Hostile fires generally occur due to one of three main causes: men, women, or children. That said, we need to recognize how we as human beings behave in our environment (**Figure 12.7**). On average, we sleep nearly a third of our lives, we work nearly a third of our lives, and we do other things for the remaining third. Our behavior in relation to hostile fire is much different, depending upon which of these three phases we are in at the time of a fire. Keep these phases in mind, reflecting on how they interact with the engineering controls we will discuss next. After we talk about engineering aspects, we will discuss behavioral issues.

Mitigate — To make less harsh or hostile or to make less serious or painful.

Figure 12.7 Stacking large amounts of combustible wood pallets next to lubricating oil within a building is not smart. People's behaviors and lack of awareness are elements of risk we must fix or mitigate. *Courtesy of the Colorado Springs, CO Fire Department*

Some basic engineering designs used in controlling fire can be divided into two types: active and passive. Some examples of each are:

- Active
  - Suppression and control (automatic fire sprinklers, hood and duct extinguishing systems, etc.)
  - Detection and alarm (smoke detectors, building alarms, etc.)
  - Fire department emergency response (last resort)
- Passive
  - Product manufacture and performance control (automatic shutoff timers on coffee makers, ignition pilot control on water heaters and furnaces, etc.)
  - Compartmentation (fire walls, fire doors, fire and smoke dampers, etc.)

Suppression and Control

Obviously, engineering or designing fire out of our environment is best. However, because we have limited financial opportunities for that degree of design, suppression and control are one of the best all-encompassing and economical methods of achieving fire protection and life safety goals. By using automatic fire suppression methods, we not only benefit from early detection that can rapidly alerts occupants for evacuation and summons emergency services, we also take an active step toward controlling a hostile fire during its incipient stage.

From an engineering perspective, many different types of fire suppression systems are available. They all work in basically the same way; it is the extinguishing agent that makes each of the systems different. Also, specific hazards require specific types of components. Some examples are:

- Automatic fire sprinklers
- Deluge systems
- Water mist systems
- Clean agent gases
- Dry chemical agents
- Foam

For more than a century, automatic fire sprinklers have protected millions of dollars in property and have, in fact, saved many lives. These systems have also provided significant alternatives to designers and builders as trade-offs for more hazardous designs, processes, or uses. This has provided greater flexibility for emerging technology and different construction products than ever before imagined.

All of these systems are relatively complex, and you should study them in detail. We strongly recommend the IFSTA manual, *Fire Protection, Detection and Suppression Systems* (5th edition) for more information and a fuller understanding of how these systems and devices are utilized and function. Water systems provide the best all-around protection; however, they have disadvantages

Figure 12.8 This photo shows a high expansion foam test in a large aircraft hangar. Special agents are ideal for particular types of hazards. *Courtesy of the Colorado Springs, CO Fire Department.*

that may warrant using a different system. For example, water-based systems in acid storage rooms may not be a prudent choice. Clean agent systems, such as carbon dioxide or FM 200, are excellent choices for these types of applications. Computer rooms or other high-tech hazards also benefit from these systems. In some applications, dry chemical extinguishing systems are the best choice for ease of installation, cost effectiveness, and reliability. Dry chemical systems are especially good for dip tanks, painting, or other coating-process hazards. Foam systems are good for protection of multimillion dollar aircraft (**Figure 12.8**). As the expert, you should become familiar with all the systems and their agents to make the best choice possible for protecting the particular hazard.

Although all of these systems do an excellent job of automatically protecting hazards during an unwanted fire occurrence, you must remember that they will be effective only if they are:

- Matched for the appropriate hazard
- Designed properly
- Tested and maintained regularly
- Tied into fuel or heat shutoff devices where appropriate

Similar to a computer database, where information technology folks say, "Garbage in, garbage out," automatic fire protection systems only work as well as they are engineered: "Bad design = bad performance."

Detection and Alarm

Fire detection and alarms are important elements of fire protection. Detection systems may initiate evacuation, fire suppression system operation, equipment shutdown, and door closure, and not to mention the notification of fire suppression forces for quicker response. The alarm may go to a geographically removed monitoring agency for notification at a corporate level, an on-site monitoring location, or simple notification of the occupants within a building or area.

The two principal means of detection are people and automatic devices.

When they are awake and in an area where they can see or smell a fire starting, people are some of the best detection devices available. For example, people are the principal means of detecting wildland fires. Although sleeping people will not generally detect the products of combustion, those who are awake generally are alert to the scents of combustion byproducts and to visible smoke and flame. In these instances, people can usually be relied upon to activate manual alarm devices such as pull stations, public announcement, or other means of notification, assuming these devices are available.

The drawback is that while humans are good at identifying hostile fire, they are not always effective at protecting lives even while people are awake and alert. Many of the larger fire loss events (dollar and life loss) occur during the hours that people are awake, at work, or out and about. While human detection appears to be fairly reliable, people's behavior does not coincide with our senses. The point being, when not asleep or incapacitated, people do a good job of discovering fires; they just don't know what to do about them.

The second means of fire detection is automatic. This may be done in any number of ways:

- Smoke detectors
- Heat detectors
- Suppression system activation by fusible links
- Ultra violet or infrared
- Sound (although still being researched, some sound-sensing security systems have detected fire)
- Air aspirating detection

These methods all provide good early warning and notification, but each is tailored to fit a relatively specific situation or fire event. Excellent resources are available that provide greater detail on system performance and design criteria. If you wish to use automatic detection as a means of hazard protection, be certain these systems are installed in accordance with the applicable National Fire Protection Association (NFPA®) standards.

Whether a fire detection and alarm system relies on people or automatic devices, its purposes are the same:

- Alert occupants of the condition requiring appropriate response (notification, evacuation, fire suppression response, or other)
- Initiate mechanical operations (smoke removal, elevator recall, fire door activation, notification appliances such as horns and strobes)
- Transmit a signal to initiate emergency fire response

A firefighter's basic training prioritizes two fundamental tasks in this order:

1. Life safety
2. Property conservation

New and upcoming notification systems that may fall on fire departments to evaluate and review are called a Mass Notification System, or MNS. These systems are becoming popular in military installations as a means of communicating various messages which not only include fire or other emergencies but weather alerts, terrorist alerts, advisories, and the like. These systems are now regulated in the same NFPA® system as fire alarms due to the need for supervisory controls and monitoring. These systems, however, require that a detailed risk analysis be conducted for the facility prior to designing these systems due to the number of systems and communication coordination involved. We encourage you to read the requirements very carefully, as this can be a very involved design and installation process.

Engineering methods must conform to the same ranked priority. If people can get out of harm's way, "stuff" can be rebuilt and or dried out. Life safety is most important and is the objective that our society and we as fire protection professionals aim to achieve above all else. Note that in our priorities of outcomes, emergency response is last. This is not to say it is not important; however, in keeping with this book's overriding theme, response after an inci-

Figure 12.9 This is a classic instance where prevention was insufficient and we are relying solely on operational intervention to protect property and suppress the fire. A little too late. *Courtesy of the Colorado Springs, CO Fire Department*

dent has occurred is our last resort and, therefore, the least preferred option. The message: Control hostile fires before you have to rely on response. It truly gives the best bang for the buck. It is far more economical to prevent a fire than to allocate resources to suppress it.

Fire Department Operations

From a fire prevention standpoint, the lowest ranked element of active protection for hazards enlists the help of our operations division or emergency responders. It is not ranked low because of resource capabilities or expected performance. It is ranked low because it comes into play only when a significant event has already occurred and we are trying to reduce a harmful or destructive situation from escalating **(Figure 12.9)**.

Conscious risk management decisions allow us to rely on reactive remedies rather than on engineered solutions. While the three "*Es*" (Engineering, Education, and Enforcement) are our motto, one additional choice is emergency response. The reality is, calling 9-1-1 to report a fire and waiting for the responders' arrival is no different than installing a very slow-acting suppression system. You lose quick response, small-fire containment, and automatic notification. However, for a community that pays several hundred thousand to a million-plus dollars a year for a fully staffed fire station, this equipment and trained staff should count for something. For that reason, it is natural to expect this mechanism to perform and do so reliably. That expectation should be clear and well defined, however, only as one part of the overall systems approach.

From a code enforcement perspective, fire prevention efforts should take into account all important aspects of fire fighting, buildings, processes, or conditions that fire prevention bureau reviews need to perform to ensure code compliance. This work should strive to minimize the risk to fire crews and other responders. Higher risk in some circumstances is unavoidable and acceptable. Firefighters do not perform their duties expecting to be totally safe and protected; however, they do expect reasonable safety while they do their job going into places that other people are fleeing. Our success as fire prevention experts not only rests on the design and installation of various systems or devices to make fire fighting easier, but also in training and educating the line fire fighter. We are the experts on those devices and processes and, therefore, should share our knowledge with those we work with and rely on to provide an important component of our fire protection system. We expect the fire suppression folks to have a fire prevention attitude. In return, we fire prevention folks must think like the fire suppression folks who will be in a smoke-filled environment at 3:00 A.M. trying to decipher the fire alarm control panel that could be inadvertently placed in a closet.

To use suppression resources properly, all fire prevention staff must be intimately familiar with the equipment, training, policies procedures, and capabilities of their emergency crews. They must continually factor integration and application of this combined resource into all engineering and educational applications of hazard protection. In a fire department systems approach, we are all in this together.

> *Example:* You go to the store and buy decorative lights for the winter holiday season. You look on the box and it has a UL stamp. You assume that it is UL approved (which it is not—UL only lists, they do not approve) and that it must be safe. However, if you were inclined to do further research in the UL directories, you may find that for this particular string of lights, UL tested only the electrical cord for electrical protection from power loss through poor insulation (which could lead to shorts), improper connections of the wires to the blades that go into an outlet, or an incorrect ratio length of a wire's length to its diameter (so as not to overheat based on the projected electrical load). However, the out-of-country manufacturer decides to install cheaper bulbs that use more power and burn hotter than initially planned. Since the bulb may not have been part of the required test, changing it may not impact the listing. So, you thought you had a good set of lights, but it is actually unsafe because the increased current flow will heat the wires and the bulbs will overheat because they are different than initially designed or tested. While the box has a listing stamp, it does not contain the product that actually was tested.

Product Manufacture and Performance Control

Products that we use and misuse every day of our lives play an important part in fire protection efforts. Thankfully, in this country, we have various independent testing laboratories like Underwriters Laboratories and Factory Mutual. These agencies test various items (power cords, heating devices, etc.) following specific test procedures and criteria. If an item meets the test criteria and passes the test as designed, it is then listed or approved (Chapter 2). A testing agency's stamp indicates the item passed whatever test it was submitted for. Do you see the problem yet? Often only one of the product's many components may have been submitted for and passed the testing process.

In addition to the numerous testing labs, many watchdog groups, such as the Consumer Product Safety Commission, regularly identify problem products, assist in issuing recalls, and verify testing procedures to be sure they match the criteria. Our point is that products are comparatively safe in America, but you must be mindful of which products are actually listed in part or whole and what that listing means. Do not assume anything! As a fire safety professional, you are the one responsible for determining whether things are safe. **(Figure 12.10, p. 302)**

Generally, the applicable code will specify using or installing only those products that are listed by some outside testing agency, at which time you, as the authority having jurisdiction (AHJ), then approve their use or installation. Learn this process well as lives and property may hinge on it. You would be surprised at what things are used for, aside from their intended or designed purpose.

Figure 12.10 Failure to properly maintain equipment in accordance with its listing resulted in an explosion and fire. *Courtesy of the Colorado Springs, CO Fire Department*

Hazard protection by product manufacture (for example, not selling children's sleepwear unless it is made entirely of flame-retardant material) and performance control (for example, designing products and instructions to meet specific safety standards) is widespread. As a result of numerous large class action lawsuits, companies in the United States are generally very careful about what they produce and sell. When people in this country are hurt physically or financially, they are quick to hold manufacturers or producers accountable, as they should. This has created a fairly reliable fire-safe market that deals in products and goods. Independent testing and certification is the cornerstone of these lawsuits, along with aggressive stances by consumer protection agencies and advocates. So while not foolproof, you can reasonably rely upon a general understanding of product control and independent laboratory testing for addressing this means of engineering control.

Compartmentation

Compartmentation — The use of passive (and in some cases active) protection features to prevent fire spread.

Compartmentation is critical to the survival of any building. It deals with our ability to engineer and install passive (and in some cases active) protection features to prevent fire spread. These features include: firewalls, fire doors, fire fire-resistant glazing, penetration protection, smoke and fire dampers, and smoke management systems. The objective of compartmentation is to restrict hostile fire to one area rather than allowing it to migrate or propagate to another location. This is done through the use of rated assemblies that meet and have passed various test criteria. An **assembly** speaks to a particular type of construction method that details specific types of materials, their specific manufacturing or installation details, and components such as specific screws, nails, ceiling tiles, and drywall (an assembly typically is viewed as a wall, a ceiling/floor, etc.). You can obtain a detailed explanation of this from the IFSTA *Fire Inspection and Code Enforcement* manual and the IFSTA *Building Construction* manual and the NFPA *Fire Protection Handbook®*. Communication and power cables are being fed through buildings from top to bottom and side to side. The holes that

Assembly — A particular construction method that details specific types of materials, their specific manufacture or installation, and components.

are punched in building assemblies to accommodate these cables violate the assemblies' integrity. Much as a chain is only as strong as its weakest link, fire barriers are only as effective as their weakest performing element. Maintaining the integrity of these assemblies is essential to providing adequate protection for evacuation and safety of occupants. Without careful adherence to requirements and specifications as spelled out in the listed assembly or wall testing criteria, you have no verified confidence in an assembly's ability to withstand a fire.

Building collapse from fire is also a major consideration. The intent of any structural assembly is to withstand thermal insult and remain structurally sound, thereby resisting collapse. If collapse does occur, it should not contribute to the additional destruction of anything more than what is already affected. The intent of compartmentation is to hold a fire in check. This must be done, relative to the design constraints, regardless of the anticipated fire's effect. Drywall is a good example of our progress over time. Because of its inherent flame retardant properties, this low-cost, easy-to-install product has done a tremendous job of containing hostile fires to their rooms of origin long enough for firefighters to arrive and extinguish them. Before drywall's invention, fires moved easily into voids in walls and ceilings, advancing rapidly unchecked throughout the building. In many cases, flames moved so fast that fire department personnel were powerless to stop their spread until the fuel was all consumed, meaning the building burned to the ground. We must consider all other products and options that facilitate the same end result. New products will come out and old ones will remain. Careful attention to how they are installed and used will provide a much greater arsenal for fire protection. Careful consideration to the manufacture's installation instructions is always paramount.

Occupant Safety

Intertwined with the proactive approach of compartmentation are certain reactive measures that play an important part in hazard protection and mitigation for occupants. These measures fall under the general category of occupant evacuation. As mentioned, protecting lives is the fire department's most important priority. To do this, people must be made aware there is a problem and then know what to do to protect themselves and others. Assuming that a building or structure was properly constructed adhering to all codes and standards, the next step is to provide a means for getting people out of harm's way **(Figure 12.11, p. 304)**.

The three principal methods for doing this are:

- Evacuating
- Providing an area of refuge
- Defending in place

Each of these methods involves specific expectations and uses. Sometimes a combination of methods may be used; however, combined methods generally are suited for specific types of occupancies. As the AHJ, you may find a need to integrate options and consider any of these methods for unique uses. Again, you are the resident expert.

Figure 12.11 Evacuation of occupants may require proper ventilation and smoke control and removal such as verified in this large-scale test. *Courtesy of the Colorado Springs, CO Fire Department*

Evacuation

As we have discussed previously, evacuation has two components:

- Notifying occupants

- Providing safe and unobstructed paths or methods of escape. **NOTE:** This can be interdependent with the mass notification systems we discussed previously.

Both of these components depend heavily upon the educational component of the three *E*s. Notification may be more important than you ever considered. After the first World Trade Center bombing, it took over eight hours for all the occupants to be evacuated from the upper portions of the buildings. Obviously the sooner people know of a hostile condition, the sooner they can leave. Accurate, prompt notification is especially important in large facilities.

After people have been notified of a fire emergency, their next critical need is a protected path of egress. Protected path of egress basically means a hallway, corridor, or stairwell that is constructed to resist fire and smoke penetration for a minimum amount of time. This minimum time should be long enough to allow anyone who could be moving through the protected path sufficient time to get from harm's way to a safe area typically referred to as a public way. This is usually accomplished by providing a minimum of two ways out.

Most exiting systems (means of egress) are protected by a minimum two-hour construction. This means that the walls, floor, and ceiling assemblies are constructed in a manner that has been proven to pass a fire severity test based on the standard fire temperature curve as detailed in ASTM E119 **(Figure 12.12)**. Keep in mind that while this test curve is a good scientific constant, it was first published in 1918. Today's phenolics, resins, and plastics were not manufactured or available then, and therefore could not be tested for their effects on the outcome of a test fire. The tests using the time temperature curve today for certifying a two-hour wall is not necessarily representative of today's fire experience because fuels are different than when the original test was conducted. However, the benefit is that it is a standardized test which provides relative comparative data on the results that can be replicated repeatedly.

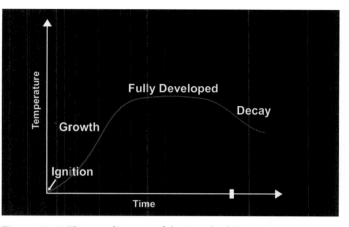

Figure 12.12 This is a diagram of the Standard Time Temperature Curve. It depicts the speed with which temperature rises over time. *Courtesy of Dan Madrzykowski, NIST*

While protecting occupants from fire and smoke exposure is important, the paths of egress must also be compatible with the occupants who must use them. If you have wheelchair-bound individuals, for example, five flights of stairs likely are not a good choice for emergency egress. Using elevators as a part of the emergency egress is not preferred, due to the compounding problems of moving people in a steel box down a chimney with fire directly below; however, in some cases elevators may be required. If a fire is above these occupants, it may not be unreasonable for them to evacuate using elevators. Understand, we are not endorsing elevators as a means of egress. We are communicating the variables from building to building and from occupant to occupant. The world is not perfect and neither will all of your fire protection solutions. Think of your options, think of the expectations, and the possible outcomes and use good practical and reasonable judgment with the tools you have available.

Code books cannot be written for every type of structure, hazard, industrial operation, or process. You will need to determine the best method currently available and exercise good judgment. Be prepared to testify in court to why you did what you did and how it met the intent of the code requirements.

Provide an Area of Refuge

Planning and designing this area requires a lot of thought. Basically, this is an area or series of areas that are separated by sufficient fire-resistive construction that we can move through or into for safety. An example could be a high-rise building where we encourage people to move above a fire floor rather than passing down through an active fire floor. This process can be contrary to many people's instincts. Another example could be a large warehouse or manufacturing facility where people may migrate to a different part of the facility to get out of harm's way.

This concept is not unusual and is widely used in warships. The U.S. Navy understands well the idea of compartmentation and isolation. If a fire is in one compartment, everyone who can do so moves to another compartment, and they close off the fire area letting it burn itself out. This happens while others are protected in areas of refuge, which may adjoin the fire compartment. Areas of refuge are also key for defending in place.

Defending in Place

Defending in place is generally thought of as a means of protecting people or processes without significant relocation or evacuation. This includes not only areas in buildings but areas in the wildfire environment as well. Terrorism is another prime example of emergencies that might require defending in place to protect from biological, radiological or toxic agents. Let's look at a wildfire scenario as an example.

Example: A large elementary school with some 500 children and around 45 staff members is nestled in a residential neighborhood in the middle of an expansive wildland/urban interface. A fire occurs downslope of the school, moving quickly uphill, uncontrolled by the available fire fighting resources. Good emergency planning on the part of the school in cooperation with the local fire department has prepared staff, students, and parents to expect either evacuation or defending in place in the event of a wildfire emergency.

Evacuation of the school's occupants is the method of choice if there is adequate time (one to two hours) to notify parents and secure transportation. Buses do not just appear; they are on strict schedules. Likewise, parents may not be able to get from work to school quickly enough to get their children. The evacuation plan also calls for vegetation mitigation around the exterior of the school so fire crews will not need to remain in place trying to protect the property.

Defending in place is the method of choice if there is an immediate threat (less than an hour), and people have insufficient time to leave the premises lest they become exposures themselves. Staff and students are instructed on how to behave. They move to interior or remote portions of the noncombustible structure to get away from windows and doors. The maintenance staff shuts down ventilation systems and closes all outside windows and doors (**Figure 12.13**).

Figure 12.13 This photo shows a grass fire adjacent to an elementary school where children were ordered to stay inside, defending in place, until the fire had been controlled. *Courtesy of the Colorado Springs, CO Fire Department*

As for evacuation, vegetation management must be sufficient to protect the structure from damage in the absence of emergency crews. A tactical operations agreement with the local fire department ensures that fire crews will respond to the school to establish an anchor for defending the site. This plan also incorporates careful education and participation of parents to understand that the children will be protected in the school and that they are not to try to get their children until advised by the local fire department. This is very important yet extremely hard to achieve.

Due to the proximity of the fire, fire crews respond to the location and set up a defensive perimeter in accordance with the plan. The staff moves the children to the core of the building, and the fire moves in proximity of the school. As expected, the heat pulse rises rapidly, consuming all vegetation downslope, throwing embers and flaming debris into the defending fire crews and the structure. Smoke begins to enter the structure through various cracks and openings, and the sound of the fire is deafening from the outside. Children begin to sob, teachers get anxious, and everyone has a harder time breathing. However, after about 10 to 20 minutes, the commanding fire officer enters the school and gives a good report that the flame front has passed. They are mopping up hot spots, but everything is okay.

This example of defending in place is applicable to many other settings, such as hospitals, nursing homes, prisons, high-rise buildings, and other locations where relocating or moving the occupants is not feasible. Defending in place requires careful preplanning and predesign work.

Communicating specific instructions on how and where to defend in place may become part of a Mass Notification System that was discussed earlier. These systems are similar to a fire alarm system, but enable greater communication and instructions of what to do and when. They may use digital phone, pager, tactile devices, or message boards to communicate messages and instructions.

To withstand a fire and its effects, various components must be included in and potentially around the structure and regularly maintained. Also, a considerable amount of training and behavior modification must take place in order to give people the necessary level of confidence that they can and will survive a fire even though it may seem to be bearing down on them.

Performance expectations, such as time of tenability or survivability, must also be considered. Designers and planners should understand a fire's expected duration and what the defended location must be able to withstand. For example, if a fire in one portion of a hospital is predicted to last approximately twenty minutes, planners could feasibly design the defended portion to remain protected a minimum of two hours. This provides a level of safety, or safety factor, above and beyond what is typically expected and allows for an unpredicted event that may last somewhat longer due to unforeseen circumstances. Looking back to the school example, we can see that by using noncombustible construction, managing the surrounding vegetation, and working with the fire department, tacticians had considered how the fire would behave and knew that an intense fire exposure of 10 to 20 minutes was survivable. In this instance, things worked out fine.

Performance-Based Design

Performance-based design is a relatively new and popular growing concept being debated and discussed extensively (see also the corresponding discussion of performance-based codes in Chapter 3). It centers on risk analysis, hazard assessment, consequence identification, and decision-making. Insofar as code books do not account for all types of structures, industrial operations, and processes that may be encountered, performance-based design is a tool to address unique or complex situations. While an in-depth understanding of performance-based design is well beyond the scope of this text, the following example offers a generalized illustration:

A prescriptive code calls for a corridor in a commercial building to be a minimum of 44 inches wide. A performance-based requirement would state: The corridor must be wide enough to allow 65 people the ability to evacuate the building in the event of a fire, within 60 seconds, without suffering any ill effects from combustion byproducts or the fire itself.

As fire modeling and research data become more available and technology allows us more quickly to calculate an exhaustive number of variables, performance-based design will find more and more applications. It frees designers from standard prescriptive constraints currently built into the codes by allowing them to design a product that performs as well or better for very specific needs or objectives.

For local application, this process is very expensive, but in a broad sense many departments employ a very simplistic variant of it every day. They think of it as "alternate materials and methods." Please keep in mind that these are different, but utilizing alternate materials and methods is slightly akin to performance-based design.

> *Example:* A designer proposes a new 15-unit, three-story apartment building. She is designing it for a developer who owns a piece of ground that is infill and landlocked. There is no good access for fire apparatus, and the water supply is limited and cannot be improved reasonably. As a progressive fire prevention manager, you listen to issues and review your local ordinances to determine all of your options. Your review of the prescriptive code shows that since the building contains only 15 units, fire sprinklers are not required.

But, because the building will be ordinary construction, your water supply requirements demand 4,000 gallons per minute, and your city can provide only 2,750 gallons per minute at the site. You step back, put on your "alternate materials and methods design" hat, and suggest installing automatic fire sprinklers and a monitored fire alarm system. Your reasoning and justification are that a fire sprinkler will work more rapidly to control any hostile fire, and early detection will alert the occupants to the fire, thereby allowing early escape. Thus, you have protected lives and property, in that order. The use of quick response fire sprinklers in lieu of slower reactive emergency response and substandard waters supplies meets the intent of the code and likely provides a better and more reliable solution. Your proposed solution saves the day for all.

You defined the risk, conducted a quick hazard assessment, and evaluated the various consequences. You then decided which alternate solution would be best. Voila! You just performed your first alternate materials and methods design process. When applying the performance-based design approach, it is paramount to strive for a solution whose results will be equivalent to or more effective than the prescribed codes. Performance-based design requires careful deliberative discussion of goals and objectives for occupants and firefighters. It requires careful performance specifications and potentially modeling to confirm. Documenting very completely the entire process and *all* of the communication is imperative. The conditions and requirements identified in the performance-based design must stay in place for the life of the structure, and fire inspectors will need to verify that they do.

Summary

Hazards are conditions that provide a source of ignition for a hostile fire or that contribute to the spread and severity of a hostile fire. In order to mitigate fire and explosion problems, you must identify and eliminate or abate these various hazards. Many texts discuss specific types of hazards such as improperly stored gasoline or exposed electrical wiring. However, our approach is more global in assisting fire officials to look at the "big picture" of occupancy type as opposed to specific individual hazards. We propose looking at the macro level and then the micro level for holistic fire prevention efforts.

The principal part of your job is to mitigate threats to your community and not just the building. While improperly stored gasoline is a threat, it would be very difficult to find every instance of that hazard in your community. We propose looking at occupancies, or risk reduction opportunities within your community that are threats to your community's life safety, property, or business and commerce. Think strategically. Look at where you are experiencing fires currently, and identify the hazards in those occupancies that aggressive fire prevention efforts can correct.

These efforts should not necessarily involve just your fire department personnel—you should also include the stakeholders who own or are involved with the occupancies. This establishes buy-in and involvement at the grass-roots level. You are not telling people what to do, but rather assisting them in taking care of themselves.

The first step is developing a strategic protection plan. Next, you must identify your resources. What can you bring to bear on the problem? What will you need to be effective? Always keep your team involved. People who are involved have a sense of ownership and do a much better job. Then constantly reevaluate your plan. Use a feedback loop to make sure you stay on top of changes in your target or modifications to your solutions. There are numerous ways to make the target protection plan succeed. Remember the Three "Es." Engineering, Education, and Enforcement. Both active and passive fire protection systems are a critical component of these three "Es." Active measures include suppression systems, detection and alarm, and emergency fire department response. Passive approaches include product manufacture and engineering controls, administrative controls, and compartmentation.

Occupant safety is our principal goal. If our occupants can get out of the structure or be isolated from hostile fire, then our principal mission of life safety is achieved. The procedures to accomplish this can vary but may include notification, evacuation, defending in place, or providing an area of refuge. A major part of these procedures is making sure occupants know of the problem and are properly trained in behavior responses.

An increasingly more popular and more difficult element of fire and life safety protection involves performance-based design. This involves a process in which an architect or designer works with the local fire authority to craft outcome objectives regarding fire and life safety. The objectives can be as simple as "all occupants are able to leave the building before a fire event causes them harm or injury." With this objective in mind, the designer and fire official establish goals by which this is to be accomplished. These goals are generally not be based on prescriptive fire code requirements. As a fire protection professional, you must apply general principles to the community and its administrators, not just to individual buildings. Identifying problems, appropriate strategies, and plans will provide a much more visible and productive approach to your overall fire prevention effort.

Chapter 12 Review Exercises

12.1 What is a hazard? _____
_____
_____
_____

12.2 What are the most common causes of residential fires nationally? Use
the Internet to contrast the causes this text identifies.
_____
_____
_____
_____

12.3 What is meant by the term *common hazard*? _____
_____
_____
_____

12.4 What is meant by the term *specific hazard*? _____
_____
_____
_____

12.5 What does it mean to identify your target hazards?
_____
_____
_____

12.6 List the seven steps of creating a target hazard identification plan.
1. _____
2. _____
3. _____
4. _____
5. _____
6. _____
7. _____

12.7 Identify the basic engineering elements used to control fire and explain how each works. _____

_____

_____

_____

_____

12.8 Explain the difference between active and passive fire protection.

_____

_____

_____

_____

_____

12.9 List in order of importance the two tasks the fire service performs.

_____

_____

_____

_____

_____

12.10 Are fire department suppression operations important? Why are they lowest in the priorities of overall protection of hazards?

_____

_____

_____

_____

_____

12.11 How does product manufacture and performance-control impact fire protection?_____

_____

_____

_____

_____

12.12 What is compartmentation? _____

_____

_____

_____

12.13 Explain the concept of defending in place. _____

_____

_____

_____

_____

12.14 What are the two principal methods involved with evacuation?

_____

_____

_____

_____

_____

12.15 What is the time temperature curve's drawback in current applications? _____

_____

_____

_____

_____

12.16 What qualifies as an area of refuge? _____

_____

_____

_____

_____

12.17 Define performance-based design?_____

_____

_____

_____

_____

Notes

1. George C. Beakley and Herbert W. Leach, *Careers in Engineering and Technology*, 2nd ed. (New York: Macmillan, 1979), p. 3.

2. Arthur E. Cote and Jim L. Linville, *Fire Protection Handbook*, 17th ed. (Quincy, Mass.: National Fire Protection Association, 1991).

Wildland Fire Mitigation

Table of Contents

Key Terms

Key Points

1. We will never be able to stop wildfire, nor should we try as it is as much a part of nature as the trees and brush it consumes.

2. The policy of extinguishing fires as quickly as possible contributes to the wildland's poor ecological health basically creating a disaster waiting to happen.

3. When wildlands can proceed through their natural cycle, the fire resistivity of the forest improves.

4. The abundance of excessive vegetation and ever-increasing populace moving adjacent to federal and state lands requires greater interaction among agencies than ever before.

5. Structural and wildland fire fighting messages must be united in a cohesive nationwide program.

6. For all practical purposes, fuel is the only factor in wildland fires that humans can control.

7. In many wildland fires, structures are more of a problem than the vegetation that surrounds them.

8. Nothing can be done with a crown fire except to watch and wait until it drops to the ground.

9. Education and awareness must compensate for our lack of resources to build a fire suppression army.

10. Once you have communicated clearly the information that adults need to make good decisions, they will honor your attempts and generally do all they can do to help themselves.

Learning Objectives

1. Identify the impact of wildland fires.

2. Identify the components of wildland and urban interface.

3. Identify agencies that take an active role in preventing wildland fires.

4. Identify methods to prevent wildland fires.

5. Define national fire problem and role of fire prevention.*

6. Identify and describe fire prevention organizations and associations.*

7. Define laws, rules, regulations, and codes and identify those relevant to fire prevention of the authority having jurisdiction.*

8. Define the functions of a fire prevention bureau.*

9. Identify and describe the standards for professional qualifications for Fire Marshal, Plans Examiner, Fire Inspector, Fire and Life Safety Educator, and Fire Investigator.*

10. Describe the history and philosophy of fire prevention.*

FESHE Objectives (USFA)

Wildland Fire Mitigation

Chapter
13

Case Study

Over the last three years, a large hillside community has been under development. This neighborhood has grown to 125 homes, all clustered among three different cul-de-sacs fed from a single, one-mile, dead-end roadway, which ties back to a much larger arterial street in Pine City. About half of the residences are brand new, with residents coming from elsewhere in the United States. While the remainders are typical Pine City residents who are moving into the area because of the majestic views, more rural setting, and their desire to get back to nature.

This development is in a heavily forested and vegetated hillside area that has elevation changes of over 2,000 feet from where the arterial roadway is. Homes are nestled right up to the vegetation; slopes are steep along the back of the lots with wooden decks hanging beautifully over the edges of most slopes.

A fire occurs downslope (downhill), which moves quickly upwards, igniting the decks of 27 homes, resulting in the loss of all 27 homes in addition to exposure damage to 13 others before the fire is brought under control.

The community's Home Owner's Association (HOA) shows up two weeks later at City Council, furious about the fire department's poor response, lack of resources, excessive response times, and extensive loss of property. Some members of the community also resent the lack of awareness given to them by the fire department about the risk of a wildfire. Much of the energy is pointed to the development process, with the City being accused of not informing the residents of the risk prior to the development being allowed to be built. The residents direct comments and questions to the City administration about how this situation is to be addressed.

1. Identify the challenges and concerns pertaining to public perception and fire prevention.
2. Which of the items previously identified could have been addressed before the fire?
3. Provide recommended solutions for each of the items identified.
4. Discuss how these solutions could be addressed with little to no additional expense to the fire department.
5. List five (5) strategic items that the fire department should initiate, develop, and manage for at least the next ten years.

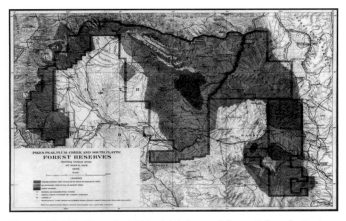

Figure 13.1 Big Burn in 1854 which burned from Colorado Springs to what is now Breckenridge, CO. Fire started in January and was caused by Native American Ute tribe herding game. *Courtesy of the Colorado Springs, CO Fire Department*

Wildfire and the Environment

Wildland fires have raged as long as the earth has had vegetation. The earth has undergone biological cleansing by fire on a regular basis. This process has been initiated by lightning, volcanic activity, and the like ever since fuel, air, and heat first combined. As man evolved and fire became a tool/weapon, wildfires — both accidental and purposeful — became a product of human activity. Humans used wildfire to herd game and wage war (**Figure 13.1**). They also used it to clear dense underbrush to make game more visible and accessible for hunting. Regardless of the man-made issues, fire is and always will be part of the natural environment, whether we humans exist or not. In many parts of the world, humankind has altered wildfire's frequency and severity, which has forced it to progress in cycles, not necessarily managed the way Mother Nature would have preferred. This alteration has occurred due to clear-cutting to make room for crops and grazing; logging for lumber; fire extinguishment policies; urban sprawl; and carelessness; to mention only a few. While policymakers at the federal, state, and local levels debate about what is right for our country and forestry issues, we need to shift from the way we have been dealing with wildfire during the recent past. As technology has improved, so have our methods of identifying and dealing with wildfire. We will never be able to stop it, nor should we try, as it is as much a part of nature as the trees and brush it consumes. Instead, we must try to understand it and work within its cycle as much as possible. The following chapter will help you understand more about where we have come from, where we are, and where we should go from here as fire service professionals.

What Is Wildfire?

Wildfire — Any hostile fire in the outdoors that is not prescribed or purposefully managed.

A **wildfire** can be described as any hostile fire in the outdoors that is not prescribed or purposefully managed. Wildfires occur in every country (**Figure 13.2**). Unfortunately, as time goes on they are occurring with greater frequency and severity in many parts of the world. Between 2003 and August of 2012, the average number of acres burned per year in the United States was 5,385,441.[1] Several of the individual fires that occurred in 2012 set records, including Colorado's Waldo Canyon Fire (**Figure 13.3**), which damaged or destroyed 346 homes, and New Mexico's Whitewater-Baldy Fire, which burned more than 297,000 acres[2] of homes. People are moving to the mountains in record numbers to escape the hustle and bustle of the city. At each event, many people are surprised at the tragedy and mayhem wildfire produces. Many people are mesmerized at the effects produced when nature unleashes her fury in wildfire events. Why? Because as we previously stated, fires have burned our countryside for hundreds of thousands of years. Driving on American highways and looking along the sides of carved-out roadways, you can see the black carbon line of historical evidence of "big burns" that occurred before the waters from floods and other geological changes covered the forest floor with silt and other debris.

Figure 13.2 Crowning wildfire in National Forest. *Courtesy of the Colorado Springs, CO Fire Department*

Figure 13.3 2012 Waldo Canyon fire just before it destroyed 347 homes. *Courtesy of the Colorado Springs, CO Fire Department*

Left to its own course, fire used to be a relatively good thing. Most forest fires were what we term *ground fires*. This means that fire generally traveled along the ground, clearing forest floor debris (dead grass, twigs, deadfall, and other noxious organic substances) and making way for new growth. Small, heavily congested trees were killed, thereby thinning potential overgrowth and allowing mature tree stands to take hold, providing cover for animals and birds. Only rare fire events were what we call *stand replacement fires* today. These fires generally occurred when the natural health of the forest ecosystem was diseased and overgrown. Stand clearing fires occurred when flames move from the ground into the crowns of the trees. They consumed everything in their path, from small ground-level plants to large ponderosa or fir stands. While not common, these fires were essential to reconstructing and realigning the vegetation that needed repair and disease eradication. While the human life span is around 75 to 80 years, the natural environment remains relatively unchanged for hundreds of thousands of years. An event such as this, while markedly tragic in our view of time, is but a small nuisance to Mother Nature. She utilizes this time to nurture, develop, and foster the ecosystem.

Our main mission in fire prevention is to protect people from the ravages of hostile fire. Wildfires raise issues that are not necessarily present in what we could term fire prevention's typical bread-and-butter operations. However, they can be addressed in much the same manner.

Historical Forest Service Issues

To begin solving our wildfire problem requires a look at its history. Fires were ravaging our forests at the turn of the twentieth century. In 1910, a Dr. Deckert noted, "devastating conflagrations of an extent elsewhere unheard of have always been the order of the day in the United States."[3] That same year, the Forest Service lost around five million acres to wildfire, which was under their jurisdiction. This stirred great debate among government officials and the Forest Service who began to throw considerable amounts of money and resources into a national fire fighting effort. Years of depression, drought, and inflation brought action on many issues. President Franklin

D. Roosevelt's New Deal presented the opportunity for a massive manual fire fighting effort. This spawned a huge program called the Civilian Conservation Corps. With it came loads of federal money that played perfectly into the hands of the Forest Service's new philosophy and policies. So perfectly, in fact, that in 1935, Chief Forester Ferdinand Silcox initiated the frequently stated, informal **10:00 A.M. Policy**, which mandated control of any fire by 10:00 A.M. the day it was reported or, failing that, control by 10:00 A.M. the day following, ad infinitum.[4] This policy led to tremendous advancements in strategies, tactics, and equipment, not to mention physical resources.

The United States Forest Service protects about 200 million acres of national forest and other lands. The Bureau of Land Management (BLM), the National Park Service (NPS), the United States Fish and Wildlife Service, and the Bureau of Indian Affairs control another 587 million acres. State and other local land under control of localized fire fighting resources make up about 840 million acres.[5]

According to a CoreLogic report, fewer than 1,000 homes per year on average were destroyed by wildfires in the 1990s. From 2000 to 2008, the average jumped to more than 2,500 per year.[6] While this seems to be a small percentage of total area, the impact is gigantic. A significant amount of this expense went toward protecting homes and businesses — structures that have been designed and built in harm's way **(Figure 13.4)**.

The 10:00 A.M. Policy has recently come under tremendous scrutiny. The public and municipal fire service agencies are asking hard questions about the issue:

- Why are we fighting fires in the middle of the forest when nothing is out there but trees and grass?

- How do we protect an ever-increasing number of structures when resources are stretched thin fighting fire on some far away mountain well away from any populated area?

- Why do so many local fire fighting agencies commit so many resources at such low federal reimbursement rates?

Figure 13.4 Homes densely nestled in the Wildland Urban Interface. *Courtesy of the Colorado Springs, CO Fire Department*

The policy of extinguishing fires as quickly as possible has led to substantial overgrowth of vegetation, which is now being scourged by disease, drought, and insects. This only contributes to the wildland's poor ecological health. Basically, in some cases, we have created a disaster waiting to happen.

Current Threats and Risks

We live, work, and play in a natural fire environment. Many believe this environment is presently out of ecological/fire balance. We are experiencing fires that grow much larger and much faster than ever before. A study published in June 2012 in the peer-reviewed journal *Ecosphere* reported that climate change is widely expected to disrupt future wildfire patterns around the world. Some regions, such as the western U.S., are expected to see more frequent fires within the next 30 years.[7] In addition to climate changes, this is also the result of the volume of thick and diseased vegetation. Communities and businesses situated in or near this combustible tinderbox greatly exacerbate the threat and risk. While the structures we protect are at great risk, the environment is equally endangered. The growing numbers of fires burning with greater intensity and the increasing thermal load of the surrounding vegetation are literally sterilizing the soil, resulting in noxious weed infestation and flooding resulting from soil erosion.

The wildland/urban interface is defined as that area of a community where forested land mixes with urbanized functions and structures. It is broken into three basic components:

- **Interface mix**. Structures that are scattered throughout a rural area where isolated homes are surrounded by undeveloped land. Usually few homes in concentration are at risk here, as the interface mix is similar to a sparsely populated mountain ranching community.

- **Occluded interface**. An isolated area of forested land or wildlands such as a large park or preserve surrounded by homes or other structures.

- **Classic interface**. Homes and other structures, especially in small dense neighborhoods, pressed directly against the forest or wildland. A perfect example is the Mountain Shadows neighborhood in Colorado Springs, Colorado, which was devastated in 2012.

These areas contain the most significant assets at risk. This is not to say that historical artifacts, such as the ruins at Mesa Verde National Park in Colorado, are not important. However, from a fire prevention standpoint, the elements of priority for our professional concerns are life, property, and resources. It is critical to understand that wildfire is not just a western phenomenon. Dense vegetation and the encroachment of developments happen in every state in America, every province in Canada, and all over the world.

Interface Mix — A rural area with structures scattered sparsely throughout.

Occluded Interface — An isolated area of forested land or wildlands surrounded by homes or other structures.

Classic Interface — An area with homes and other structures, especially in small dense neighborhoods, pressed directly against the forest or wildlands.

Outside of developed communities, the major natural elements that are directly impacted by wildfire are:

- Water
- Wildlife
- Air
- Plants
- Timber
- Soil

Watersheds (drainage basins), or land that provides critical water from rain and snow runoff, are becoming more critical to increasing developments, agricultural needs, and recreational use, not to mention the sustainability and growth of communities. These waterways are being polluted by wildfire, often for extended timeframes, costing millions of dollars to clean up and restore. The water pollution created from wildfire is extensive and complex. Smaller organisms and the fish they support become poisoned and suffocated. Streams and creeks become obstructed and contaminated. Silt and sediment diminishes storage capacity in supply reservoirs and lakes. In addition to water issues, the destruction of forest canopy due to crown fires eliminates wildlife habitat for smaller birds and animals on which larger predators feed. The overall ecological impact is significant.

Wildfires frequently produce thousands of pounds of airborne particles (smoke) that affect not only human health, but the weather as well. Toxic gases, such as carbon dioxide, carbon monoxide, nitrous oxides, and other organic gases not only harm animal life, but also can contribute to global warming. Big fire events can paralyze the lumber industry, which impacts construction, paper, and other industries. Trees do not reach harvesting size overnight and, therefore, must be established for many years before the product can be properly harvested. Forest soils, which once could be cleared of down and dead vegetation by relatively minor ground fire events, are now being turned to glass, resulting in an inability to soak up moisture or rapidly spawn new vegetation (**Figure 13.5**).

Figure 13.5 Hayman Burn in Colorado reveals how severely a crown fire devastates the soil and landscape for years to come. *Courtesy of the Colorado Springs, CO Fire Department*

Causes of Wildfire

The same conditions that cause most of our fire problem also cause wildfires. However, one more element, while not a direct cause, is certainly a significant contributor to catastrophic wildfire: Interruption of Nature's fire cycle. We have clearly recognized recently the negative effects of removing fire from the natural cycle. This intervention is detrimental and disruptive to a wildland's ecology. Just a few of the many factors that influence a wildland's ecosystem are:

- Climate
- Topography
- Geography
- Vegetation
- Animal life
- Pollution

Defining a local ecosystem is complex, as it requires an examination of many elements. It depends upon events, objects, processes, and all the details of how they interact. An example could be a typical western mountain forest. If the forest is permitted to grow without fire, the vegetation will gradually become very dense and overgrown **(Figure 13.6)**. Animal life will dwindle as the vegetation becomes too hard to navigate. Sunlight is reduced, thereby restricting the types of vegetation that will grow there. Plants, just like animals, are limited by the amount of nutrients they receive. Aggressive competition then becomes destructive. Disease and insects become common, as they are among the few organisms that can survive this predicament. The insects further stress the vegetation, making it even more susceptible to disease and parasites. The bark beetle is a good example. This

Figure 13.6 Densely overgrown vegetation is not beneficial to the overall environment. *Courtesy of the Colorado Springs, CO Fire Department*

beetle burrows into the bark of a tree, its host, and lays eggs. This damages the tree's natural protection (bark) and causes the tree to lose life-giving moisture due to the network of tunnels the beetles created as they spread throughout the tree. These numerous tunnels soon fill with a bluish fungus that blocks nutrients from feeding the tree. Where drought is not present, a healthy tree will form a plug of pitch or resin that in the colder months, when the beetle is inactive, will bubble outward forcing the beetle out and thereby saving the tree. However, when the tree is stressed and competition limits its moisture supply, the tree cannot protect itself and eventually succumbs to the beetle infestation, which it passes on to the next tree. Fire, in this example, would play a beneficial role by:

- Influencing plant community composition
- Interrupting and altering succession
- Changing the amount, kind, and size of various vegetation types in a given area
- Regulating vegetation (fuel) accumulation
- Influencing and improving the nutrient cycles and energy flow

- Affecting the wildlife habitat
- Interacting with and controlling insects and disease
- Influencing the productivity, diversity, and stability of the ecosystem

When wildlands can proceed through their natural life cycle, the fire resistivity of the forest improves, the balance of detrimental insects and disease is maintained, and wildlife interaction benefits the overall symbiotic relationship.

Factors Determining Risk

The high intensity wildland fires of the last few years have forced a significant shift in federal directives giving priority to structures and community infrastructure over natural resources. This shift to address the "larger" risk is likely to quickly drain federal reserves and force more interdependence with state and local resources. As national economic issues strain federal resources, funding for suppression will become more difficult. As budgets tighten, the federal government will quickly be put in the position of forcing local communities to spend more money protecting them rather than depending on federal assistance.

Political and sociological factors also play a big part in the wildland/urban interface fire problem. Society is moving from a philosophy of "taking care of ourselves" to always depending on someone else. Our "don't-blame-me" culture causes extraordinary dependence upon governmental resources, whether federal, state, or local. This cannot be sustained and therefore is forcing all fire protection experts to foster leadership and cooperation or be left in a lurch when a disaster occurs.

From a fire prevention standpoint, this interagency cooperation is very important. The fire prevention message is difficult enough to convey in general fire safety issues, but is compounded in wildlands because Americans in general do not perceive a relevant risk from wildfire. The property owners who typically build in or adjacent to forested areas generally choose to do so because they want a more rugged lifestyle or desire a location that is "away from it all," not to mention to "have a view." This attitude is typical for those people who tend to show more independence and also who typically do not like being told what to do.

Much like those folks who live in hurricane alley, these people also have a belief that insurance companies will pay for their losses and therefore do not show much concern. In our experience, people who have survived or been closely involved with wildfire, exhibit excellent fire-prevention attitudes, but sometimes this is short lived. After some time has passed, many of these same people develop an attitude that "It won't happen again" or that it was a "Once in a million years event." The amount of publicity surrounding wildfire events also puts a lot of focus on the fire and forest services. This is a major barrier to getting our fire prevention message across to the public we serve because there is a false sense of security with folks believing there will be a fire truck in every driveway or the firefighters will protect us.

The factors that determine risk in the wildland/urban interface are relatively consistent throughout the country:

- What are the potential loss impacts to the community?
  - Direct
  - Indirect
  - Life loss
- What is the community's assessment of which losses are and are not acceptable?
- What communities have really done this? What is our ability to mitigate identified hazards?
  - As previously discussed, historically, the fire and forest service spends money on response (reaction), not prevention or mitigation.

To effectively deal with this risk, the fire service must determine what level of emergency they can reasonably contend with and then remedy those bigger emergencies in some other manner.

Wildfire Management

Once a community has determined its acceptable risk, elements of prevention and suppression should be readily identifiable to help target specific wildfire management schemes. This can involve a number of elements such as:

- Interagency cooperation
- Fire and life safety education
- Prevention tools
- Prevention and mitigation
- Suppression

Interagency Cooperation

Fire protection problems in the wildland/urban interface are very complex. The numerous political and bureaucratic barriers that must be dealt with include but are not limited to:

- Legal mandates such as water quality laws
- Zoning regulations
- Fire and building codes
- Fire protection infrastructure
- Insurance grading and rating systems
- Environmental concerns
- Cooperative agreements with affected agencies (Bureau of Land Management, Forest Service, Water Purveyors, etc.)

The wildland fire problem has increased steadily over the years. With an abundance of excessive vegetation and an ever-increasing populace moving adjacent to federal and state lands, the interaction among various agencies is unlike anything in the past. Federal agencies are being relied upon more heavily than ever before because of their expertise at running campaign fires.

Federal officials, however, face the dilemma of ever-dwindling resources. Expenses for war, homeland security, Social Security, health care, etc. are higher than ever before. Tax cuts loom. Resources are taken for the funding of various government programs in addition to federal stimulus packages. Federal managers have been placed in an interesting predicament due to the massive retirements in the forest service command cadre coupled with limited funds to train and develop younger fire management officers. How does this impact fire protection professionals? What are we going to do without an effective and competent army to respond to and suppress hostile wildfires? It is reasonable to think that the federal government may need to take a lesser role in protecting communities that already fall under the purview of other governmental fire fighting agencies.

We need to maximize our local efforts and resources to provide the best prevention and protection possible. Most wildland/urban interface departments do not have the resources to respond at a level sufficient to battle a serious wildfire. The ability of a single agency to provide the necessary leadership in these catastrophic events is becoming more and more difficult. Local communities typically do not have the resources to combat large campaign events. As a result, many smaller communities have to muster alternate resources to combat these large conflagrations, while still subordinating command and control responsibilities to federal or state agencies. Conflicts between state and federal boundaries, not to mention those of municipalities and private land, all create huge logistical complications that should be worked out well in advance of a looming wildfire. The best method for doing this is through proper fire prevention efforts, which will reduce the need for operational support.

Attempts to foster cooperation between the "structure guys and gals" and the "yellow shirts" or forest service folks are ongoing. Typically, these two groups have had very different missions and training. Wildland firefighters generally have not had tactical structural fire fighting training and structural firefighters typically have not been "red carded" (a wildfire training process equivalent to Firefighter I or II certification). While this effort at networking and building working relationships between the groups is important, much more needs to be done to unite the messages of fire prevention in a more cohesive nationwide program.

Fire and Life Safety Education

The forest service started a massive public education campaign with Smokey Bear in 1950. Smokey was an injured bear cub found by a fire fighting crew in the Lincoln National Forest in New Mexico. The young cub was nursed back to health and placed in captivity, launching one of the largest forest fire prevention campaigns ever seen. This campaign is still in use today and is credited with preventing thousands of fires. In fact, considering the number of fires started by careless campfires or outdoor fire use (very small) vs. the number of people who recreate outdoors each year (very high), the number is amazing. If only our success was that good for building fires caused by unattended cooking!

In addition to fire prevention campaigns using Smokey, many more partnerships have been initiated to keep the message current and visible. Many communities have teamed Sparky with Smokey to emphasize the importance of fire prevention. The emphasis is that as fire service professionals, we need to keep current with the latest trends, the latest marketing strategies, and the latest fire problems to make sure we are adequately communicating our issues and solutions.

Prevention Tools

Wildfire cannot easily be explained; however, it can be experienced. Nothing quite resembles the awesome power of these events. Three elements dictate how a wildfire will behave and move:

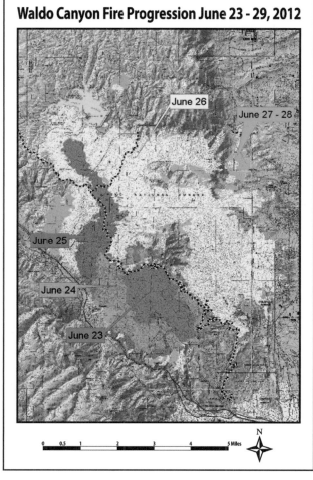

Waldo Canyon Fire Progression June 23 - 29, 2012

- Fuel
- Weather
- Topography

Understanding these three elements is crucial to understanding how to prevent wildfires. Humanity can modify only one of them. To some extent, we might change a site by modifying its topography, but most communities or regulatory agencies prohibit any excessive ground disturbance that would impact the aesthetics of a hillside. For all intents and purposes, we will consider fuel as the only factor we can control.

Weather is a natural phenomenon that has baffled mankind since our creation. It can and will do mysterious things. The Hayman fire in Colorado in 2002 made a 19-mile run in a single afternoon as a result of extreme fire weather. While our science and technology have allowed us to learn much about the weather, we have very little control, if any, over how it will behave **(Figure 13.7)**. We can do a reasonable job of predicting what will be happening within a few hours or a day; however, changing the weather is out of our reach (since cloud seeding does not really pertain to the fire service, we'll leave its effectiveness for others to debate).

Figure 13.7 Waldo Fire Progression Map that reveals how rapidly fire can move based on weather, topography and fuel. *Courtesy of the Colorado Springs, CO Fire Department*

Wind obviously has a major effect on the severity of a fire. We also know that large fires will create their own weather. El Niño, la Niña, and other weather patterns all have their effects, whether they be drought, flooding, lightning, or winds. Weather is a force to recognize and to prepare for, but we also must realize we can do nothing about it.

Topography is another significant factor. Just as tipping a match up or down as you hold it can affect the aggressiveness of the fire's behavior in relation to your fingers **(Figure 13.8, pg. 330)**. What happens when you light a wooden match and tip the flaming head downward? Ouch, right? Well, what happens when you build a house on the top of a steep slope? Ouch, again! Fire moves rapidly uphill. Although science can explain that, no one should need

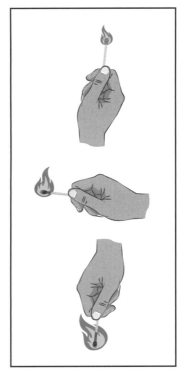

Figure 13.8 Disregarding wind, fire and the heat it generates will move much faster uphill than on flat ground or in a downhill direction, as depicted by the heat generated from this match.

much explanation as it also follows the basic understanding that fire moves upward. So, flat land in Florida will not burn as severely as it would if it had steep, similarly vegetated hills and slopes, but make no mistake; it still burns aggressively and can be grossly impacted by weather.

In comparison to the Rocky Mountains, Florida has to contend with fuel and weather but not topography. Large mountains with steep slopes and drainages that are heavily vegetated can create massive firestorms that, without significant initial weather influence, can create their own weather.

From a fire prevention standpoint, it would not do us much good to recommend that people not live somewhere based on the weather. People generally choose weather as a principal reason *for* living where they do. Telling someone not to build a home at the top of a steep ridge overlooking a scenic valley or open meadow with beautiful pines and mountain mahogany surrounding them can be quite a struggle as well. Why do you think they paid so much for the property on that ridge top? The view! Why would they build down low where they could not see anything? They won't, if they don't have to.

What about fuel? Can it be modified or mitigated? Yes, very easily in many cases. Does this mean that all vegetation needs to be clear-cut and removed from around a house or structure? Absolutely not (**Figures 13.9 and 13.10**)!

Figure 13.9 Overgrown vegetation. *Courtesy of the Colorado Springs, CO Fire Department*

Figure 13.10 Mitigated plot shown in Figure 13.9 AFTER mitigation to reduce fuel loading. *Courtesy of the Colorado Springs, CO Fire Department*

Wildland fire models classify available fuels into 13 basic types of vegetation according to fire behavior and heat-release rates. We propose there are actually 14 types of fuels, with structures being the 14th. In many instances, structures prove to be more of a problem than the vegetation that surrounds them.

If you study the Cerro Grande fire that struck Los Alamos, New Mexico, or the Waldo Canyon fire that bumped into the city of Colorado Springs, Colorado in 2012, you will see that structures in many instances were more of an issue than the vegetation involved in the forest fire. Structures also possess a tremendous amount of mass and in terms of mass-to-air-ratio and are a significant source of heat and ember generation contributing greatly to conflagration outcomes. In fact, while much has been written about the Oakland Hills fire, we propose that it was not as much a wildland/urban interface fire as it was a conflagration or structure-to-structure fire. The way the structures blended into the surrounding vegetation made them as significant a fuel model or fuel package as the eucalyptus surrounding them. Assuming for a moment that this is true, let's consider how to mitigate vegetation and then apply the same methodology to structures.

Vegetation can be grouped into three components for fire prevention or mitigation purposes:

- **Ground fuels (flash fuels)**. Grasses or very light fuels that are easily ignited and spread fire rapidly; duff and roots
- **Surface fuels (intermediate fuels, brushy fuels)**. Mountain mahogany, gambol oak, chaparral, or small trees
- **Crown fuels**. Branches, tree tops, high brush

In fire prevention, we work very hard to educate people to trim or interrupt ladder fuels. **Ladder fuels** are those fuels configured so that a ground fire can become a surface fire, and a surface fire can become a crown fire.

Ladder Fuels — Fuels that are configured so that a ground fire can become a surface fire and a surface fire can become a crown fire.

Fires can be extinguished when they are on the ground or on the surface; however, when fire moves into the crowns, fire crews have to stand back and watch just like everyone else. Nothing can be done with a crown fire except to watch and wait until it drops to the ground. These fires, which can be termed *stand replacement* fires, cause tremendous damage. Their heat release rate and heat output are so intense that hand crews, engines, or aircraft have no extinguishment effect. Provide education to urge people to manage their vegetation to interrupt these ladder fuels. Remove or strategically clump the brush and large surface fuels, and the ground or small surface fire will not get into the canopy starting a crown fire. This does not stop the fire. However, it alters or mitigates it and prevents the fire from becoming impossible to combat or helps to keep it relatively benign. Remove or greatly thin one of the two lower fuels, or prune low-hanging branches from large trees or brush. Doing this will not allow fire to be drawn into the canopy as easily, thus preventing crowning.

Considering homes or other structures, we can do similar things. Why would we treat a home or other building in the wildland urban interface any differently? If we view the structure as a fuel source or fuel package, then we can address it like we do other fuels, managing it or mitigating it. As research physical scientist Jack D. Cohen points out, "Understanding how homes ignite during wildland/urban fires provides the basis for appropriately assessing the potential for home ignition and thereby effectively mitigating wildland/urban fire ignitions. Fires do not spread by flowing over the landscape and high intensity fires do not engulf objects, as do avalanches and tsunamis. All fires spread by meeting the requirements for combustion—that is, a sufficiency of fuel, heat and oxygen."[8]

Colorado Springs, Colorado, which is said to have the largest wildland/urban interface for a single jurisdiction in the state of Colorado, takes the approach of factoring this knowledge into home building. Developers have teamed with the fire department to provide several key elements that vastly improve new homes' survivability:

- Class A roofing ordinance (excluding solid wood material)
- Hardening the structure (noncombustible siding and soffits)
- Metal screens on all exterior vents
- Composite decking materials
- Double-pane glazing (energy requirement)
- Vegetation (fuels) management
- Fire engine access for manual suppression hose reaches on all exterior portions of buildings

While this approach does not guarantee preventing a home from catching fire, it certainly reduces the possibility and or severity. Communication with the residents in those areas suggests a 50 percent improvement in a structure's chances for surviving a wildfire. An important part of this philosophy relates to economies of scale. If we can improve the ignition resistance of structures, we reduce the number of ignitions. If we reduce the number of ignitions, the more effective our suppression forces will be at protecting or extinguishing fires that do occur. This will further reduce other structure ignitions, thereby reducing the number of forces and equipment needed to further suppress and extinguish. It is akin to falling dominoes but in this case, all in our favor.

Our objective as fire safety professionals should not be to prevent wildland fires in total. Remember, fire is part of the natural environment. Since we choose to build homes in that environment, we should expect fires to threaten them periodically. This being the case, we should then engineer out what fuel sources we can and allow fire to move into and through a neighborhood without destroying homes along the way. Jack Cohen refers to the area around a home as the ignition zone. He writes, "Given low ignition potential and enough time, homeowners and or firefighting forces can make significant reductions in the little things that influence ignition potential before wildfire encroachment. Then, if possible, homeowners and firefighting resources can suppress small fires that threaten the structure during and after the wildfire approach."[9]

Prevention and Mitigation

Fire prevention and mitigation are the keys to any structure's long-term survivability. The fire service alone cannot achieve this. Again, using Colorado Springs, Colorado, as an example, over 34,000 addresses are in the revised wildland/

urban interface. The City of Colorado Springs has a total of 21 fire stations. This provides an on-duty force of over 110 firefighters. Assuming there are no other emergencies, training, or public assistance requests to deal with, the department could throw 110 people per shift against a major fire. However, the department decided to utilize its emergency resources for what they are paid to do: respond to fires and other emergencies. Proactively then, they enlisted those who live in the affected areas to help by managing fuel (vegetation), thus keeping those fires that do occur small and relatively benign **(Figure 13.11)**. This potentially created an army of 34,000 people or more. If only 10 percent actually participate, that still yields a minimum of 3,400 people. Which army is better equipped, staffed, and able to communicate the issues? One hundred firefighters or 3,400 motivated citizens? Thousands of motivated citizens will win every time!

Figure 13.11 A CSFD chipping crew helping homeowners to mitigate fuels around their homes. *Courtesy of the Colorado Springs, CO Fire Department*

The key is for the fire service to stop telling people what is good for them. This is an adult education issue and requires a different approach than Stop, Drop, and Roll. We suggest following these guidelines:

1. Assume adults living in large homes in forested areas are educated, understanding, and responsible.
2. Assume that these same adults do not know much about the natural fire environment or their actual risk.
3. Assume that they want to do the right thing if they are informed, if they have the capability, and if they can measure progress by themselves, for themselves.
4. Be factual, be truthful, and make clear all expectations about survivability and fire suppression capabilities. *Do not* exaggerate the issues.

Once you have communicated clearly the information that adults need to make good decisions, they will usually honor your attempts and generally do all they can to help themselves. In Colorado Springs, neighbors motivate neighbors. Why would you want to act as safely as possible, protecting your property and family, while living next door to a person who has not learned about the issues and therefore does not care? You are likely going to go have coffee with that neighbor and share a little insight and maybe a little peer motivation **(Figures 13.12a and b, pg. 334)**.

Remember, think of this as adult education, not as government telling little people what to do. Government should provide factual information about problems, solutions, and resources, but this does not mean the government can provide those solutions and resources. Government just has to help answer questions that the ordinary person cannot. Also remember that this takes time...lots of time. Be patient. Earlier in the chapter we spoke of the perception of people. Until they live through a tragic or scary event, they typically cannot grasp the true reality. Do your best to provide them up-front information but realize they may not get it *until they get it*. They don't know what they don't know. This is why your communication must be clear, factual, and repetitive.

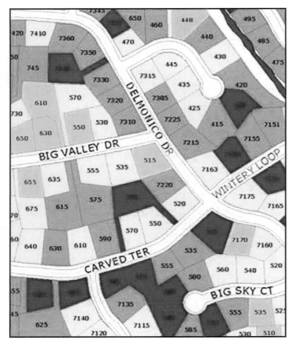

Figure 13.12a A wildfire risk map. *Courtesy of the Colorado Springs, CO Fire Department*

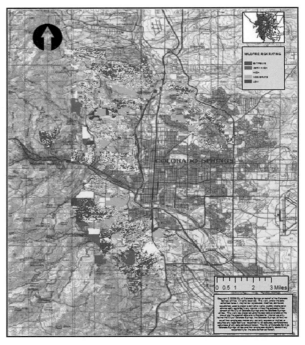

Figure 13.12b Wildfire Risk Map of Colorado Springs, CO. *Courtesy of the Colorado Springs, CO Fire Department*

Figure 13.13 An Interface Engine surrounded by vegetation. *Courtesy of the Colorado Springs, CO Fire Department*

Suppression

As with all good fire prevention programs, a systematic approach to suppression is critical. Engineer, educate, and enforce those things that facilitate the best emergency response capabilities. Even in the best of worlds, we are giving structures only a fifty-fifty chance of survivability on their own if and when a wildfire does occur. But we are also helping people to help us. By providing good, survivable space and smart prevention/mitigation measures, we are also giving our fire suppression crews an opportunity to interrupt a hostile fire's advance on a structure. We give our crews the best chance possible for protecting the property because we already designed and factored in effective prevention and mitigation measures. Ultimately, we must use a systems approach. Without limitless funding, prevention and mitigation alone are not enough. We need to rely reasonably on fire suppression forces. Practicing these methods will go a long way toward ensuring the system's overall effectiveness **(Figure 13.13)**.

Paradigm Shifts in Wildfire Management

In this chapter we have called for several paradigm shifts in the ways we think about and deal with wildland/urban interface fires. Where these shifts already have been made, they have proven successful. They include:

- Be aware that structures provide as much fuel or more than the trees surrounding them.
- To build a fire suppression army, provide education and awareness to compensate for our lack of resources.

- Quit telling everybody what is "good" for them. Give them the information and help them to understand it. Then, they will make good, informed decisions.

- Accept decisions with which you may disagree. Remember, you work for the public and not the other way around. Sometimes people want to accept greater risk. As long as they know the potential outcomes, why not let them? Do not put firefighters in harm's way. And, of course, if someone's behavior becomes a fire risk to others who do not share the same beliefs and values, we have to do something to address it. This is why they pay us (**Figure 13.14**).

Figure 13.14 CSFD Mitigation Crew emphasizing everyone needs to "Share the Responsibility" for a safer wildland urban interface. *Courtesy of the Colorado Springs, CO Fire Department*

Summary

Wildfire is any hostile fire in the outdoors that is not prescribed or purposefully managed. Wildfires are becoming a significant cause of annual fire loss throughout our country. As our forests and wildlands become more dense and overgrown, and as more and more developments encroach within that space, the losses will continue to rise.

The philosophical and scientific basis for forest and vegetation management and wildfire control has a long history. However, most agree today that our aggressive posture of extinguishing wildfire has prevented much of the natural vegetation management that occurred from "frequent" fire. This has led to overgrown vegetation and to dead and diseased forests threatening our communities with huge unnatural fuel loading that contributes significantly to our wildland disasters.

Watersheds are also becoming a major exposure due to wildfire. As more and more people move and populate the west, water is becoming a more vital and valuable resource. The ponds, lakes, and estuaries from which we receive our water are easily destroyed or disrupted for many years if a wildfire event moves through them. This causes significant drought and water shortage issues for our communities, not to mention the death and destruction of the aquatic life and ecological environment these areas support.

Wildfire is generally impacted by three major components:

- Fuel
- Weather
- Topography

The issues affecting and determining risk are hugely varied. These issues depend greatly on the community where you live, as various values on the risk are readily apparent. Obviously, people moved into the mountains to be in the wildland in some sense and likely to be away from regulation or at least for the perception of being away from regulation. Thus, the issue of how to mitigate wildfire risk can stir controversy.

Various prevention measures are available and depending upon the area of the country, some are emphasized more than others. Those we believe to be most important are:

- Noncombustible roofing
- Hardening of the structure
- Vegetation management
- Defensible space

We will never have enough fire suppression capability to completely protect our communities from wildfire. Those departments that have this problem must be smart in applying prevention and risk management strategies that maximize their efforts by involving the community. No matter how you divide it, you will always have more citizens than firefighters. Use the public to your best advantage to build coalitions. They are inexpensive, and if you motivate and engage them, they will do most of the work for you.

Chapter 13 Review Exercises

13.1   What is wildfire? _____
_____
_____
_____

13.2   Explain why preventing wildland fires from occurring is or is not a worthwhile fire prevention function. _____
_____
_____
_____

13.3   Identify and define the three components of the wildland urban interface.

1.   _____
_____
2.   _____
_____
3.   _____
_____

13.4   List six natural elements that wildland fires impact.

1.   _____
2.   _____
3.   _____
4.   _____
5.   _____
6.   _____

13.5   Identify those agencies that need to take an active role in preventing wildland fires. _____
_____
_____
_____
_____
_____
_____
_____

13.6 What are the two key methodologies to a structure's long-term survivability and why?_____

_____

_____

_____

13.7 Name at least two methods of modifying fuel.

_____

_____

13.8 Explain how fire is part of the natural fire environment.

_____

_____

_____

13.9 What is the "10:00 A.M." policy, and why is it significant?

_____

_____

_____

13.10 What does it mean to interrupt ladder fuels? _____

_____

_____

_____

13.11 List five key resources that are impacted by wildfire and explain how.

1. _____

_____

2. _____

_____

3. _____

_____

4. _____

_____

5. _____

_____

13.12 What three main factors determine how wildfire behaves?

1. _____

2. _____

3. _____

13.13 What are the three main factors that determine risk in the wildland/urban interface?

1. _____

2. _____

3. _____

13.14 Why are collaborative interagency relationships important?

_____

_____

_____

_____

_____

13.15 Which of the three wildfire components can human efforts reasonably mitigate?

1. _____

2. _____

3. _____

13.16 List and describe the three main fire fuel substances.

1. _____

_____

2. _____

_____

3. _____

_____

13.17 Define a stand replacement fire. _____

_____

_____

_____

_____

13.18 List the four key guidelines for providing adult education about wildfire mitigation.

1. _____

_____

2. _____

_____

3. _____

_____

4. _____

_____

NOTES

1. http://www.nifc.gov/fireInfo/nfn.htm

2. Ibid.

3. Stephen J. Pyne, *World Fire: The Culture of Fire on Earth* (Seattle: University of Washington Press, 1995), p. 187.

4. Ibid., p. 195.

5. *Fire Protection Handbook*, p. 8-221.

6. CoreLogic Wildfire Hazard Risk Reprort, Residential Wildfire exposure Estimates for the Western United States 2013; Howard Botts, Ph.D, Thomas Jeffery, Ph.D., Steven Kolkm ACAS, Sheila McCabe, Bryan Stueck. http://www.corelogic.com/research/wildfire-risk-report/2013-wildfire-hazard-risk-report.pdf

7. http://www.newscenter.berkeley.edu/2012/06/12/climate-change-global-fire-risk/

8. Jack D. Cohen, "Wildland-Urban Fire: A Different Approach," Missoula Fire Sciences Laboratory, Rocky Mountain Research Station, Forest Service, U.S. Department of Agriculture, 2002.

9. Ibid.

Using Technology to Improve Fire Prevention Efforts

Table of Contents

Key Points

1. How we use the data we collect will determine the effectiveness of our fire prevention efforts.

2. Coordinating all data collection activities and sharing access to that data benefits the customer as well as the fire department.

3. Creating coalitions to collect and manage data should be no different than for any other area of fire prevention.

4. Collect those data that will help to evaluate the effectiveness of your fire prevention efforts or identify new areas that need attention.

5. The fullest potential for the use of GIS information has yet to be seen.

6. For the fire protection profession to advance, we must be willing to accept new technology and the changes associated with it.

Learning Objectives

1. Identify technological advancements that can improve fire prevention and community risk-reduction efforts.

2. Understand the elements needed for technology to be a beneficial tool.

Using Technology to Improve Fire Prevention Efforts

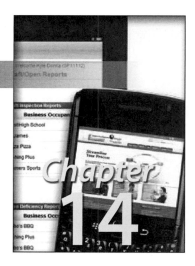

Case Study

The fire department's fire prevention bureau is an organization operating smoothly and serving a city that has been working hard to bring in new business, which includes semiconductor manufacturing and a very competitive commercial broker/realtor business attraction program. The brokers and realtors are working to sell older, vacant properties and office suites, but based on the influx of new businesses, there is a boom occurring with commercial office buildings. Unfortunately, your community has not had the resources to invest in technology. This lack of investment has resulted in antiquated permitting and inspection processes that are principally paper driven. System inspection testing and maintenance records are received by mail and then either filed or handed to an inspector. Customers must come into your fire prevention bureau and make cash or check payments for permits, call or come in to schedule appointments for inspections, and generally have to schedule meetings to get any information regarding code requirements or the inspection processes. The community has had an aggressive fire sprinkler ordinance in place for a number of years, resulting in an enormous amount of paper inspection testing and maintenance (ITM) reports from service providers.

These reports have resulted in a need for several clerical people, as they remain available for handling most questions while inspectors are out in the field. When the clerical staff is not working with customers, they enter data of manual inspection reports on a paper form. This staff also files incoming ITM reports and in-house fire inspection reports. In addition, the clerical staff schedules inspections and processes invoices and collections. The fire prevention staffing levels have decreased in the fire prevention bureau in the past five years. The fire chief does not have the support to fill the vacant positions and feels that a self-inspection program for select occupancies is something to consider. This creates another demand for records management.

1. Based on this scenario, what processes can you identify that could be improved upon (i.e. handling of ITM and inspection reports, self-inspections, improved permitting process, etc.)?

2. How would you propose making improvements or changes in the areas you identified?

3. What types of technology or solutions would you recommend to your chief? Should we purchase or lease? Are there other options to consider?

4. How can we cost effectively maintain the ITM reports to ensure all of the life safety systems are operable?

Technology and Change

Our environment is constantly changing, both at home and in the workplace. How we deal with those changes will determine how we survive productively in the workplace and in our home life. Some changes are occurring so fast that it is difficult to keep up, especially in the world of technology. Fifteen years ago, who would have thought that fire inspectors would carry computers that were also phones; have mobile data terminals that plot out inspection routes, locations of hydrants, and other infrastructure information? As well, did we ever think that cameras could provide ultra-high resolution and video documentation of various fires, violations, or incidents. Is all of this stuff really necessary, or is it just a collection of toys and gadgets?

Let's take a look at how we can use technology to actually make our job as fire protection professionals easier and more efficient. Efficiency is very important, particularly today given our push to be as customer and business friendly as possible. We have learned that fire departments will continue to do more with less. One way to compensate for that is to work smarter and more efficiently. Having to work harder is a given because there is a good chance that you will face less than adequate staffing to accomplish the mission.

Today's workforce is more accustomed to electronic and digital technology, individual problem solving, and independent work processes. This paradigm shift forces managers and supervisors to be more nimble and adaptive to different and faster ways of accomplishing and managing tasks.

Figure 14.1 The nature of fire inspections can generate enormous amounts of paper.

A critical function of managing and working in fire prevention bureaus is record or data management. The nature of the work in fire prevention bureaus is conducive to creating tons of records, which were notoriously paper (**Figure 14.1**). Essentially, fire prevention bureaus and their staff are data collectors, managers, and analyzers. How we use or don't use data and our ability to share data with our stakeholders to improve our community risk reduction efforts will determine the success or failure of our fire prevention efforts. Think about what goes on during typical fire prevention activities that we have covered in the previous chapters in this book. Let's analyze tasks of a fire inspection. The inspector conducts an inspection, identifies deficiencies, and generates data on items that must be corrected. In addition to violations or deficiencies, the inspection data may also consist of a number of other items such as the following:

- Business name
- Business license
- Occupant contact
- Emergency contact

- Fire alarm system information (ITM records)
- Fire suppression system information (ITM records)
- Other fixed fire suppression system (ITM records)
- Fire pump (ITM records)
- Key Box
- Building square footage
- Hazardous material information
- Violations noted
- Address
- Construction document review comments
- Acceptance test data
- Utility shutoff information
- Dates for follow-up or compliance dates. Guidance for additional information including contact numbers etc.

Fire prevention bureaus may also be involved in developing preincident plans used at emergency incidents. Even if the task of gathering the preincident data is performed outside of the fire prevention bureau, the data collected are very similar, if not identical. Data typically collected as part of the preincident plan include:

- Occupant contact information
- Address
- Building construction details
- Building dimensions
- Hydrant location
- Available water supply
- Hazardous materials
- Fixed fire suppression (i.e., sprinklers)
- Key Box information
- Fire lanes
- Fire department access
- Utility shutoff

For discussion purposes, we will consider a fire department in the process of conducting a hazard risk assessment similar to the one we examined in Chapter 6. Some of the information gathered in this process would include the following:

- Occupancy classification
- Building construction details
- Building dimensions
- Number of building stories
- Occupant load
- Available water supply
- Hydrant location
- Fixed fire suppression (i.e., sprinklers)

Now compare the data collected in each of these tasks. **Table 14.1** clearly illustrates a redundancy. This redundancy will not just be for these three tasks. Consider other tasks, such as construction document reviews and fire investigations. Information submitted by third party ITM inspection companies will repeat similar information. Finding ways to be more efficient rather than redundant is very important for fire prevention managers and supervisors. Interruptions to businesses are costly. Making multiple visits can be very annoying to the owner and portray a perception that the fire department is wasting resources and time. In addition, failing to share the data with the operations side of the fire department or other regulatory agencies does nothing to enhance our community risk-reduction efforts.

The more data we can collect and utilize across all lines, the better and more efficient our communication and resolution of problems will be. As multiple organizations and regulatory agencies are stretched thin also, remember that the data for the occupancy previously noted may not pertain only to fire prevention activities. Some could be used as dispatch information for emergency response, health and safety code enforcement, law enforcement information, etc. Coordinating all data collection activities and sharing access to that data would benefit the customer, the fire department and whoever else may have use for it. Fire departments are now placing computers in their apparatus, or even more commonly, in their Tablets, Slates, or Smartphones to collect and manipulate data **(Figure 14.2)**. Global Positioning System (GPS) devices are in vehicles and Smartphones and are a mainstay for delivery operations. These devices and others are used by Incident Commanders and company officers, both while en route to and at the emergency scene. The amount of information that can be provided to the responding units is limited only by the department's ability to define and collect it. Real-time photos can be taken and relayed to the Command Post, the emergency operations center, or across the country. The responding units can be provided with specific directions of how to get to the scene as well as a map. The communications system also can provide plan review data if it has been collected and managed properly in the beginning.

Figure 14.2 Having occupancy data available at incidents can benefit fire department operations.

Table 14.1
Data Collected during Fire Prevention Activities

Data Collected	Activity		
	Inspection	*Preplanning*	*Risk Survey*
Address	X	X	X
Business name	X	X	X
Business license	X		
Emergency contact	X	X	
Occupant contact information	X	X	
Building construction details	X	X	X
Building square footage	X	X	X
Building dimensions		X	X
Number of building stories		X	X
Occupancy classification	X	X	X
Building dimensions		X	X
Hydrant location		X	X
Available water supply		X	X
Hazardous materials (chemical inventory)	X	X	X
Fixed fire surpression (i.e. sprinklers)	X	X	X
Fire pump	X	X	
Fire alarm system information	X	X	
Knox Box information	X	X	
Fire lanes		X	
Fire department access		X	

Many businesses and governments are utilizing some means of Geographic Information Systems (GIS). This is a very detailed and complex system that can provide a wealth of information from parcels, tax information, addresses, water and gas lines, centerline road data, right-of-ways, topography, etc. This data can be invaluable in viewing and combining various layers of information for management, communication, awareness, emergency incidents, and the like.

It is one thing to collect and maintain data, but it is another thing to manage that data. Basically, our imagination is the limit of our use. However, while we can put all the data in the world into a fire prevention database, what are we really going to do with it? How do we intend to use it? Answering those questions dictates what we actually need to collect. This is critical because having information at our disposal is good. Having gigabytes of useless data just makes us hoarding practitioners of the infamous "garbage in vs. garbage out" paradigm. It is important to research what data is already available somewhere else in another format. Do not collect superfluous or incorrect data.

Referring to Table 14.1, we emphasize that if all of these program managers kept separate databases, they would be performing redundant and inefficient tasks. Do not limit your questions about who is collecting what data to just your fire department. Some of the information you need might have been collected or be maintained by another municipal department, municipality, county, or emergency services department in the community. An engaging stakeholder process discussed early would be very beneficial to help identify other participating partners. Maybe the department responsible for water billing or business licenses is also collecting and maintaining similar data. What about the building department? Are they keeping the same construction project information as the fire prevention bureau? Throughout this text we have emphasized the need to build coalitions and seek ways to be collaborative. For example, are inspection testing and maintenance service providers sharing the inspection results with your business owners (i.e. coalition member)? Creating and using teams to collect and manage data should be no different than how we team up on fire and life safety education programs.

Most fire departments do not have a full-time person assigned to information technology (IT) or management information systems (MIS). But this is not to say that no one in the department is performing those tasks. In many fire departments, there are resident experts who have had previous jobs, degrees, or trades in technological processes just before hiring on the job. Seek out these experts, as they can be a wealth of assistance in a multitude of tasks, if not just to provide quality assurance to various processes. If you are in a large department, we recommend developing an in-house IT group to both prospectively design business plans as previously discussed, but also to develop strategic plans for improving if not just keeping pace with technological developments in various arenas. With the advancement of web-based systems, the need for specific software and people to manipulate data is dwindling. There are APS and web-based solutions that are very user friendly to manipulate and manage data in a cost-effective manner.

Because of their size and scope of responsibility, smaller departments may not warrant a full-time information technology (IT) position or a management information systems (MIS) position. However, both are critical functions that can enhance the fire department's operation, efficiency, and customer service. If the functions are not managed correctly, the department's operation can be hindered and animosity may be created among the work group, your clients, or the public. So, should all fire departments begin to eliminate one line firefighter or officer and hire a full-time IT or MIS person? *Maybe not.* Experience has shown that within any community, a number of individuals are more than happy to devote volunteer time to assist a fire department in accomplishing its mission. Remember, team building is the key to fire prevention. Engage and solicit help, and likely you will find it. Obviously, the best situation is to have a person who is trained in fire department operations and functions and also has a formal education or at least a strong background in MIS or IT. Any MIS person hired to fill that position from outside the fire service will need some time to become familiar with fire department operations. It may even be feasible to use or contract with another fire department's MIS person if the size of your departments does not justify a full-time position. This is another instance where coalitions and agreements can pay off. This move can be significant if future financial situations prove to be challenging for one or the other department or both. Leveraging strengths provides huge economies of scale. Think outside the box.........look for web-based systems and or APS.

The "typical" data fields that we have discussed are only a small sample of the data that can be collected. The operations or functions your department is undertaking should be the driver to the data you collect.

The data that was previously noted discussed the deficiencies recorded during the inspection or the results of fire investigations. These are actually the type of data that are most useful in tailoring the fire prevention program to the community. As discussed in previous chapters, an essential component in our community risk reduction efforts is coalition building and utilization of outside organizations and agencies resources. Who else in the community is a stakeholder, and who else is collecting data that you can use in your community risk-reduction efforts?

For example, consider the additional data coming from an outside stakeholder, such as the service providers or inspection companies, within our communities. This entity, in some cases, is more visible to our commercial or business sectors because it actively inspects, tests, and maintains the fire protection systems on a regular basis. Almost every jurisdiction across the country requires some sort of proof from these companies that the fire systems in place are compliant. Again, most of these letters of compliance are paper copies and sent via mail or fax, or even printed from emails. The goal of our fire prevention efforts and the bureau's responsibilities is to reduce the impact of fire for our firefighters and community. However, at what point does the data and information management of not only our inside efforts, but the incoming inspection testing and maintenance (ITM) become a priority? Or is it simply too much? Consider a coalition of building owners ITM providers and the AHJ. All have a common goal... compliance. All need the data. All should share the data.

The development of Tablet devices and Smartphones has provided another means for fire prevention bureau personnel to collect data. As mentioned, fire departments historically have collected data on paper forms. They may be stored in their original paper format or through duplicate efforts, entered or scanned onto a server, and stored electronically. The electronic data may or may not be systematically placed in a database format for retrieval and statistical analysis. Again, these discussions must be held before determining what type of data collection and filing you intend to use. Fire departments are finding that the sooner they can format the electronic data for retrieval and report generation, the more accurate and useful they are. Simply storing scanned documents is good for space saving, but very difficult to use for research and analysis. Since computing technology has improved our means to retrieve data quickly, we need to ensure that we can enter them at an efficient and productive pace. Nothing is more frustrating than collecting data by hand, only to find out that your data entry is a year behind.

The best method to keep your data current is to enter them directly during the inspection, emergency incident, or survey. This can be done now wirelessly via mobile services, data transmissions, or via Wi-Fi hot spots (**Figures 14.3 a and b**). This provides instant or nearly instant data management that ties directly to the server and network that the rest of the community or department is working on. The data would be entered at nearly real time into the handheld device during the fire inspection, preplan inspection, or risk survey. It can then be available to all personnel in a matter of minutes instead of weeks or months. Emergency response personnel could use the information during an emergency incident in real-time. This type of ready access and accounting can also be used to establish trends, patterns, or targets that can be worked through the fire prevention bureau. Other data such as systems out of service could play an important role in the company officer's tactical decisions.

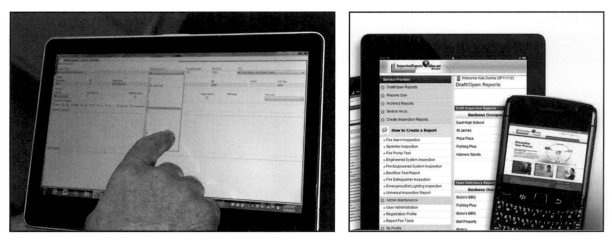

Figures 14.3 a&b Many fire departments use Tablets or Laptops for inspection data gathering.

Another possible fire-prevention use for computer technology is cataloging and tracking plan submittals through the use of Quick Response (QR) codes. These are becoming an important tool for cataloging and referencing many sets of data from identification of items and tracking to automatically going to website quick links for additional information. Since the advent of Smartphones, everyone can have a QR reader right in their pocket.

A number of commercially available software products can enhance the fire department's ability to use the data after they have been collected. The key is to ensure that the correct data are being collected for the intended outcome. For example, if you are collecting data pertaining to the color of exit signs, it is important that this information be used at some time. Never collect useless data. All it does is take up room, and then, like your junk drawer, the database will need to be cleaned.

Web-Based Solutions

In addition to the software choices already mentioned (GIS, QR, etc), a new trend has emerged giving departments yet another option for effectively and efficiently using technology: Web-based Inspection Report Management Systems. The web-based solution allows access from any device with internet connection. This option is eliminating many of the costs and requirements of traditional software programs, such as in-house storage and devices, license fees, upgrades and maintenance fees, and overhead costs such as the need to copy, print, mail, fax, and even store paper copies.

Currently, the development of web-based systems focuses on inspections; specifically the receiving, managing, following-up, and filing ITM reports within a jurisdiction, but some have expanded to include self-inspection and an in-house inspection program for fire departments/prevention bureaus (See Chapter 11 for self–inspection discussion). These systems are providing a means of streamlining the inspection process and providing and active platform for the stakeholders involved (fire department, inspection company, and the commercial property owner/occupant) **(Figure 14.4)**.

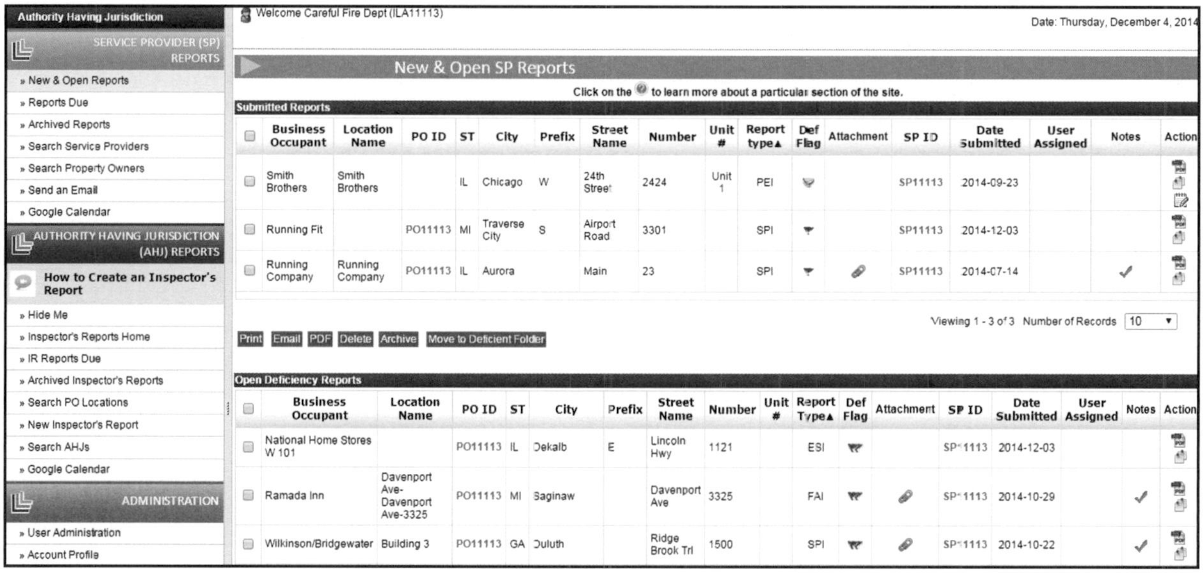

Figure 14.4 Web-based solutions can offer real time data to both operations and prevention personnel. *Courtesy of Inspection Reports on Line (IROL)*

We've already mentioned some of the key issues with current processes implemented in our fire prevention efforts, such as redundancy of incoming information, lack of staff to manage data, too much paperwork, etc. The inspection process is no different and in fact, offers additional challenges such as:

- Budget
  - Software is expensive and until now, each stakeholder never had the opportunity to purchase a program linking and engaging each other.
  - Departments, let alone property owners, do not have the money to spend on programs or systems.

- Time
  - With everyone doing more with less, the ability to manage incoming inspections or in-house follow-through with deficiencies has become increasingly harder to accomplish. In some cases, departments are unable to respond to deficient systems in a timely manner.

- Record Management and Retention
  - It is not unheard of for departments to have off-site storage for in-house inspections, but also for the ITM reports submitted to their jurisdictions.
  - When in the field, inspectors are finding it is harder to track down specific information regarding a single location because the data or information is either not easily accessible or cannot be found.

Web-based systems were developed to provide a comprehensive plan incorporating industry-proven strategies that also reach the 3 E's of fire prevention: Education, Engineering and Enforcement. The centralized and secure web platform links the stakeholders and provides a means of fire and life safety responsibilities, minimizing issues/deficient systems on all ends (education). Fire departments can utilize state-of-the-art tools to manage the inspection, testing, and maintenance of fire protection systems or other inspection processes. This levels the playing field and increases voluntary compliance, thus reducing the impact of fire (engineering and enforcement). Never before has technology given us the ability to be in the field with access to needed information at the touch of our fingertips, not to mention the ability connect with other entities and increase our partnerships and communication.

What are some of the key features of these systems?

- Organized in-boxes, giving the end user the ability to quickly identify and track information such as business occupant, address, type of system inspected, and status of inspection (deficient, critical or clear).

- Real and direct communication similar to a chat between entities allowing the sharing of inspection specifics and attachments such as documents, videos, etc.

- Built-in automatic notification capabilities, such as critically deficient reports and upcoming and past-due inspections

- Access to system using any device with internet connections (Tablet, Laptop, Smartphone, etc.)

- One central and secure location for data management, filing, and data retention, thus eliminating the need for paper storage, copies, mail-ins, faxes, and even e-mails.
- Built-in self-assessment capabilities.

Behind the scenes, some Web-based systems are more adaptable and flexible than software, which is one reason many departments are going in this direction. The cost to implement this solution is typically zero on the fire department's end. Many of the companies offering a web solution place a pay-as-you-use submittal fee on the inspection company. This, too, is another major benefit of incorporating this type of technology in our fire prevention bureaus. The cost is absorbed by another entity.

A number of Web-based systems are available. Evaluate them to ensure they exceed your expectations and have no hidden costs. An example of a highly reputable Web-based system with ITM and self-inspection capability used by a number of fire departments large and small throughout the United States is *InspectionReportsOnline.Net* (IROL). This program utilizes a state-of-the-art web-based system with cloud technology, which allows Authorities Having Jurisdiction (AHJ) and Commercial Property Owners (PO) to receive, manage, and consolidate Inspection Reports submitted electronically by Service Providers/Contractors (SP).

IROL allows AHJs and Commercial Property Owners to receive and manage Fire/Life Safety Inspection Reports from any and all service providers at **NO Cost to the AHJ** from one central location. In addition, the site assists fire departments in performing, completing, and submitting their own in-house inspection reports. Our experience has shown that this method increases communication and code compliance between all three entities, streamlining the entire inspection process. (For more information visit *www.InspectionReportsOnline.Net)*.

Data Collection

What data should be collected? That depends on the fire prevention efforts determined in the strategic planning process (see Chapter 4). The data should help to evaluate the effectiveness of your fire prevention efforts or identify new areas that need attention.

Geographic Information Systems

Geographic Information Systems, as previously mentioned, have become critical and are a mainstay of analysis and communication. Google Earth is a format of this data and is a classic example of the technology available to anyone with an internet connection. Beware, communities suffering disasters are receiving nearly real-time shots of damage and destruction of their neighborhoods and homes before they even have a chance to survey the damage.

These systems are an extremely powerful and unique way to share information with not only fire department staff, but also the public we serve and the policymakers who provide the resources to accomplish our mission. The kinds and levels of analysis available now to researchers, policy advisors, and decision makers were only dreamed of as recently as a decade ago. With every passing day, more and more information and potential use come to light.

GIS works by taking base data of given parameters and laying those data over other layers of data. An example is gathering information on fire hydrant locations, which are specifically cataloged by geo-based coordinates. They are placed into a map format called a layer. A second layer of data can be created—a satellite photo of the community, for instance—that reveals the exact footprint of each structure and all the street centerlines within the study area. By combining these two layers, one over the other, the footprint of the structures along the streets with specific hydrant locations can be placed on a map that provides great detail and visibility.

The base data come from any number of sources, such as private contractors, survey companies, public and private utilities, government agencies, and the military. These data can be in the form of infrared, topographical, photographic, or other images. Tabular data are easily placed into map layers that are graphical images. For example, census data can be layered over postal zip code tracts to provide an accurate picture of demographic information for use in public information programs. Similarly, fire incident data can be transferred to fire demand zone (FDZ) layers.

The fullest potential for this type of information has yet to be seen. When the tragic events of 9/11 unfolded, GIS information was used to show accurately where buildings and open spaces used to be versus where they were after the disaster. The information was shared with all of the emergency response teams, and the photos could even be seen over the Internet. This provided invaluable information for search and recovery operations. Another example is the use of satellite imagery during the severe wildfire season of 2012 and 2013 in Colorado. Infrared satellite images of fires, such as Waldo Canyon and Black Forest, provided real-time pictures of where the flame fronts were advancing; yielding accurate understanding of time relationships to decision points that had been established for initiating community evacuation orders.

Are We There Yet . . . Is the Technological Advancement Over?

Technology will continue to advance. The progressive fire protection professional should be situationally aware of those advances and constantly explore how to incorporate them into his or her work. There are wristwatch Smartphones currently available that were only a dream in the old Dick Tracy comics of days gone-by. Technology should be used whenever possible if it enhances your job performance or improves the way you do business.

The use of digital photos for fire inspections and fire service training is common. These digitized images are useful not only for in-house applications, but also for board of appeals hearings, city council meetings, and other venues. More fire prevention bureaus are using digital cameras to document fire safety deficiencies or existing conditions during inspections. Almost every

Smartphone has a camera. For example, restaurants often modify their cooking appliance arrangement to accommodate their changing menus. These modifications can alter the effectiveness of the automatic extinguishing system. Taking a digital picture of the approved cooking equipment during initial installation allows the fire inspector to compare what he or she sees in the restaurant a year later to the digital photo. The digital photo can be accessed along with other photos of the occupancy. Other uses could be that the first responders would have a map of where they are going on their vehicle laptops, and they may even have a photo of the building.

In some jurisdictions it is also acceptable to share photos during self-inspection programs or send photos of corrected violations to inspectors to eliminate additional trips and time delays. This has also proven to be helpful in gaining collaborative cooperation with business owners who can share in the overall code enforcement and inspection process by exhibiting trust and confidence that all are working together to fix dangerous conditions.

For the fire protection profession to advance, we must be willing to think ahead of the power curve. We must anticipate our needs and the needs of our customers and seek opportunities to meet those needs through advancing technologies, products, or processes. We do not want to be left on the corner, as an opportunity to improve the way we prevent and mitigate fires passes us by. Businesses will not tolerate that and neither should we.

Cautions and Drawbacks

The digital age, while exciting and beneficial, is fraught with challenges. As a manager or supervisor of any fire department or division within the fire department, the aspects of social media and Smartphone photos has proven to be an adventure. Develop and cultivate policies and practices that manage the use of photos from incidents, horseplay, or other activities that can be compromising, illegal, or just foolish. For example, a fire department arrives on the scene of a tragic vehicle accident to provide aid to the victims. One of the crew members takes a photo of the scene and immediately posts it on Facebook or tweets it. The carnage of that act becomes instantly detrimental to the department and the community—let alone the family of those involved in the accident. Having this same type of thing occur on a fire that is determined to be arson causes immediate complications, such as release of confidential crime scene information and opportunities for other issues.

Manage digital information carefully, educate and coach staff and others properly in its collection and use, and ensure policies are in place to help everyone. The adage that, "Once it's on the internet, it's there forever," is very true. Be thoughtful in your plans, develop policies and training, and adjust regularly as technology changes and advances.

Summary

Technology continues to play an important role in fire prevention. The use of a multitude of handheld devices for data collection and fire prevention bureau management is just one of many advancements that have assisted fire departments. The advent of the personal computing and communication devices and their continued evolution has allowed further advancements, such as geographical information systems (GIS), QR coding, real-time access to important information, and more. This allows fire departments to access information that may range from the street location of the incident and the nearest hydrant to topographical features and drainage locations or more.

Other important information, such as topographical details, can be essential during a wildland fire or hazardous materials incident. Other advancements like Smartphones have enabled fire departments to improve efficiency and reduce costs. This is critical to justify purchase of such equipment. Technology will continue to advance, and fire departments must constantly monitor the environment for opportunities to use the latest technology to handle hostile fires through suppression and prevention to provide the best possible customer care and support we can. Web-based systems are more adaptable and flexible than software and are considered the optimal solution for many fire departments. The cost to implement this solution is typically zero on the fire department's end. Many of the companies offering a web-based solution place a pay-as-you-use submittal fee on the inspection company. The ability to use a web-based system and build a coalition with ITM service providers is a big step forward in building your departments community risk-reduction program.

Chapter 14 Review Exercises

14.1 How can the use of technology support fire prevention efforts?

_____

_____

_____

14.2 Explain what technological advances in the past 10 years have impacted the fire department's ability to prevent fires? _____

_____

_____

_____

_____

14.3   Explain how you would present a request to purchase an item(s) such as a Tablet computer for use in the fire prevention bureau?

_____

_____

_____

14.4   How can global information systems benefit a fire department?

_____

_____

_____

14.5   Provide examples of where the technology available today needs to advance to a higher level to improve the way fire departments prevent fires._____

_____

_____

_____

_____

14.6   To keep up with the advances of technology, should fire departments purchase the latest products as soon as they become available? Why or why not? _____

_____

_____

_____

14.7   What do the following acronyms abbreviate: GIS, IT, and MIS?

_____

_____

_____

14.8   What are some of the benefits of using Laptops or Tablets on routine inspections?_____

_____

_____

_____

14.9   Do courts allow digital images to be used as legal evidence?

_____

14.10   Discuss ways technology could be used in the field of fire prevention. _____

_____

_____

_____

_____

Appendix A

Websites Listed in Fire Prevention Applications 2nd Edition

(Current as of this printing)

1. http://www.nfpa.org/research/reports-and-statistics/the-fire-service/fire-department-calls/fire-department-calls

2. http://www.strategicfire.org

3. www.IFSTA.org

4. www.usfa.fema.gov

5. http://www.nifc.gov/fireInfo/nfn.htm

6. http://www.corelogic.com/research/wildfire-risk-report/2013-wildfire-hazard-risk-report.pdf

7. http://www.newscenter.berkeley.edu/2012/06/12/climate-change-global-fire-risk

8. www.InspectionReportsOnline.Net

9. www.usfa.gov

Appendix B

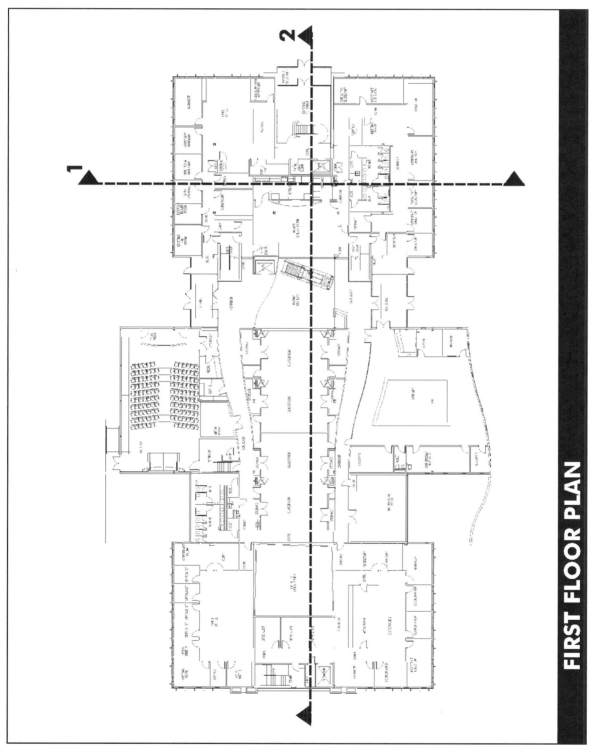

FIRST FLOOR PLAN

Figure 10.3 Floor plans are a type of architectural plan.

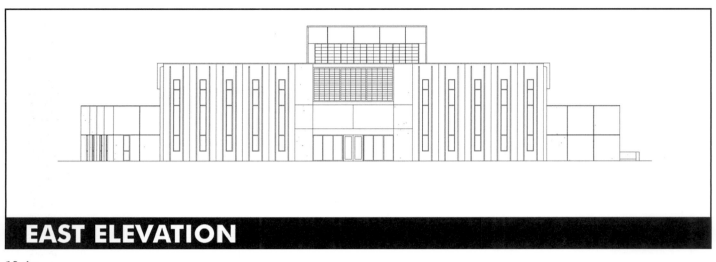

EAST ELEVATION

10.4 a

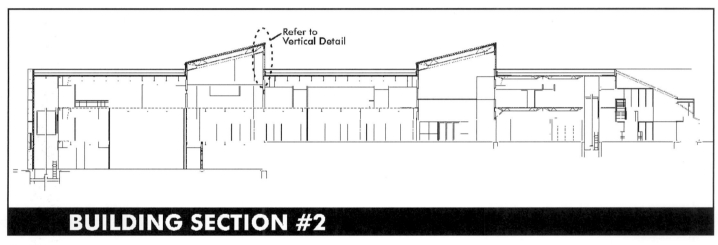

Refer to
Vertical Detail

BUILDING SECTION #2

10.4 b

Figure 10.4 a-c Three types of construction drawings are (a) elevation views, (b) section views, and (c) detail views. *All three courtesy of C.H. Guernsey & Company*

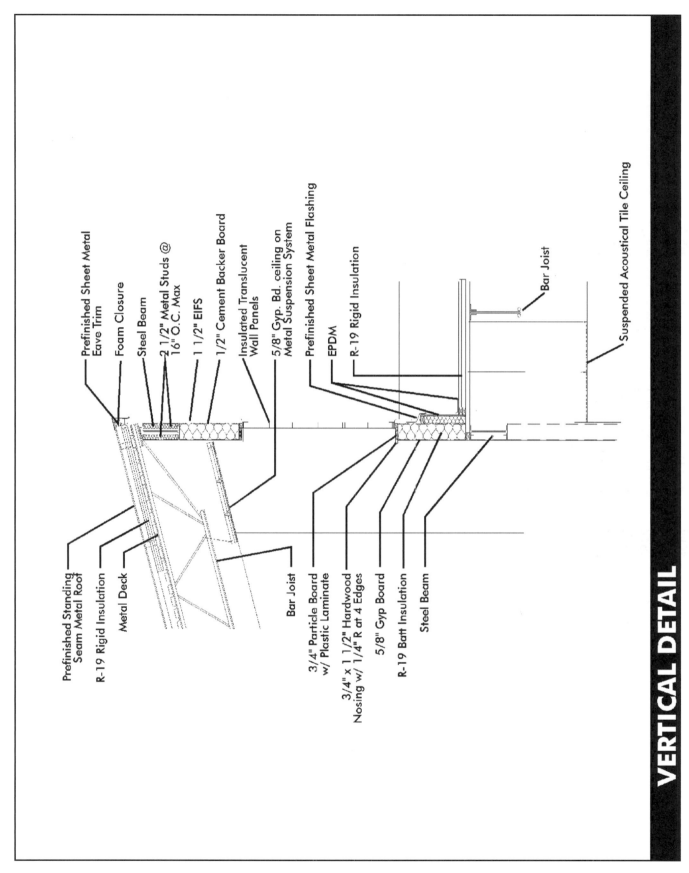

Prefinished Sheet Metal Eave Trim

Foam Closure

Steel Beam

2 1/2" Metal Studs @ 16" O.C. Max

1 1/2" EIFS

1/2" Cement Backer Board

Insulated Translucent Wall Panels

5/8" Gyp. Bd. ceiling on Metal Suspension System

Prefinished Sheet Metal Flashing

EPDM

R-19 Rigid Insulation

Bar Joist

Suspended Acoustical Tile Ceiling

Prefinished Standing Seam Metal Roof

R-19 Rigid Insulation

Metal Deck

Bar Joist

3/4" Particle Board w/ Plastic Laminate

3/4" x 1 1/2" Hardwood Nosing w/ 1/4" R at 4 Edges

5/8" Gyp Board

R-19 Batt Insulation

Steel Beam

VERTICAL DETAIL

10.4 c

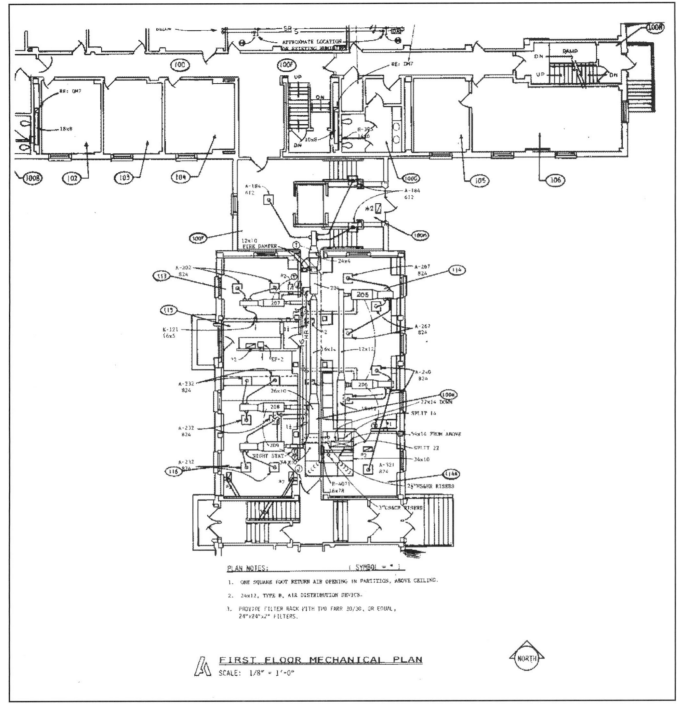

Figure 10.5 A typical mechanical plan.

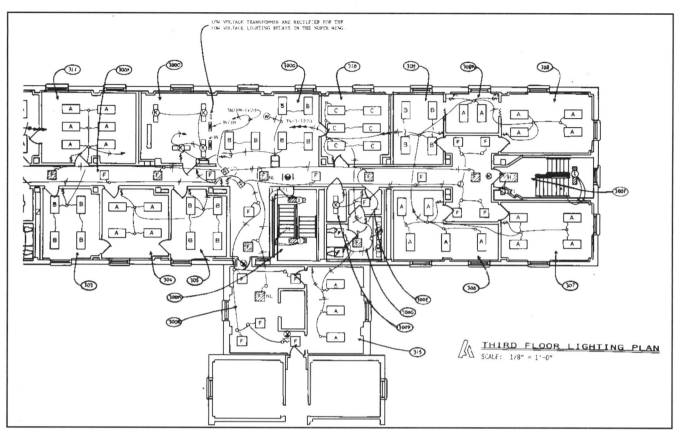

Figure 10.6 A common electrical system drawing.

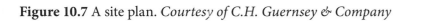

Figure 10.7 A site plan. *Courtesy of C.H. Guernsey & Company*

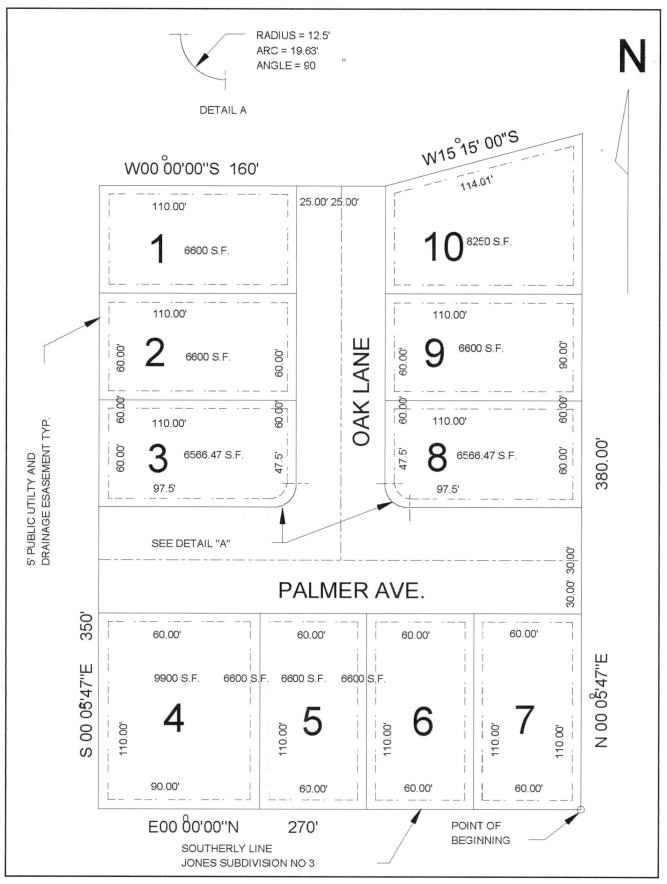

Figure 10.8 A typical plat.

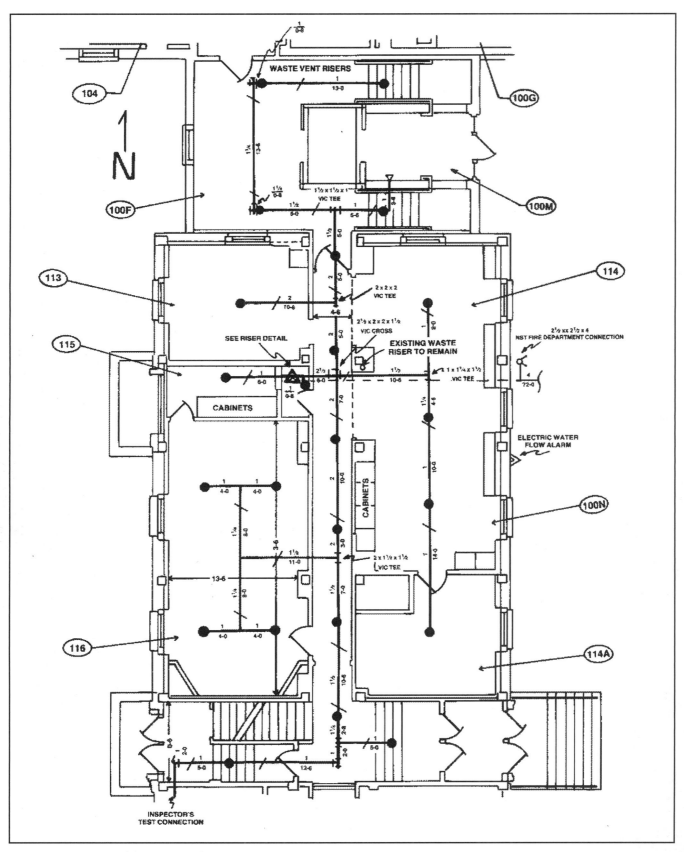

Figure 10.9 A sprinkler system drawing.

Fire Prevention Applications

Chapter Review
Exercise Answers

Chapter Review Exercise Answers

CHAPTER 2

2.1   An understanding of the relevance of historical fire events lays the groundwork for how we got where we are today in the field of fire prevention. Fire prevention professionals will at times be asked to explain why we need to do something or what is the purpose or intent of the code requirement. If fire professionals understand what led to the development of fundamental code requirements, then they will be capable of explaining the purpose of or reasoning for code requirements. Fire protection professionals can also use historical events to gain insight into what contributed to a fire loss and apply those lessons in preventing similar losses.

2.2   A tragic event creates media attention. The public tends to show an interest in preventing the event from occurring again. Many times the codes are changed as a result of tragic fires. This is sometimes perfect timing for code changes at the local level even before they are modified by the model codes.

2.3   In 1189 the first lord mayor of London mandated that houses be built of stone. Some buildings were required to have party walls with rings for hooks.

In 1608, a fire in Jamestown, Virginia, led to requirements for thatch roofs and wood chimneys.

2.4   The Iroquois Theater fire provided a number of lessons:

Stages need to be protected by automatic sprinklers. At the time of the fire, the building had none.

Draperies, such as the stage curtain, need to be fire resistive.

Permitted occupant load limits must be obeyed.

Employees need to be trained in evacuation procedures.

2.5   One of the most significant findings was that a building's means of egress must include a sufficient number of stairs in fire-resistant shafts with rated fire doors at each opening.

2.6   At the National Fire Protection Association's annual meeting in May 1911, R. H. Newborn presented a paper on exit drills and educating factory workers. A year later, Newborn's paper became the NFPA's first safety-to-life publication under the title "Exit Drills in Factories, Schools, Department Stores, and Theaters."

2.7   The lessons learned from the Our Lady of Angels incident included:

Automatic sprinklers with water flow alarms would have reduced or eliminated the loss of life at this incident.

The open stairways created vertical passageways for the smoke to travel. Like the other fires, this event again demonstrated the need for enclosed stairs.

The dangers of transoms over doors and combustible finishes were also identified.

2.8   The building had undergone many additions and had no automatic sprinklers. The interior finishes along the walls and ceiling in many areas of the building were combustible. The facility's occupant load was exceeded. Exits were obstructed or locked. Some exits were not clearly identified.

2.9   Both fires occurred in an industrial occupancy that resulted in large life loss. The Triangle Shirtwaist fire was the basis for the establishment of what later became the *Life Safety Code®*. The Food Processing Plant fire occurred 80 years later at a time when the *Life Safety Code®* was already established and when technology was on the rise. In both fires, contributing factors to the loss of life included the lack of automatic sprinklers and blocked or obstructed exits.

2.10   Vison 20/20 was created under the auspices of the Institution of Fire Engineers (US Branch)— a group of fire service and related professionals joined together to conduct a national strategic planning process. The result was the Vision 20/20 project which created a National Strategic Plan for Loss Prevention.

2.11   *Prevention and Public Education.* Emphasizes the need to expand this resource.

Training and Education. Addresses managers increasing their professional and leadership roles in order to remain credible to policymakers, administration, staff, and the public.

Fire and Life Safety Systems. This issue concerns the need for adopting and supporting more codes and standards that mandate these protective systems use.

Strategic Partnerships. Participants explored the need for the fire service to reach out and enlist the support of other individuals and groups in accomplishing the overall mission of fire protection and emergency service response.

Data. Measurable data are crucial to understanding where we have been and where we are going.

Environmental Issues. The need for the fire service to comply with local and federal laws both in mitigating incidents and in providing for the safety and welfare of our employees and partners.

2.12   *America Burning* is a document that explains America's overall fire problem going into the 1970s and the issues that needed to be addressed.

2.13   The nation needs to place more emphasis on fire prevention.

The fire services need better training and education.

Americans must be educated about fire safety.

In both design and materials, the environment in which Americans live and work presents unnecessary hazards.

The fire protection features of buildings need to be improved.

Important areas of research are being neglected.

2.14   The standard test was developed using wood and other combustibles. It is used to compare how things burn and often is applied today as a benchmark to draw a comparison of a product used in construction.

2.15   There needs to be more research on providing cost-effective fire protection features for residential occupancies such as the single-family home. This is just one of the many areas that need to be addressed to reduce fire deaths.

2.16   The National Fire Protection Association®. This organization promulgates codes and standards.

2.17   Underwriters Laboratories and Southwest Research Institute

2.18   The functions may vary from state to state but typically include: code enforcement, fire and arson investigation, construction document review, inspections, fire data collection, fire legislation development, fire and life safety education, fire service training, and licensing.

2.19   The United States Fire Administration (USFA) administers the federal data and analysis program and serves as the primary agency to coordinate arson control programs at the state and federal levels. This agency also administers a program concerned with firefighter health and safety.

2.20   The tragic fires identified in this text all resulted in multiple loss of life. The facilities had some form of exit deficiency such as blocked or obstructed exits. The facilities lacked the installation of automatic sprinklers throughout. Installing automatic sprinklers and providing adequate exits could have altered the outcomes. An aggressive fire safety education program combined with the installation of the built-in fire protection features are also important components of altering the outcome of the tragic fires studied in this text.

CHAPTER 3

3.1   Buildings or structures collapsing. Buildings constructed poorly. Pressure from the insurance industry and loss of life.

3.2   A body of law systematically arranged to define requirements pertaining to the safety of the general public from fire and other calamities. The purpose of codes is to establish minimum requirements for life safety.

3.3   ISO provides an analysis of the community's fire protection system and grades it based on recognized standards.

3.4 One.

3.5 Ten percent.

3.6 Standards are rules that dictate how something is to be done.

3.7 Codes typically are thought of as written documents that answer the questions who, what, when, and where concerning various requirements and their enforcement. A code will tell when something is required. A standard will tell you how to do it. The building code requires a sprinkler system. NFPA® 13, *Standard for the Installation of Sprinkler Systems*, indicates how the sprinkler system is to be installed.

3.8 Appendices in model codes or individual standards contain a great deal of information ranging from background material, history, simplified tables, or interpretations to good fire protection practices. However, they are not enforceable unless specifically adopted.

3.9 Those that govern the construction and occupancy of a building when it is being planned

Those that govern the construction and occupancy of a building when it is being constructed

Those that regulate activities conducted within a building once it has been constructed

3.10 The building code specifies how to construct a building to prevent the spread of fire by construction features or hazard arrangement.

3.11 A maintenance code manages and "maintains" the safe conditions in a building as it ages and as processes are changed or modified.

3.12 Highly skilled fire prevention staff

Computer applications

Web-based communications

3.13 These codes differ from the prescriptive model codes in that they have no standards for reference but utilize an engineering approach to determine protective measures based on goals and objectives.

3.14 Performance-based codes are typically used in large, irregularly shaped or uncommon facilities with complex industrial processes that may not be addressed in the normal prescriptive code process.

3.15 The performance-based codes provide design flexibility for the building designer.

3.16 The performance-based design is based upon the conditions of the building at the time of construction as well as the anticipated use of the facility. Once the building is constructed and occupied, the building may undergo changes that impact the performance-based design. The fire official is continually responsible for the building and must monitor changes that impact the performance-base design. Additionally, the use of the building may be severely restricted.

3.17   Technical committees or groups formulate drafts of new codes as requested. A standards council or other similar group generally issues the approval for certain codes or standards to be drafted and or revised if it is not already on a set printing schedule. Technical committees also may hear numerous appeals throughout a given code cycle. Once the assigned committee or group makes drafts or revisions, drafts are put out for public comment. The technical committee receives the comments and then makes changes. In some cases, staff makes these changes rather than a technical committee. The revised proposal is then republished for public comment. Often, hearings are held to discuss the proposals and potentially to receive comments from the floor. In other instances, residents submit comments again, and the technical committee makes one last revision. The code is then ratified, accepted, and printed.

3.18   Code renewal cycles repeat approximately every four years. The renewal process offers an opportunity for code committee members as well as the general public to submit requests for changes.

3.19   The federal government's facilities are exempt from compliance with local ordinances. However the municipality may request compliance in order for the facility to get water or emergency service protection from the municipality.

3.20   The level of authority is actually situational. There are cases where the local authority does not have jurisdiction and the state does. The level of authority will also vary from state to state. In some states, the community can adopt codes and ordinances stricter than the state codes. The most stringent codes would then apply.

3.21   The appeals process must be spelled out in a form of ordinance or part of the adopted model codes. The process begins with a written appeal of a fire official's code interpretation. For example, the adopted appeals procedure may require the appeal to be made to an appointed board of professionals or to an individual such as the city manager.

3.22   The level of authority will be determined by the structure of the two entities. For example, there are municipalities where the fire department is responsible for the building requirements and vice versa. If, for example, the fire marshal manages the fire prevention bureau and the building department, then he or she will have more authority. The authors of the text believe that the building department and the fire department should have equal authority and strive to work together to address all areas of life safety.

3.23   Codes cannot be written and approved at the same pace that technology may advance. While the codes are in the revision cycle, technology is already creating new products or commodities that may not be in the codes.

3.24   The authority having jurisdiction.

CHAPTER 4

4.1 Most government jobs are being scrutinized, and taxpayers are not always in favor of the cost associated with the services provided. Citizens are paying for the fire service professionals employment; therefore, they need to be treated like any other paying customer in a business establishment selling services.

Empathy for clients

Try to be more of a customer advocate. Understand that your organization likely knows far more about fire code issues than customers do. Be their resident consultant regarding your policies, procedures, and laws.

Remember that every time you open your mouth or put pen to paper while doing code enforcement work, you are likely costing somebody a lot of money.

Be mindful of community economic conditions and try to be flexible. Meet the intent of the code, not necessarily the rigid prescriptions listed in the document. Your job is to help business, commerce, and quality of life in your community, not to stifle it.

4.2 3–5 percent

4.3 Change is inevitable and economics is a large driver. The fire service will likely be pricing itself out of a job if it does not begin to consider better and more efficient ways of providing a necessary service for a lower cost. While we are not a business, we should try to be more businesslike.

4.4 Privatization
Enterprise
Cost recovery/fee based
Revenue offset

4.5 To expedite and better control their interests. The less regulation on their industry, the better their performance in their world.

The result, whether good or bad, depends solely on the local community and what they desire and get for their money.

It can free up general fund monies to sustain other fire department operational needs or help improve other city financial concerns.

It prevents governmental agencies from squandering fire prevention operational funds.

It typically is cheaper to operate and run than traditional Civil Service employee options.

It enhances consistency in operation and congruency in code enforcement due to reduced employee turnover.

4.6 Which method is best depends solely upon local conditions, organizational culture, community demands, and policy decisions. The key is to have flexibility, technical knowledge, economies of scale, and economically sound business practices.

4.7 Familiarity with operational procedures

Organizational promotional opportunities

Experience with organizational culture and policies

Familiarity and respect among sworn ranks

4.8 High costs

Frequent turnover

Limited experience

Limited flexibility with talent

4.9 Low cost

Consistency

Good training investment (they stay after you pay)

Tenure

4.10 Requires careful search and preemployment screening for qualified candidates

Potentially limited knowledge of local policies and procedures

Organizational respect issues

4.11 Upward organizational mobility

Consistency

Educational opportunities

4.12 Our external environment (business, community and governmental) changes constantly.

Fire and injury problems change.

Employee personality types change (X Gen vs. XY Gen types).

4.13 *Any three of the following:*

Help people get rid of the old stuff before moving to the new.

Provide answers to questions while change is taking place.

Involve people in the process.

Don't just talk the talk, but walk the walk.

Communicate, communicate, communicate.

4.14. Who are we?

What basic social or political needs do we exist to meet?

What do we do to recognize, anticipate, and respond to problems?

How do we respond to our stakeholders?

What are our philosophy, values, and culture?

What makes us distinctive or unique?

4.15 Answer will be based on local issues.

4.16 Much of what a fire prevention division does impacts not only community injury but response times, organizational costs, resource planning, and overall community risk management.

4.17 *Any five of the following:*

Proactive education programs at nursing homes

Low-hazard inspections reduced from annually to every five years

Fewer vehicles responding to selected types of incidents

Public education programs for EMS

Smoke detector blitzes in key areas such as housing projects

Aggressive inspection programs

Public education and awareness programs

Ambulance service providers taking nonemergency medical calls

4.18 *Any four of the following:*

Defines the purpose and objective for specific hazards based on organizational mission and values

Creates an environment where your entire department has ownership

Creates synergy by pushing everyone to be his or her best all at the same time

Communicates goals and strategies to your clients and the policymakers

Helps make sure you are using your department's resources to their fullest

Provides a benchmark or baseline from which to measure progress

4.19 Yes. Dead-end jobs are not always attractive. Promotional opportunities and enrichment should be attempted and provided whenever possible. This not only satisfies the employee but helps the organization hold onto valuable resources.

CHAPTER 5

5.1 The different methods of staffing the fire prevention bureau include sworn personnel, civilian personnel, or a combination of both, as well as either full-time or part-time employees.

5.2 The differences between using sworn and civilian personnel will vary from department to department and from person to person. Generally speaking, however, sworn personnel can offer an understanding of firefighters' needs as they relate to built-in protection systems and fire department access. They have an extensive knowledge on how things burn in a fire. They can use these skills to mitigate and prevent fires. Civilian personnel may offer some technical training, such as fire protection engineering, that most sworn personnel do not have. Some civilian personnel can be hired at a lower cost than a sworn firefighter.

5.3 Outsourcing is sending work outside the organization for completion.

5.4 Outsourcing can reduce workloads and use specifically trained or educated individuals to perform certain tasks; for example, outsourcing construction document reviews to a fire protection engineering firm.

5.5   Education, engineering, and enforcement. Education takes place mainly in fire and life safety education presentations. Engineering aspects are found mainly in the construction document review. Enforcement includes ensuring compliance with codes.

5.6   The position is typically an assistant chief or battalion chief who reports to the chief.

5.7   Prevention or suppression

5.8   Organizational structure

        Staffing options (i.e. civilian vs. sworn)

        Number of employees

        Level of service to provide

        Budget

5.9   Span of control is the number of people a person can directly supervise effectively.

5.10   Construction documents related to zoning

        Building construction (architectural)

        Roadways (civil)

        Fire suppression and fire detection

5.11   Historically, fire prevention bureaus were staffed with individuals who did not fit the mold of a firefighter. Many fire departments have seen the need to have competent people perform the roles of fire prevention who meet national fire protection professional qualification standards.

5.12   Performance designs allow the builder to select construction methods and materials as long as they can be shown to meet the performance criteria through an engineering analysis.

5.13   Forcing firefighters to work in fire prevention bureaus may cause turnover if they are reluctant to perform fire prevention tasks. The firefighter may have entered the fire service with suppression tasks in mind. The civilian employee may become frustrated with the pay structure and lack of promotional opportunities.

5.14   Options include using a third party for construction document review and inspections.

5.15   Those departments that choose to contract out will lose some degree of ownership as they will not have direct control over all the aspects they otherwise could. In many cases, there will be a lack of quality control regarding a true vested interest in the overall fire department's mission. More importantly, suppression personnel within the fire department are among the customers supported. It is imperative that all fire prevention bureau personnel have the interests of fire suppression personnel in mind when they are performing their fire prevention duties.

5.16 The engineering section is responsible for construction document review and inspection of fixed systems. The fire and life safety education section is responsible for conducting fire and life safety education presentations. The inspection sections conduct routine fire inspections. The investigation section is responsible for investigating the cause and origin of fires. The public information section serves as a primary media contact at emergency incidents in the capacity of public information officer.

5.17 NFPA® 1031, *Standard for Professional Qualifications for Fire Inspector and Plan Examiner*.

5.18 Types of inspections can include business licenses, complaints, and inspections of existing occupancies.

5.19 Preincident planning is the process of identifying specific occupancies, buildings, or locations likely to require special treatment or operations should an emergency occur.

5.20 NFPA® 1033, *Standard for Professional Qualifications for Fire Investigator*.

5.21 Policies and guidelines for conducting fire investigations

Fire origin and cause determination

Data collection methods

Relationships with various legal authorities within the jurisdiction

5.22 The prevention and mitigation of wildland fires

CHAPTER 6

6.1 Risk is the exposure to possible loss or injury.

6.2 Provides strategic planning, communicates, and clarifies what issues are impacting your community and your department and provides a basis for addressing the problems.

6.3 Credible and accurate data for strategic planning

Communicate better with your public and elected officials

Identifying the right targets

6.4 The answer will depend upon local area and culture.

6.5 Mitigation is simply the modification of an unwanted outcome to a level that is sustainable, survivable, or nearly nonexistent due to protection measures.

6.6 Mitigation changes the outcome or severity of an event, where prevention actually stops the event from ever occurring.

6.7 A typical risk could be a simple single-family residence that your department routinely handles time and time again. No special effort, equipment, or training is required to deal with the event. Another typical risk could be a small office building that is easily addressed by a department's existing resources.

6.8   A measure of the severity and probability of an unwanted event

6.9   The actual value of the risk. This is basically the public's opinion of the risk.

6.10   So that policymakers, citizens, and elected officials will be better able to understand, agree, and communicate priorities for dealing with the right issues and in the best time frame.

6.11   The act of engaging the public with factual information concerning issues that are important to them and your department while learning from them "what they want and can't live without."

6.12   Accepting what the public feels is important, even if it seems not to be in alignment with what the department thinks.

CHAPTER 7

7.1   The effectiveness of Fire and life safety education has been documented to reduce injuries and losses due to fire. Elimination of the fire department's fire and life safety education program will result in an increase in injury and loss due to fire.

7.2   The goal is to focus on human behavior modification to reduce deaths and losses from the effects of fire.

7.3   Fire and life safety education is a proven method to motivate the public to act in a fire-safe manner. It is also an opportunity to teach people what to do.

7.4   Fire departments have expanded their role in educating the public to act safely in a number of areas. The inclusion of additional prevention topics led to the term *fire and life safety education.*

7.5   Fire departments have expanded their emergency response from just putting out fires to other services that include emergency medical services, swift water rescue, ice rescue, hazardous materials, and other specialty teams. Because of the increased services, fire departments were witnessing citizens being injured from a variety of activities. Fire departments then began to incorporate prevention topics for these activities.

7.6   Topics may include: water safety, bike safety, seatbelts, car seats, ice safety, and babysitter training

7.7   Vision 20/20 National Forum gathered together in Washington DC for the development and support of a national strategic agenda for fire loss prevention. Many free programs as well as community risk reduction material is made available free to fire departments.

7.8   The effectiveness of a fire safety education program should be measured by its impact on the "bottom line." In this case, the bottom line would be the reduction or elimination of the community's fire incident, injury, and death rates.

7.9   The following can be obtained from the United States Fire Administration (www.usfa.fema.gov). Funding for grants, juvenile fire setter intervention handouts, after-the-fire brochures, reports on residential sprinklers, *America Burning Revisited,* and Beyond Solutions 2000 publications.

The following are available from the National Fire Protection Association at www.nfpa.org. Fire statistics in the United States, firefighter deaths, fire reports for specific occupancies, Fire Prevention Week information, and fire and life safety education articles.

An online **Community Risk Assessment Guide** (also available as a PDF) to help fire departments and other organizations to conduct a basic or more complex assessment of risks within their community. Ultimately, the results of the risk assessment can be used to develop a CRR plan.

The **Community Risk Reduction Planning Guide** takes you through the remaining four steps of the CRR planning process.

The Consumer Product Safety Commission offers information on product recalls that may have contributed to fires.

7.10   Programs can include a fire engine ride to school for children who successfully accomplish an academic goal determined by the school principal (i.e., reading a set number of books), having the fire chief read a fire safety related book out loud to the children, sponsoring a fire safety education poster contest, producing a fire safety education calendar, having a save-our-senior day with fire prevention topics for the seniors, and inviting a senior housing complex to the fire station for a cookout and conducting a presentation.

7.11   Sign boards or message boards can be used for a variety of ages. Web pages with interactive fire and life safety messages are good for children. Stickers and coloring books are effective for preschool age children. Newsprint is a medium for adults and seniors. A variety of social medias exist.

7.12   A firefighter speaking to the group works well with seniors; puppets or clowns work well with younger children. Children in kindergarten through grade 5 learn well with videos, TV, and compact discs. Social media and internet based interaction works well with most children and young adults.

7.13   Answer will depend on local circumstances.

7.14   The report indicated that fire prevention could not be effective without public education.

7.15   The report should address the recommendations that focus on targeting the high-risk groups by improving fire safety in the areas of egress, early alarm notification, and fire sprinkler protection. It also should explain in detail how each of the areas of improvement will impact fire safety for the targeted groups.

7.16   Answer will depend on local circumstances.

7.17 Community Risk Reduction (CRR) is defined by Vision 20/20 as a process to identify and prioritize local risks, followed by the integrated and strategic investment of resources (emergency response and prevention) to reduce their occurrence and impact. Much of the current literature and training materials suggest that Community Risk Reduction programs use a six-step approach towards development.

7.18 The Vision 20/20 planning guide identifies the following six steps.

CHAPTER 8

8.1 NFPA® 1035, Standard on *Fire and Life Safety Educator, Public Information Officer, Youth Firesetter Intervention Specialist and Youth Firesetter Program Manager Professional Qualifications*

8.2 To communicate with the public those things that will maintain the department's image, to inform the public of emergencies, to communicate events, and to keep audiences motivated in fire and life safety behaviors

8.3 Community relations

Media relations

Writing

Public speaking

Audio/visual presentation

Fire department operations and functions

8.4 Who works, invests, and maintains a high profile within the community

Awareness of organizational relationships and partnerships within the community

Community views regarding emergency preparedness and their political opinions and volunteer involvement

Cultural and religious values and beliefs

8.5 Answer will depend on local circumstances.

8.6 Answer will depend on local circumstances.

8.7 Answer will depend on local circumstances.

8.8 Important things are not important unless they are different.

Emotional messages are better than factual messages.

Keep information in small bites.

Create shelved "canned" messages for immediate release.

Keep information easy to get and easy to give.

Treat the media with respect by answering their questions directly.

Repeat messages just as in teaching.

Never make stuff up.

Never speak for another agency or official.

Don't give personal opinions.

Assume you are always on camera or tape.

If you don't want something on the national news, don't say it.

8.9 Worksheet should include at least the following:

Cause of event

Place for eyewitness accounts

How many and what type of resources responded to the incident

Statistics (money, loss, deaths, injuries, etc.)

Past history of similar events

8.10 Answer will depend on research quality and instructor preferences.

8.11 *Any four of the following*:

Looking sloppy

Holding onto a podium when speaking

Pacing from side to side

Reading to any audience

Using hypnotic filters

8.12 Answer will depend on local circumstances and instructor preferences.

8.13 Could vary but:

a) Releasing information that the fire is arson before details or witnesses have been identified and interviewed.

b) Reporting that an employee is accusing another of creating a hostile work environment before it has been investigated.

c) Releasing to the media a factual statement that the Chief plans on closing a fire station before he has had a chance to inform his boss or council.

CHAPTER 9

9.1 Fires are investigated to determine:

Origin and cause

Code deficiencies

Data collection

Training deficiencies

9.2 Fire prevention bureau personnel can use the fire investigation infomation to try to prevent similar events. One excellent method of using this information is through fire and life safety education. The information gained from a fire investigation can and should be used to educate and inform the public of potential fire causes or modifying code elements to prevent future occurrences.

9.3 An investigation to determine where (origin) and how (cause) fires started by analyzing the area of origin, the heat of ignition, the combustible or flammable materials involved, and the actions of the occupants at the time of the fire

9.4   Some of the most significant mistakes investigators make on fire scenes include:

Not taking enough photographs

Failing to dig and scrape to the bottom of the fire scene

Failing to dig and scrape with care

Not documenting where the evidence samples were taken

Failing to get and document accurate statements from witnesses

Assuming that something has happened without making certain

Not writing a comprehensive report

Failing to seek the assistance of outside agencies or experts

9.5   Arson, in the laws of the United States, is the act of willfully and maliciously setting fire to a house, building, or other property.

9.6   Arson can be costly to the taxpayers for the investigation as well as result in the loss of tax revenue if a business is destroyed. Arson can be lethal.

9.7   NFPA® 921, Guide for *Fire and Explosion Investigations*

NFPA® 1033, *Standard for Professional Qualifications for Fire Investigator*

9.8   Just as conducting a fire inspection, determining the origin of a fire requires a systematic and consistent approach such as outside-in, etc.

9.9   The data collected by the cause of fires can be used to develop a code that would prevent the incident from occurring again. For example, if the use of charcoal grills on wooden balconies in large combustible apartment complexes causes a number of fires, the code could prohibit charcoal grills on wooden balconies in multi-family residences.

9.10   The list should include at least the following: address, deaths, cause of fire, material ignited, time of fire, names of witnesses, names of suppression crews, time and day of fire, injuries, insurance company, dollar loss, and occupancy classification.

9.11   The list should include at least the following:

- Personal protective equipment
  - Helmet
  - Coat
  - Gloves (heavy and light)
  - Boots
  - Respirator or self-contained breathing apparatus (SCBA)
  - Coveralls
- Flashlight
- Cell phone or portable radio
- Pump can (for hotspot touchups and cleaning the floor after debris removal for better examination)
- Portable electrical lights

- Hydrocarbon sampling device
- "Get after it and dig for it" equipment
  — Trowel
  — Scraper
  — Shovel
  — Small saw
  — Chisel/hammer
  — Wire cutters
  — Multipurpose scissors
  — Needle-nose pliers
  — Vise grips
  — Gooseneck pliers
  — Pry bar
  — Garden tools
  — Battery-powered reciprocating saw
- Measuring tape
- Ruler
- Camera and film (digital work well)
- Barrier tape
- Paper
- Pen or pencil (pencils always work in a variety of climates!)

9.12   The answer should include at least the following:

Day of the week fires most often occur.

Time of day fires most often occur.

Most common type of fires.

Most common ignition sources causing fires.

Most common locations where fire occurs.

Trend identification is also important for justifying requests for more resources.

9.13   Extreme climate conditions, the potential for falling objects, the instability of the structure, and poor air quality. Risk of cancer due to chemical exposure is also significant.

9.14   Answer should include at least the following:

Develop good policies and procedures for everyone to follow.

Integrate fire investigation functions into the mission of the fire department.

Train everyone on the job about what it means to do good fire investigation.

Get management's commitment to this process.

Provide proper training and equipment.

Develop arson task forces where appropriate.

Gather and track good data.

Develop an arson early warning system.

Develop a progressive fire and life safety program citing all aspects of fire.

Partner with community organizations.

Develop relationships with law enforcement agencies and prosecutors.

Support advances in fire investigation technology.

CHAPTER 10

10.1 The construction document review is an opportunity for the fire department to begin building a fire prevention coalition with the owners of the building. This is an opportunity to identify potential hazards and risks that will impact the safety of the occupants and the firefighters who may be called there in the middle of the night to mitigate an emergency incident.

10.2 Building officials have a responsibility to enforce the building code that contains elements that protect firefighters.

Insurance companies want to ensure their level of risk is minimized through adequate construction methods.

State fire marshals have jurisdiction to enforce state codes in some construction document reviews.

10.3 Firefighter safety

Occupant safety

Enforcement of codes

Preincident planning

Property conservation

10.4 The construction document review package will include:

Architectural drawings: The fire department will be concerned with egress and passive fire protection features.

Structural drawings: The fire department will be concerned with how the building is put together and how it may fail during a fire.

Mechanical drawings: Fire department concerns include how the mechanical systems will affect firefighter operations.

Electrical drawings: The fire department concerns pertain to the emergency lighting, generators, fire pump wiring, egress lighting, and location of transformer vaults.

Site, landscaping, civil, utility drawings: The fire department will focus on access, hydrants, and vegetation management.

Plats: The fire department will review elements that may impact fire protection access.

Fire protection drawings:

- Sprinklers
- Fire detection and alarm
- Standpipes

The fire department's concern is compliance with the applicable standard as well as the impact the systems will have on fire department operations.

10.5 Preconstruction meetings provide the architect, other design professionals, and owner the opportunity to ask the construction document review team specific questions regarding the project.

10.6 The preconstruction meeting is an excellent opportunity for everyone involved in the project to meet all of the individuals from the design team and construction document review team. The architects and designers benefit from such a meeting because they will reduce or eliminate design mistakes, and the construction document review team benefits because ensuring their concerns are addressed in the initial design will most likely shorten the review time.

10.7 Conceptual designs are preliminary documents and are not intended to be used for construction or permit approval. Conceptual drawings typically serve as a means to facilitate a meeting of the construction document review team with the owners and their design professional.

10.8 A certificate of occupancy is issued to the occupant when the building has been determined to be safe to occupy and has been inspected to meet the code requirements.

10.9

1. The deficiency. What is the problem with the document under review?

2. The code and code section of the deficiency. Where can the design professional go to find more information?

3. What is needed to correct the deficiency? What must the design professional do?

10.10 Fire protection drawings include a variety of different systems and elements pertaining to the building's fire protection systems.

10.11

1. The need is determined.

2. A design professional is contacted.

3. Conceptual designs are prepared.

4. Construction documents are prepared.

5. Construction documents are reviewed.

6. Construction document review comments are completed.

7. Construction document revisions are submitted.

8. Construction document revisions are approved or denied.

9. Permit is applied for and issued.

10. Construction begins.

11. Construction is inspected.

12. Construction is completed.

13. Certificate of occupancy is issued.

14. Occupant moves in.

15. Business license is approved.

16. Periodic fire inspections continue.

Chapter 11

11.1 Research

Interior inspection

Exterior inspection

Documentation

Explanation to owner(s)

11.2 Yes. A fire inspector must be credible for fire prevention and code issues to be accepted and trusted by an owner or occupant.

11.3 To provide a reasonable degree of safety to the occupants of commercial and public buildings. Americans typically rely on safe buildings just as they rely on fire engines or ambulances to be available in an emergency.

11.4 Research the occupancy and any past history.

Contact the owner/manager to schedule an appointment.

Be punctual.

Be prepared with all your necessary equipment.

Examine the exterior for various requirements such as access, fire hydrants, etc.

Begin the interior inspection in a top-down or bottom-up procedure (just be consistent).

Document the inspection findings, both positive and negative.

Close the inspection with a verbal summary of the findings.

11.5 Ensuring that a building or process is safer than when an inspector entered it

11.6 Occupants conduct their own inspections and provide a copy of the inspection report to the fire department.

11.7 *Advantages*:

More people involved in the inspection and accountable for their own safety

Inexpensive

More businesses are inspected at a greater frequency

Disadvantages:

No guarantee the inspection will be done

No professional guidance

Lack of quality control and limited technical knowledge

Potential that occupant may be dishonest

Self-inspections can be limited to certain occupancies or inspection types (i.e fire doors, exit lights etc.)

11.8 Because the history of the findings and the end results are important not only for future reference but possible legal defense in civil or criminal litigation. Documentation is critical to any fire prevention work.

11.9 Communication

Research capabilities

Code knowledge

Fire department knowledge

Familiarity with hazards and processes

Interpersonal skills

Conflict resolution skills

Salesmanship (being persuasive to convey the importantce of fire prevention practices, not a used car salesman approach)

11.10 Conduct research.

Be familiar with the occupancy and processes.

Gather tools and equipment.

11.11 Electronic method for organization (phone, Tablet, etc.)

Laptop or Tablet computer

Clipboard

Pen

Paper

Rulers or tape measure

Flashlight

Cell phone

Radio

Forms

Camera

11.12 Answer will depend on local circumstances.

11.13 A reasonable time frame as established by the inspector and the business owner. However, the inspector has the final say based on the severity of the violation and hazard.

11.14 To provide guidance to the inspection staff and to provide and promote continued focus on mission and purpose

11.15 To provide a safe environment for the public

11.16 Let the occupant know how and why a violation is a problem and why it must be fixed. Provide good information as to the risk and the exposure and how that relates to the occupant's business operations and to firefighters as well.

11.17 Identify your fire problem. Those with the most significant potential losses (life and then property) are your targets. Place them in descending order and you will have a prioritized list. However, do not forget about large significant risks that may not have suffered a loss yet. That future single event may cause more damage than all the others combined.

11.18 *Any ten of the following*:

Fires

Fire alarms

Request for service

Outstanding fire code violations

Construction permits/alterations

Occupancy changes

Process uses

Written correspondence

Permit activity

Fire alarm test results

Suppression system inspection /test results

11.19 They are equally important as one must have the technical knowledge to understand and communicate the violations and then be skilled enough to communicate the needs and problems so as not to offend the owner/occupant.

11.20 Domination

Compromise

Integration

11.21 No method is necessarily better as long as the inspector is consistent.

CHAPTER 12

12.1 A condition or element that provides a source of ignition for a hostile fire or that contributes to the spread and severity of a hostile fire.

12.2 Heating and cooking. While some years the statistics may vary, these two generally lead the list. Smoking used to be very prevalent, but over the years this cause has begun to decline.

12.3 A hazard that is common among occupancies or locations.

12.4 A hazard, like concentrated hydrogen peroxide storage, is isolated to specific operations or locations — not a common sight in most homes.

12.5 Selecting those occupancies where your most commonly occurring fire incidents take place or where extreme life risk exists.

12.6 Create a team.

Create a mission statement.

Establish strategic goals.

Establish strategies to achieve the goals.

Sign a contract document affirming the goals and strategies.

Implement the plan.

Evaluate.

12.7 Suppression systems: extinguish the fire in its smallest incipient phase.

Smoke control systems: evacuate smoke and/or pressurize certain compartments to keep smoke out.

Detection and alarm systems: detect fire in an early phase to alert occupants and/or the fire department.

12.8 Active fire protection systems do something to suppress, warn, or control fire when it occurs. Passive fire protection involves design features that do not actively attack a fire but by their mere existence prevent fires or prevent them from getting worse or spreading, such as fire walls or barriers.

12.9 Life safety

Property protection

12.10 Yes, very! This is the last and least effective fire protection and suppression measure because of time delays, reliance on human beings, and dependence on various factors.

12.11 Typically, flame-retardant properties can be incorporated in the manufacture of various materials making them inherently flame resistant. If something will not burn, it does not have to be protected. Temperature limit switches can also be installed in various appliances to kill power to a device that may have an abnormal increase in temperature.

12.12 Compartmentation is the physical separation of one area from another, which keeps fire or toxic gases from moving between the areas.

12.13 Defending in place involves protecting people or processes without evacuation or significant relocation.

12.14 Notification

Safe and unobstructed paths of egress or methods of escape

12.15 This curve represents a standard test fire but does not accurately reflect the type of fire that commonly occurs nowadays.

12.16 An area or series of areas separated by sufficient fire-resistive construction that occupants can move through it or into it for safety

12.17 A process of risk and hazard assessment that identifies goals, objectives, consequences and decisions that dictate the design of a building or process.

CHAPTER 13

13.1   Any hostile fire in the outdoors that is not prescribed or purposefully managed.

13.2   While it will help prevent disasters from occurring to some extent, it also prohibits nature from "taking care of business" by cleaning the forest of noxious weeds, insects, and overgrowth. This lack of cleaning increases the severity of the problem so that when an event does occur, it likely will be well beyond anyone's capability to stop or defend.

13.3   Interface mix: structures scattered throughout a rural area surrounded by undeveloped land

Occluded interface: isolated area of forested land surrounded by homes or structures

Classic interface: homes or structures pressed directly against the forest or wildlands

13.4   Water

Wildlife

Air

Plants

Timber

Soil

13.5   Answer will depend on local circumstances. Agencies may include:

USDA

USFS

State forest service

Local fire department

Neighborhood associations

County

Land trust owners

City and county parks

13.6   Prevention and mitigation.

13.7   Hand or mechanical vegetation management

Control burns

13.8   Fire has always been a major method of reclamation and restoration for old, dead, and dying forests. It has been the mechanism for rejuvenating new growth and clearing out overgrowth as well as controlling disease and insects. It has been around as long as forests and, in fact, probably did the best job of managing forests until we intervened.

13.9   It called for all fires to be extinguished by 10:00 A.M. the day after their discovery. This caused an environmental imbalance by allowing vegetation to become too dense and overgrown, which weakened some species allowing additional insect infestation to also impact various forested areas.

13.10 Reducing or eliminating the possibility of fire communicating from lower fuels like grass into brush and then brush into trees.

13.11 Water – Flood and debris in watershed areas contaminates drinking water and incapacitates reservoirs.

Air – High volumes of smoke and ash exacerbate diseases in chronic patients and impact clean air over communities.

Wildlife – Fire destroys animal habitat and in some cases kills wildlife. It forces relocation of various animals and can greatly impact domestic animals such as horses, cattle, and other livestock.

Timber – Stand clearing fires can destroy timber and eliminate access to lumber harvesting areas.

Soil – High temperature fires can virtually sterilze the soil, removing organic material causing it to become hydrophobic delaying regrowth of understory and grasses.

13.12 Fuel

Weather

Topography

13.13 What are the loss impacts to the community?

What is the community's assessment of which losses are and are not acceptable?

What is our ability to mitigate hazards that have been identified?

13.14 Because wildfires cross many boundaries and no one department has the resources to deal with a large wildfire situation once it occurs.

13.15 Fuel

13.16 Ground fuels

Surface fuels

Crown fuels

13.17 A large wildfire moves into the crown of large vegetation. This event creates enormous amounts of heat release, generally burning so hot that it sterilizes the soil underneath. Obviously, little vegetation can survive these extreme temperatures.

13.18 Assume adults living in large homes in forested areas are educated, understanding, and responsible.

Assume that these same adults do not know much about the natural fire environment or their actual risk.

Assume that they want to do the right thing if they are informed, if they have the capability, and if they can measure their progress.

Be factual, be truthful, and make clear all expectations about survivability and fire suppression capabilities.

CHAPTER 14

14.1 Technology can allow you to work smarter and accomplish more with less. Technology can also improve efficiency.

14.2 The advancement of the personal computer has enabled fire prevention bureaus to work more efficiently.

The global information system permits fire departments to share information in layers. An example would be fire hydrant location with a second layer of streets or topographical information.

Digital cameras can provide a cost-effective means to document conditions that need to be corrected.

Tablet computers can be used by fire prevention bureau inspectors to conduct inspections and reduce paperwork.

Database software is an affordable means to track fire loss data and determine the areas on which to focus fire prevention efforts. The advancement of web based inspection programs enhances the inspector's ability to work efficiently in the field.

14.3 Answers will vary. In any case, the item needs to show a benefit either of reducing cost or improving efficiency. For example, the digital camera feature on the tablet can reduce photo-processing costs and eliminate the time it takes inspectors to take film to a lab for processing. The customer can instantly be emailed inspection results

14.4 Global information systems can determine hydrant locations, locate streets for emergency response, and identify topographical information that may benefit the wildland mitigation and prevention efforts. The inspector can use many web based maps to view the facility prior to the inspection as well as for diagraming any needed report details.

14.5 Answers will vary. One possible answer is given below:

The fire deaths in the United States continue to be the highest in our homes. It is important that technology advance to reduce residential fire deaths. Progress has been made toward the installation of residential sprinklers. Residential fire safety improvements are just one area to be addressed by technological advancements.

14.6 It is not always wise to purchase products as soon as they become available because of the cost associated with the first generation of the product. Typically the second generation of the product corrects any product deficiencies and further improves upon the product's first generation.

14.7 GIS: Global Information Systems

IT: Information Technology

MIS: Management Information Systems

14.8 Answers may vary. One example may include the identification that Laptops and Tablets have become very cost-effective tools used by many fire prevention divisions. In many cases, the use of web-based inspection systems provides the inspector "real-time data." The inspector also has the ability to send an email of the inspection results or print while at the facility. Many other tools, such as contacts, calendar etc. also enhance the inspector's ability to organize his or her inspection process.

14.9 Yes

14.10 Technology can be used to identify a community's fire problem by tracking the frequency and causes of fires through a computerized database. The information of the property where an inspection is being conducted can be sent to the fire inspector's Laptop in his or her vehicle. Technology can provide the inspectors with real-time data while they conduct their inspection.

Fire Prevention Applications

Bibliography

Bibliography

Amdahl, G. (2001). *Disaster Response: GIS for Public Safety*. ESRI Press, Redlands, CA.

Arno, S. F. and Steven Allison-Bunnell (2002). *Flames in Our Forest: Disaster or Renewal?* Island Press, Washington, DC.

Babcock, Chester, and Rexford Wilson, "The Chicago School Fire," NFPA Quarterly (January 1959).

Beakley, George C., and Herbert W. Leach, *Careers in Engineering and Technology*, 2nd ed. New York: Macmillan, 1979, p. 3.

Best, Richard L., "Tragedy in Kentucky," Fire Journal (January 1978).

Best, Richard L., Investigation Report on the MGM Grand Hotel Fire National Fire Protection Association, Quincy, Mass., 1982.

Bryson, J. M. (1995). *Strategic Planning for Public and Nonprofit Organizations*. Jossey-Bass, New York.

Cohen, Jack D., "Wildland-Urban Fire: A Different Approach," Missoula Fire Sciences Laboratory, Rocky Mountain Research Station, Forest Service, U.S. Department of Agriculture, 2002.

Coleman, R. J. (1997). "It's the Fire Service, Not the Fire Business." Fire Chief, (April)

Colorado Springs (Colorado) Fire Department Wildfire Mitigation Plan, 2001.

Cote, A. (1991). *Fire Protection Handbook®*, 17th ed. National Fire Protection Association®, Quincy, MA.

Cote, R., ed. (2000). *Life Safety Code®* Handbook, 8th ed. National Fire Protection Association, Quincy, MA.

Covey, Stephen R., *Principle Centered Leadership* (Summit Books, 1991), pg. 173.

Custer, R. L., and B.J. Meacham, (1997). *Introduction to Performance-Based Fire Safety®*. National Fire Protection Association, Quincy, MA.

Dungan, K.W. (2001). "Risk-Based Methodologies." Fire Protection Engineering (Spring 2001), Society of Fire Protection Engineers, Bethesda, MD.

Employment Policy Foundation (2000). The American Workplace, 1998. Retrieved February 21, 2000, from the World Wide Web: http://www.opf.org/labor98/98intro1.htm.

Favreau Donald F., *Fire Service Management* (New York: Donnelley, 1969).

Federal Emergency Management Agency (2001). Public Fire Education Planning: A Five-Step Process (brochure). Federal Emergency Management Agency, Emmittsburg, MD.

Federal Emergency Management Agency, United States Fire Administration (1998). Strategies for Marketing Your Fire Department Today and Beyond (brochure). Federal Emergency Management Agency, Emmittsburg, MD.

Illinois State Fire Marshal (2000). Media and Budgeting Resource Guide for the Fire Service (brochure). Illinois State Fire Marshal, Springfield, IL.

International Association of Fire Chiefs, et al., The Fire and Emergency Services in the United States, Wingspread IV, October 23–25, 1996.

International Fire Service Training Association (1998). *Fire Inspection and Code Enforcement*, 6th ed. Fire Protection Publications, Stillwater, OK.

International Fire Service Training Association (2000). *Fire Investigator*, 1st ed. Fire Protection Publications, Stillwater, OK.

International Fire Service Training Association (1997). *Fire and Life Safety Educator*, 2nd ed. Fire Protection Publications, Stillwater, OK.

International Fire Service Training Association (1993). *Fire Service Orientation and Terminology*, 3rd ed. Fire Protection Publications, Stillwater, OK.

International Fire Service Training Association (2000). *Public Information Officer*, 1st ed. Fire Protection Publications, Stillwater, OK.

Kiurski, T. (1999). *Creating a Fire Safe Community: A Guide for Fire Safety Educators*. Fire Engineering, Saddle Brook, NJ.

Machlis, G. E. and Artley, D., eds. (2002). Burning Questions: A Social Science Research Plan for Federal Wildland Fire Management. National Wildfire Coordinating Group, Washington, DC.

Mathis, Mark, *Feeding the Media Beast* (West Lafayette, Indiana: Purdue University Press, 2002).

McEwen, Tom, *Fire Data Analysis Handbook*. Washington, D.C.: U.S. Fire Administration

National Fire Protection Association (1994). NFPA *Inspection* Manual, 7th ed. National Fire Protection Association®, Quincy, MA.

National Fire Protection Association® (1998). *Professional Qualifications for Fire Inspector and Plan Examiner* (NFPA 1031). National Fire Protection Association, Quincy, MA.

National Fire Protection Association (1998), NFPA 1033, *Professional Qualification for Fire Investigator National Fire Protection Association,* Quincy, MA.

NCFPC, *America Burning*: Report of the U.S. National Commission on Fire Prevention and Control (Washington, D.C.: U.S. Government Printing Office, 1973).

North American Coalition for Fire and Life Safety Education (January 2002). Beyond Solutions 2000. Retrieved February 1, 2003, from the World Wide Web http://www.usfa.fema.gove.

Osborne, D., and T. Gaebler, (1992). "Anticipatory Government." In Reinventing Government (pp. 223–226). Addision-Wesley Publishing Company, Reading, MA.

Pages from the Past, "Flammable Decorations, Lack of Exits Create Tragedy at Cocoanut Grove," Fire Engineering (August 1977).

Pages from the Past, "Theater Was 'Fireproof' Like a Stove but 602 Persons Lost Their Lives," Fire Engineering (August 1977).

Powell, P., and M. Appy, (1997). "Fire and Life Safety Education: The State of the Art." In Ron Cote, ed. *Fire Protection Handbook®*, 17th ed., pp. 2–55. National Fire Protection Association, Quincy, MA.

Pyne, S. J. (1997). "World Fire: The Culture of Fire on Earth." University of Washington Press, Seattle.

Robertson, J. C. (1979). *Introduction to Fire Prevention,* 2nd ed. Glencoe Publishing Co., Inc., Encino, CA.

Teague, Paul E., "Case Histories: Fires Influencing the Life Safety Code," in Ron Cote, *Life Safety Code Handbook®* (Quincy, Mass.: National Fire Protection Association, 2000), pp. 931–933.

TriData Corporation (1987). *Overcoming Barriers to Public Fire Education in the United States.* TriData, Arlington, VA.

TriData Corporation (1990). *Proving Public Fire Education Works.* TriData, Arlington, VA.

Wallace, William H., *Community Risk Issue: Structure Fires,* Colorado Springs, Colorado: Colorado Springs Fire Department, 1997.

Wallace, William H., Colorado Springs Fire Department, Summary of Survey Responses, page 25, January 3, 2003.

Wolski, A. (2001) "Risk Perceptions in Building and Fire Safety Codes." Fire Protection Engineering (Spring 2001), Society of Fire Protection Engineers, Bethesda, MD.

Fire Prevention Applications

Glossary

Glossary

A

adoption by reference a local jurisdiction's formal decision to follow state laws exactly as drawn

arson the act of willfully and maliciously setting fire to a house, building, or other property

assembly a particular construction method that details specific types of materials, their specific manufacture or installation, and components

C

classic interface area with homes and other structures, especially in small dense neighborhoods, pressed directly against the forest or wildlands

code a body of law systematically arranged to define requirements pertaining to the safety of the general public from fire and other calamities

common hazard found among many occupancies or locations

community injury any significant loss of property or monetary value, as well as physical injury or death

compartmentation the use of passive (and in some cases active) protection features to prevent fire spread

compressed workweek scheduling system that permits full-time employees to perform the equivalent of a week's work in fewer than five days

compromise method of resolving conflicts in which each side gives up something in order to reach an agreement

conceptual design preliminary document that is not intended to be used for construction or permit approval

D

defending in place protecting people or processes without significant relocation or evacuation

domination method of resolving conflicts in which one side is the victor and one side is the loser

E

electrical drawing reflects the details of the building's electrical system

enabling act method of adopting state regulations that allows the local jurisdiction to amend them based on local needs or preferences

enterprising running a service organization or a division of the organization so that it operates fully on a cost-recovery basis

F

fire protection drawing indicates the systems and elements pertaining to a building's fire protection systems

G

grading (fire suppression rating schedule) method of evaluating fire suppression capabilities and crediting them to individual property fire insurance rates

H

hazard a condition or element that provides a source of ignition for a hostile fire or that contributes to the spread and severity of a hostile fire

hostile fire any unwanted or destructive fire

I

incendiary fire any fire that is set intentionally

injury personal injury, monetary impact, job loss, or aesthetic impact

integration method of resolving conflicts in which a mutual solution is found and both sides achieve their goal to some degree with neither being wrong or bad

interface mix rural area with structures scattered sparsely throughout

L

ladder fuels configured so that a ground fire can become a surface fire and a surface fire can become a crown fire

M

maintenance code details how to properly safeguard the activities or operations in a building

mechanical drawing indicates the elements of the building's mechanical systems, such as plumbing, heating, air conditioning, etc.

mission statement a description of the purpose of the organization

mitigate to make less harsh or hostile or to make less serious or painful

mitigation the prevention or reduction of severity of an undesired event

O

occluded interface isolated area of forested land or wildlands surrounded by homes or other structures

P

paradigm shift a change from old ways of thinking that opens the door for new insight, new methods, and different views

performance-based code allows designers to determine how best to meet an individual building's unique fire protection needs

plat legal document illustrating the legal description of a property as well as any legally binding easements

preconstruction meeting people involved in a building project review the conceptual designs

preincident planning the process of identifying specific occupancies, buildings, or locations that will likely require special treatment or operations during an emergency

prescriptive code lists specific design requirements, such as number of exits, fire separation, construction type, and fire suppression systems

privatization the transfer of functions or duties previously performed by a government entity to a private organization

public information provided to the general public so they remain informed about events, become better educated and/or prepared for various situations, and remain motivated in fire- and injury-prevention behaviors

public process the act of engaging the public with factual information concerning issues that are important to them and learning from them "what they want and can't live without"

R

risk the exposure to possible loss or injury

risk assessment process of analyzing the risk impact and the risk perception and then combining the results

risk impact a measure of the probability that something will occur and the severity of its results

risk perception the actual "value" of a risk

S

service delivery deployment, response times, and service level objectives of the overall fire service system

site drawing indicates a variety of details concerning topography, landscaping, and civil engineering details

specific hazard isolated to particular operations or locations

standard document that details how something is to be done in order to comply with the applicable codes

Standard Fire Test published results of research that determined time/temperature measurements of a typical large fire in the early twentieth century

structural drawing provides details on how a building is put together

T

target hazard location or building that is different than those that are "typical" throughout a jurisdiction

10:00 A.M. policy mandates that any fire be controlled by 10:00 A.M. the day it is reported or, failing that, by 10:00 A.M. the day following, ad infinitum

V

value statement a summary of the ethical priorities for everyone's behavior when working on the mission

vision statement a brief description of how the fire department or more specifically how the fire prevention bureau will operate

W

wellness being physically fit, mentally prepared, and emotionally healthy

wildfire any hostile fire in the outdoors that is not prescribed or purposefully managed

Fire Prevention Applications

Index

Index

occupancy hazards, 268

Fire Prevention Applications for the Company Officer, 27, 105, 265

    mission statement, 32–33

Fire suppression and alarm systems, wildfires, 334

Fire suppression rating schedule (FSRS, grading), 43

Fire testing laboratories, 32

Fires

    Apollo space program, 204

    Beverly Hills Supper Club (1977), Southgate, Kentucky, 16, 20–21

    Big Burn (1854), Colorado, 320

    Black Forest, Colorado, 354

    Cerro Grande (2012), Los Alamos, New Mexico, 331

    Chicago fire (1871), 42

    Cocoanut Grove (1942), Boston, Massachusetts, 16, 19, 204

    code changing fires, 16

    Cook County Building (2003), Chicago, IL, 23

    Crowning wildfire, National Forest, 321

    decreasing incidence of, 70

    DuPont Plaza Hotel (1986), San Juan, Puerto Rico, 207

    food processing plant (1991), Hamlet, North Carolina, 22

    Happy Land Social Club (1990), Bronx, New York, 21

    Hayman Burn (2002), Colorado, 324, 329

    hostile, 289

    Iroquois Theater (1903), Chicago, Illinois, 16, 17

    Jamestown fire (1608), codes following, 15

    Los Alamos, New Mexico, wildland fire, 111

    MGM Grand Hotel (1980), Las Vegas, Nevada, 16, 21

    One Meridian Plaza (1991), Philadelphia, PA, 16

    Our Lady of Angels School (1958), Chicago, Illinois, 16, 19–20

    power production facility, 202

    Rhode Island nightclub fire, 3–4

    Station Nightclub (2003), West Warwick, Rhode Island, 16, 22–23

    Triangle Shirtwaist (1911), New York, New York, 16, 17–18

    Waldo Canyon (2012), Colorado

        citizen preparedness, 146

        GIS information used for data collection, 354

        homes destroyed, 150, 320, 321

        progression map, 329

        structures destroyed, 331

    Whitewater-Baldy Fire, 320

Five-Step Planning Process to Community Risk Reduction, 156, 157

Flash fuels, 331

Flextime, 74

Floor plan, 240

FM 200 clean agent systems, 298

FMNA (Fire Marshals Association of North America), 32

Follett, Mary Parker, 271

Food processing plant fire (1991), Hamlet, North Carolina, 22

Forest service. *See* United States Forest Service (USFS)

FPP. *See* Fire Protection Publications (FPP)

Franklin, Ben, 6

FSRS (fire suppression rating schedule), 43

G

Gambol Oak removal, 137

General Services Administration (GSA), 47

Geographic Information System (GIS)

    Google Earth, 353

    layers, 354

    occupancy data collection, 348

    for preincident planning, 109

    risk analysis, 134, 136

    for sharing data, 353–354

GIS. *See* Geographic Information System (GIS)

Global Positioning System (GPS), 346

Google Earth, 353

GPS (Global Positioning System), 346

Grading [fire suppression rating schedule (FSRS)], 43

Ground fires, 321

Ground fuels, 331

Groups, engineering section, 101–102

GSA (General Services Administration), 47

Guide for Fire and Explosion Investigations (NFPA® 921), 210, 213, 214

H

Hammurabi, King, 41

Happy Land Social Club fire (1990), Bronx, New York, 21

Hayman Burn (2002), Colorado, 324, 329

Hazard risk assessment, 345

Hazardous Material Inventory Statement (HMIS), 277

Hazardous Materials Incident Worksheet, 183–184

Hazards

    case study, 289

    common hazards, 289

    defined, 289

    hostile fire, 289

    performance-based design, 308–309

    solutions for control, 296–307

        active fire control, 297

        area of refuge, 305

        compartmentation, 302–303, 305

        defending in place, 306–307

        detection and alarm, 298–300

        evacuation, 304–305

        fire department operations, 300–301

        mitigation, 296

        occupant safety, 303–304

        passive fire control, 297

        product manufacture and performance control, 301–302

        suppression and control, 297–298

    specific hazards, 290

    target

Intermediate fuels, 331
International Association of Arson Investigators (IAAI), 32
International Association of Black Professional Fire Fighters (IABPFF), 32
International Association of Fire Chiefs (IAFC), 32
International Association of Fire Fighters (IAFF), 32
International Fire Code® (ICC)
    fire investigations, 203
    fire sprinklers in apartment buildings, 44
    Hazardous Material Inventory Statement (HMIS), 277
    model fire codes, 53
    older buildings, 58
    preincident planning, 107–108
    references to codes and standards, 45
International Fire Service Training Association (IFSTA)
    Building Construction, 302
    Essentials of Fire Fighting, 27
    Fire and Emergency Services Orientation and Terminology, 192
    Fire and Life Safety Educator, 104, 154, 174
    Fire Inspection and Code Enforcement
        assembly, 302
        hazards and issues, 291
        inspector's level of training, 264
        occupancy hazards, 268
    Fire Investigator, 210, 218
    Fire Protection, Detection and Suppression Systems, 297
    as research organization, 32
    website, 27
Internet for providing public information, 107
Investigations. *See* Fire investigations
Iroquois Theater fire (1903), Chicago, Illinois, 16, 17
ISO. *See* Insurance Services Office, Inc. (ISO)
IT (information technology), 348–349
ITM (inspection testing and maintenance) reports, 343–345, 349

J
Jamestown fire (1608), codes following, 15
Jensen, Gary, 100
Junior Fire Marshal Program, 148
Jurisdictions
    code adoption, 45
    federal laws and properties, 46–47
    state laws and statutes, 47

L
Ladder fuels, 331
Landscaping drawings, 244–246
Laws to prevent fires
    America Burning report
        causes of ongoing fire problem, 15–16, 34
        fire code delays, 54

    enabling act, 48
    federal laws and properties, 45–47
    fire and life safety education, 148–149
    history of, 14–16
    Jamestown fire, codes following, 15
    local laws and ordinances, 48
    party wall, 15
    state laws and statutes, 47
Leadership issues, Wingspread Conference issues, 25
Learn Not to Burn, 104
Learning environment, 163–164
    elementary children, 163
    older adults, 164
    preschool children, 163
Legal issues and public information officers, 192–193
Level of fire prevention services determination
    determination of, 72–73
    staffing levels to provide services, 89–90
License, business, 253
Life Safety Code® (NFPA® 101)
    building codes, 19
    exit requirements, 265
    means of egress, 18, 22
    older buildings, 53
    performance-based codes, 95
    reentry from stairwells, 21
Life safety educator, 92
Life safety group, 101–102
Life safety systems, Wingspread Conference issues, 25
Life threats, 134–135
Liquid oxygen tank hazards, 290
Lloyd's of London, 42
Local laws and ordinances, 48
Local research organizations, 33–34
Los Alamos, New Mexico, wildland fire, 111

M
Maintenance code, 48, 235
Malaysian jet, flight 370, disappearance (2014), 175
Managed care issues, 24
Maps
    Waldo Canyon progression map, 329
    wildfire risk, 334
    wildland urban risk map, 111
MAQs (Maximum Allowable Quantities), 49
Marketing issues, 24
Mass Notification System, 307
Mathis, Mark E., 176, 179–180
Maximum Allowable Quantities (MAQs), 49
Mayor's office and construction document reviews, 236
McEwen, Tom, 207
Means of egress. *See also* Exits
    elevators, 305